HYDROELASTICITY OF SHIPS

Hydroelasticity of ships

R. E. D. BISHOP

Kennedy Research Professor in the University of London and
Fellow of University College

and

W. G. PRICE

Lecturer in Mechanical Engineering, University College London

CAMBRIDGE UNIVERSITY PRESS

CAMBRIDGE

LONDON · NEW YORK · MELBOURNE

CAMBRIDGE UNIVERSITY PRESS
Cambridge, New York, Melbourne, Madrid, Cape Town, Singapore, São Paulo

Cambridge University Press
The Edinburgh Building, Cambridge CB2 2RU, UK

Published in the United States of America by Cambridge University Press, New York

www.cambridge.org
Information on this title: www.cambridge.org/9780521223287

First published 1979
This digitally printed first paperback version 2005

A catalogue record for this publication is available from the British Library

Library of Congress Cataloguing in Publication data

Bishop, Richard E. D.
Hydroelasticity of ships.

Bibliography: p.
Includes index.
1. Ships–Hydrodynamics. 2. Hydroelasticity.
3. Naval architecture. I. Price, W. G., joint author.
II. Title.
VM156.B534   623.8′144   78-67297

ISBN-13 978-0-521-22328-7 hardback
ISBN-10 0-521-22328-8 hardback

ISBN-13 978-0-521-01780-0 paperback
ISBN-10 0-521-01780-7 paperback

Contents

Preface

Oh will you swear by yonder skies,
Whatever question may arise,
'Twixt rich and poor, 'twixt low and high,
That you will well and truly try?
Trial by Jury

Just as with aircraft, it was quite inevitable that the dynamics of ships would initially be founded on the assumption of rigidity. We now possess a valuable body of theory constructed on that basis. But a ship is not really a rigid structure and, as we show in this book, this has some profound consequences which cannot easily be ignored – particularly when one considers the stresses and strains of the hull in a confused sea.

That a ship is really a flexible structure and might be treated as an elastic beam is not a new idea. This was presumably the underlying theme of a paper published nearly fifty years ago.† But in years gone by the point could not be pressed. The beam being non-uniform, the calculations for the structure were virtually impossible to make before the advent of electronic computers. There was then very little understanding of the mechanics of non-conservative systems. The excitation of the beam by the sea could not be specified with any certainty. Indeed there was no acceptable way of describing the response, let alone of determining it. Even the far more pressing problem of aircraft distortion in flight had not, at that time, given rise to anything like a coherent body of adequate theory.

This book is written in the belief that a return to the fundamentals of ship dynamics is no longer out of the question. Advances in linear structural dynamics, in oceanography, in random process theory, in marine hydrodynamics and in computing appear to have made a big difference. What is more we have concluded that revaluation of existing theories need not lead to a mere collection of proofs, studies and solutions, all produced *ad hoc* as a result of practical demands. It should be possible (we think) to put a thread through this whole subject of Ship Hydroelasticity.‡

† Inglis, C. E. (1929). Natural frequencies and modes of vibration in beams of non-uniform mass and section. *Trans. INA*, **72**, 145–66.

‡ Hydroelasticity is that branch of science which is concerned with the motion of deformable bodies through liquids.

This question of a 'thread' is more important than it may appear. In this book we have given a general treatment, adapting contemporary techniques of structural theory, hydrodynamics and statistical theory for use in it. Obviously progress will be made in these fields, but that progress will not render the general approach useless. New theories too, can be adapted if need be.

This book is essentially one on naval architecture. But it is radically different from any existing book in that field and, in fact, much of the material has appeared only recently as published research. In other words the techniques presented have not yet withstood the test of time. In the circumstances we urge our readers not to regard this as a book of recipes but, on the contrary, to subject it to as many tests as possible. While we have left no intentional errors, it would, we are afraid, be a miracle if none remain. Indeed, having been moved to work in this field almost solely by sheer interest in the subject,† we should regard it as surprising if commercial and/or contractual pressures fail to produce worthwhile improvements.

The fact that this book is concerned with a rapidly developing subject has made it difficult to decide where to draw the line. Our decision has been in favour of presenting only basic theory, together with some specific results for actual ships. In the few months before the book actually appears, studies will be completed on a number of topics. The superposition of slamming responses on steady wave-induced responses has been investigated, additional results have been found for antisymmetric response and several other matters have been examined. In other words, we have decided to restrict this book to the presentation of basic ideas and not to develop those ideas as fully as one might.

The results that are given place greater emphasis on symmetric responses than on antisymmetric. Admittedly this has been dictated by the availability of adequate structural data and by the shortcomings of present-day techniques of estimating antisymmetric fluid loading. Even so this emphasis on symmetric responses hardly makes for artificiality since they are largely responsible for structural damage – notably when slamming occurs. Existing literature on antisymmetric responses is far smaller than that on symmetric responses; but it is to be expected that its sparseness and its known shortcomings will be remedied before long if only because antisymmetric motions are of great importance in questions of safety, and in particular of capsizing.

The origins of this book stem from the decision, reached about ten years ago, to start teaching naval architecture in the department of which we are members. This decision had been reached because it was

† The exception is chapter 9. The Ministry of Defence (Procurement Executive) commissioned the studies on slamming and we gladly acknowledge our gratitude for that support.

agreed that (among others) entrants to the Royal Corps of Naval Constructors should henceforth receive their specialist training in University College London. The RCNC being that body of engineers which is responsible for the Royal Navy's ships, it became necessary to think about the design of high-performance vessels. The then Head of the Corps, the late Sir Alfred Sims, hoped that such thinking would be aimed essentially at the longer term and the writing of this book is one outcome of our attempts to meet the challenge.

We really must say some words of thanks. The exploratory nature of this work has meant that our research assistant, Dr P. K. Y. Tam and, also, in the last stages, Mr P. Temarel and Mr Ö. Belik, have had to get used to our never being quite satisfied, to our perpetual wish to try something else and to our wanting all computer results 'yesterday'. We enjoyed working with Mr C. V. Betts on the particular subject of damping. Jenny Price must sometimes have wondered if the typing of drafts at breakneck speed was entirely compatible with running a particularly lively household and Jane Saffin must occasionally have wondered if the flow of tracings to be made would ever dry up. We are very grateful indeed to them all.

The Ministry of Defence (Procurement Executive), Yarrow (Shipbuilders) Ltd, Ocean Fleets Ltd, Howaldtswerke-Deutsche Werft A.G. and the British Ship Research Association have all helped us by supplying data on actual ships. Finally we wish to acknowledge our indebtedness to the Editors of the *Transactions of the Royal Institution of Naval Architects*, of the *Journal of Sound and Vibration* and of *International Shipbuilding Progress* for permitting us to reproduce material that was first published by them.

University College London   R.E.D.B.
April 1978   W.G.P.

1 Ship response

Fanned by a favouring gale,
    You'll sail
Over life's treacherous sea
    With me,
And as for bad weather,
We'll brave it together . . .
Ruddigore

1.1 The effects of waves on ships

When a ship proceeds through waves it executes motions that it does not perform when it moves through flat calm water. These parasitic motions raise a number of serious questions and it will be helpful to discuss the motions and their effects in general terms at the outset.

1.1.1 *Increased resistance*

A ship suffers increased resistance in rough sea. That is to say, for a given propulsive thrust its speed is lower, and the higher the waves the greater is this involuntary loss of speed. This behaviour is represented in a rough and ready way in fig. 1.1, though it is really a matter for conjecture what happens to ship resistance when the significant wave height is great because effects other than mere increase of resistance begin to intervene. The loss of speed is then a matter of choice, being ordered by the ship's master so as to lessen other ill effects of a more pressing nature.

Fig. 1.1. A sketch illustrating the effects of waves on ship performance. To the left of the broken line, the curve represents involuntary loss of speed due to increased resistance. To the right, the loss of speed is voluntary, depending largely on the judgement of the ship's master.

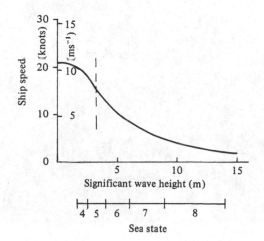

Resistance in waves is not a well-understood subject. Several attempts have been made to calculate it, and a number of theories exist by which estimates can be made.

1.1.2 *Deck wetting*

If a ship's master attempts to drive his vessel too hard in heavy seas, he may cause it to suffer damage by 'digging in'. The bow may dip beneath the surface so that water breaks over the upper works. The ship is said to be 'shipping green water' or to suffer 'deck wetness'. Numerous films have been made of warships shipping water in this way since, for them, the need to maintain speed in heavy seas is a matter of concern and therefore of research. These films make it very plain that deck wetting can be extraordinarily severe. (It is not unknown for the forward end of the flight deck to dip beneath the waves when a large aircraft carrier is driven hard in heavy seas.) When severe deck wetting occurs, no one could survive on the ship's forecastle and serious damage may be done to the deck fittings. Some idea of the ferocity of the conditions can perhaps be gained from fig. 1.2, which shows a frigate with its bow dug well into the sea.

Fig. 1.2. A frigate on trials experiencing severe deck wetting. (Photograph by courtesy of the Royal Navy.)

1.1.3 Slamming

Fig. 1.3 shows a frigate with its forefoot lifted clear of the water surface. Provided the relative velocity of the ship's bottom and the sea surface is large enough, the vessel will 'slam' when its forefoot re-enters the water. That is to say, an impulsive loading will be applied which will make the ship shudder and may well cause damage. The bottom plating of a steel ship may fatigue, local damage may be done due to overstressing and equipment (particularly sonar domes) may suffer as a result of the shock loading. Some ships are left vibrating significantly for comparatively long periods of time after one of these impacts. Fig. 1.4 shows a strain gauge signal taken from the hull plating of a ship following a severe slam.

The possibility of frequently repeated slamming in a heavy sea is a very real one and a ship's master could well jeopardise his ship by allowing it to happen since the effects of slams are in some degree additive. For this reason, too, it is necessary to slow down when conditions require it.

There are two related phenomena, neither of which is well documented yet both of which are worthy of mention. First, when a ship's

Fig. 1.3. A frigate with its forefoot lifted out of the water. When the bows re-enter, the impact of the sea may cause severe stressing of the hull. (Photograph by courtesy of the Royal Navy.)

bows have a marked flare (as with some high-speed vessels or the forward end of the flight deck of an aircraft carrier), rapid immersion can severely strain the vessel. Secondly, it is known that some ships may suffer slamming at the stern in a following sea. Although this behaviour is not as common or as violent as slamming at the forefoot, it is still a serious matter since a ship is most vulnerable at its stern as that is where the propeller and rudder are located.

1.1.4   Vertical acceleration

It is not known with any precision what causes sea-sickness. Many factors play a part (including lack of fresh air). The same can be said of the other ways in which human performance may become impaired in heavy seas. Perhaps the best practical guide is vertical acceleration; at all events this is commonly used in deciding which motions in a seaway are acceptable if severe loss of human performance is to be avoided. A ship's master may very well have to reduce speed in heavy seas to prevent conditions from becoming virtually intolerable for the crew or for a fragile (possibly live) cargo. In conditions like those of fig. 1.5, however, severe loss of human performance is probably quite unavoidable in a small vessel like the one shown.

1.1.5   Stressing

A ship is a more or less elastic body and, quite apart from slamming, waves cause it to distort. In other words, time-dependent strains are set up in the hull which are superimposed on those that are present in still water. Fig. 1.6 shows the stern of a tanker which split because stresses in the plating were excessive. (Tankers have been lost as a result of explosion in a tank, but there was no suggestion from the survivors of this ship that an explosion occurred.) This book is very much concerned with this aspect of wave response.

Leaving aside slamming for the present, it would be misleading to imply that a ship's master reduces speed in heavy seas in order to limit stresses in the hull. In fact he has no accurate means of judging what the stresses and strains are, let alone how serious they may be. If, then, stresses are to be limited by reducing speed, the naval architect (in the capacity of ship designer or ship surveyor) should place restrictions on operating conditions, or else furnish the master with some sort of monitoring device.

Fig. 1.4. A strain gauge signal taken from the hull of a merchant ship showing response to a slam. The high-frequency oscillations have a frequency of about 0.9 Hz, while the low-frequency component is of approximately 0.1 Hz.

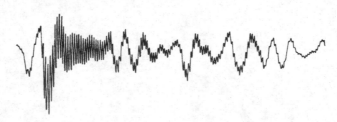

Fig. 1.5. The conditions in this ship as it weathers a severe gale must be difficult and normal working must be quite impossible. (Photograph by courtesy of the Royal Navy.)

Fig. 1.6. The stern of the 153 m tanker *Gem* which split in early March 1962 off Virginia. Survivors reported that the ship broke in two 'with a cracking sound'.

1.2   Response to wave excitation

The various phenomena that we have mentioned provide the ship designer with some difficult problems. Briefly they reduce to that of so designing a ship that its responses to waves are as small as possible. The responses may be bodily motions of the hull as a whole (as if it were rigid) or they may be distortions of some form or other. In other words he must be able to estimate responses for different wave- and operating conditions by investigating the way in which suitable parameters vary.

1.2.1   *Motion responses and distortion responses*

In the past it has been the custom to distinguish sharply between the 'motion responses' and the 'distortion responses' of a ship. Motion responses of a rigid ship have been dealt with in the theory of 'seakeeping' on which there is a large literature. Hull stressing on the other hand has been investigated in a semi-empirical fashion, stresses being held to be of two sorts – those that would be set up in a rigid ship, and those which must be added, if necessary, by way of a correction for something called 'springing'. An approach of that nature must eventually be discarded (as its predecessors were) and refuge will presumably be taken in more basic principles of the dynamics of a deformable body.

It is a return to first principles that we shall attempt to accomplish in this book, and our approach will be that of 'linear modal analysis'. This approach allows us to do two things:

(*a*) strength analysis can be rationalised, and
(*b*) seakeeping theory can be reconciled and combined with the strength analysis.

Note particularly that this does not mean either that we can produce, ready made, a working set of design rules which will immediately supplant those that are already in existence, or that all existing theory is valueless. A tremendous amount of practical knowledge is reflected in the present rules and the need now is to sift that knowledge carefully with a view to making it more reliable and more useful in ship design. That will take time.

1.2.2   *Selection of coordinates*

The structural dynamicist may be forgiven if he assumes that the choice of coordinates for investigation of ship response presents no problems. The computer will permit him to select hundreds of coordinates if he so wishes. Unfortunately things are not that easy. At some stage it becomes necessary to estimate hydrodynamic actions and that is virtually impossible to do with precision and economy without *convenient* measures of structural motion. Unless the naval architect is prepared to invest an unrealistic amount of time and effort in some

Fig 1.7, Outlines of some typical hulls. The hulls are drawn with the same length to differing scales so as to show the differing degrees of realism involved when regarding a ship's hull as a 'thin beam'.

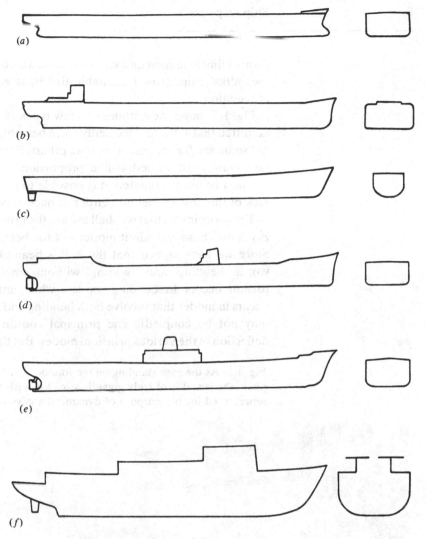

(a)

(b)

(c)

(d)

(e)

(f)

Ship	Weight loaded MN (tonne-f)	Approx. dimensions in metres			Type
		LOA	Beam moulded	Depth moulded	
a	1871 (190 800)	326	47.2	23.7	Tanker
b	382 (38 940)	201	25.9	14.4	Bulk cargo carrier
c	41 (4 184)	130	13.6	10.3	Frigate
d	640 (65 200)	224	31.7	15.7	Tanker
e	176 (18 000)	156	19.5	12.3	Cargo ship
f	5.43 (554)	55.1	9.5	5.3	Pilot vessel

form of finite element analysis, he is almost forced to think in terms of a 'beamlike' ship. How reasonable this is, is very much a matter for speculation.

Fig. 1.7 shows the outlines of a few types of vessels and it has to be admitted that some are decidedly more beamlike than others; that is to say some are far more slender than others. Paradoxically perhaps, the large tanker of cathedral-like proportions (fig. 1.8) is in fact the thinnest beam. Fortunately it is possible to make some allowance for lack of thinness of the 'hull girder' if necessary.

The coordinates that we shall use are the principal coordinates of the *dry* hull. These will admit motions of the beam as well as distortions. Since we have agreed that the hull is beamlike we can employ the words 'bending' and 'twisting' without ambiguity. Symmetric distortion occurs in bending modes, while antisymmetric distortion occurs in modes that involve both bending and twisting (which may or may not be coupled). The principal coordinates are measures of deflection in the various principal modes. But there is a drawback – it is

Fig. 1.8. As the men standing on the fore deck of this large tanker make plain, the vessel is of truly gigantic size. Yet such a ship is very well represented for the purpose of dynamical analysis as a *thin* beam.

not possible to find an unsupported dry hull, let alone to observe its behaviour.

Other sorts of coordinates can be used to specify ship response, but the only serious alternative to those we have mentioned seems to be the set of principal coordinates of the wet hull. At least the behaviour of a wet hull can be observed. Unfortunately these latter coordinates suffer from serious disadvantages; thus

(a) the deflection corresponding to unit value of any particular coordinate is dependent on the hydrodynamic theory that is chosen to represent fluid actions if modes are to be defined for the wet hull;

(b) the precise meaning to be assigned to the 'principal coordinates of the wet hull' is by no means obvious, any more than the properties of the relevant modes are.

1.2.3 *Estimation of response to waves*

Having seen why one should wish to estimate ship responses and how they might be specified, we have now to think about how those responses might be obtained. Since that is the object of this book we shall briefly outline the approach we shall use.

The dynamical 'system' with which we are concerned is a dry hull. In other words all fluid actions – be they due to hull motion or to waves – are to be thought of as 'applied'. Our first task, then, is to examine the hull and to derive appropriate equations of motion.

We shall discover that the fluid actions can be regarded as being of two distinct sorts. Some fluid forces applied to the hull are those which would be due to the actual motion of the hull if that motion were performed in flat calm sea. The second type of fluid actions are those that would be applied by the waves to the hull if the *hull* had no motion. The wave actions are found, ultimately, from the observations made by physical oceanographers. We shall therefore deal with the business of putting the entries in a wave atlas to use in ship design. With this task completed we shall have assembled our 'input' data.

The 'output' is a motion, a bending moment, a shearing force, a stress, or some other quantity whose variation is of interest, and in a sense its derivation completes the task before us. Unfortunately there remains the difficult task of assessing the results of these calculations. We must try to assess the motions and stresses that are set up in a hull.

1.2.4 *The assessment of responses*

The assessment of ship response is not at all an easy subject. The analyst can try to estimate the incidence of deck wetting or some other form of output for given conditions. But the absolute and relative degrees of importance to be assigned to these results really lies in the province of the ship designer.

The design of a ship is an enormous undertaking and the profit to be derived from piling more and more responsibilities on the designer is perhaps open to question. But, particularly with large flexible ships, he is providing himself with some hitherto unexpected problems. We can all agree that the size of the galley, the layout of the bridge, and so forth, are important, so 'architecture' matters a great deal. But where the integrity and performance of the vessel is concerned, the technology is (or should be) that of hydroelasticity, and ship designs have to be judged in the light of it.

To take an example, over the last few years it has become apparent that hull fatigue can be a serious matter. Does it follow that, because the vast majority of ships do not suffer serious damage from this source they are grossly 'over-designed'? This is not a question that can be answered with much confidence – rather little is known about the fatigue properties of steel in sea water – so how can hull distortions be assessed even if they can be estimated? The fact that this is an area of very considerable difficulty, so that recourse has to be had to empirical rules, is a very poor reason for failing to try to get at the truth. We must learn to estimate the output responses of various sorts *and* to assess their importance.

Even at sea in an actual vessel, the assessment of response to waves is in some respects exceedingly difficult. When a ship is in heavy seas, the master has to weigh up conflicting requirements. His first concern is for safety, but safety can only be bought at a price and he must decide to what extent he must exercise caution. He will not want to miss tides at his destination, he may have a tight sailing schedule, he may in times of war have other very pressing reasons for wishing to maintain speed. Different captains will inevitably reach different decisions in the same circumstances, because they have different temperaments if for no other reason.

But things are not really as simple as this. The master's decision to slow his ship down (to lessen slamming for instance) is based on his own observations. When he feels violent shuddering after a slam he is more likely to slacken speed than when he does not. If his bridge is located at an antinode in an appropriate mode he may be unduly cautious, but if it is at a node he may not slacken speed enough. Again the captain can under- or over-estimate the ill effects of waves on his crew for comparable reasons.

So one can go on. What, for instance, should be regarded as an acceptable incidence of deck wetting? Personal judgements play a vital role in ship operation and crucial decisions are reached with the wisdom born of experience. This is an area where we need much more information on the so-called 'man–machine interface'.

1.3   Excitation by machinery

It seems largely to have been overlooked in the past that excitation of a ship by waves is closely related to excitation of the hull by mechanical means. The subjects appear to be completely separated in the literature of naval architecture. Now vibration 'at blade rate' due to propeller rotation and vibration due to unbalanced machinery are two of the major problems in contemporary ship design. It is therefore natural to enquire whether or not the techniques to be described in connection with wave excitation can be employed in an attack on them. The one obvious difference between wave excitation and mechanical excitation is that the former passes along the hull whereas the site of the latter remains fixed in the ship. It turns out, however, that this is not, of itself, a serious obstacle provided care is taken. Nor does the fact that wave excitation is a random process produce any real difficulties.

The main difficulty is one of detail rather than principle, and it will be very familiar to the dynamicist. It has to do with the driving frequency. Serious mechanical excitation in a ship is of higher frequency than the range of dominant frequency in a wave spectrum. In other words, machinery produces resonances in modes of higher order than those with which we shall be concerned. And those modes are likely to be very different – sufficiently so as seriously to call into question our (probably perfectly legitimate) assumption that a hull can be treated as a free–free beam whose section does not distort when it bends or twists.

To put the matter another way, while mechanical excitation of ships demands use of the basic principles of linear dynamics that we shall invoke, it is likely to throw emphasis on some very different aspects. In particular it is likely to raise some much more difficult problems of structural dynamics.

2 The dry hull

Let us grasp the situation,
Solve the complicated plot –
Quiet, calm deliberation
Disentangles every knot.
The Gondoliers

2.1 Types of deflection

A ship hull, at rest floating in flat calm water, is acted upon by gravity
and buoyancy forces with the result that it deflects, adopting a particular attitude and distortion. These deflections depend on the loading
of the ship, diurnal temperature variations, salinity and so forth. They
provide the configuration about which time-dependent fluctuations
take place.

The still water configuration can be of very considerable practical
importance. The handling characteristics of a ship that is trimmed by
the stern may be very different from those of the same ship trimmed
'bows down' since she will be more stable (in the directional sense) in
the former case. Again, rules have to be laid down governing the
distribution of ballast water in unloaded tankers since there is a very
real possibility otherwise, that dangerous bending moments will be set
up amidships. Nor are still water deflections of importance confined to
symmetric responses; mal-distribution of load can easily cause too
large an angle of heel or impart a dangerous twist in the hull, for
instance.

It should be remembered that the datum configuration from which
the still water deflections are reckoned is by no means easy to define.
The meaning assigned to the datum *attitude* is a matter of arbitrary
definition; usually the ship will be supposed to have its plane of
port-and-starboard symmetry vertical, and then decisions have to be
made about an acceptable waterline. Since it is not possible to con-
struct a strain-free hull, the distortions too have to be defined with
respect to an arbitrarily chosen datum. Thus axes $OXYZ$ are imagined
fixed to the hull with OXY horizontal and OZ pointing vertically
upward while the ship is dry and in its datum state.

The frame $OXYZ$ is to be thought of as proceeding uniformly in
the direction OX as the ship moves ahead. The deflections (both 'still
water' and time-dependent) occur relative to these 'equilibrium axes'.

In other words static and dynamic deflections, be they bodily motions or distortions, are to be regarded as 'parasitic'; the axes define the datum configuration. Remembering that henceforth we shall usually place bodily motions and distortions on the same general footing we may sum the matter up as in fig. 2.1.

2.2 Symmetric response of a ship

For reasons that have already been explained we shall treat a hull as a 'beam', using linear theory. First of all we shall discuss symmetric deflections. That is to say we shall now discuss bodily motions of heaving and pitching and distortions in which motions are confined to a vertical plane, preserving port and starboard symmetry. In this chapter the distortions will be assumed to conform to the simple Bernoulli–Euler theory of beam analysis; a less restrictive theory will be employed in chapter 3.

2.2.1 *Symmetric modes of a non-uniform beam*

Fig. 2.2 shows a short length of a non-uniform beam that occupies the region $0 \leqslant x \leqslant l$, the section $x = 0$ being at the stern. The shearing force and bending moment applied to it are V and M respectively. The upward force $Z(x, t)$ per unit length applied to the slice includes contributions from weight, buoyancy and all other fluid forces.

Motion of the slice of the beam in the vertical direction is governed by the equation

$$V_1 - V_2 + Z(x, t) \Delta x = \mu(x) \Delta x \, \ddot{w}(x, t),$$

where $\mu(x)$ is the mass per unit length of the beam and $w(x, t)$ is the upward deflection. Hence

$$\frac{\partial V}{\partial x} + Z(x, t) = \mu(x) \ddot{w}(x, t). \tag{2.1}$$

If rotatory inertia of the beam is neglected,

$$M_1 - M_2 + V \Delta x = 0,$$

Fig. 2.1. The forms of deflection with which we are concerned. One might, thus, speak of 'SWBM', 'WBM', 'PBM' referring to bending moment amidships in still water, due to waves and due to propeller excitation respectively.

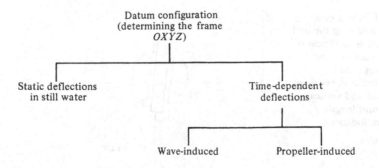

so that

$$V = -\frac{\partial M}{\partial x}. \tag{2.2}$$

According to elementary beam theory

$$M = EI(x)\frac{\partial^2 w(x, t)}{\partial x^2} + \beta(x)\frac{\partial^3 w(x, t)}{\partial x^2 \partial t}, \tag{2.3}$$

where $EI(x)$ is the flexural rigidity and $\beta(x)$ represents viscous structural damping. It follows that

$$V = -\frac{\partial}{\partial x}\left[EI(x)\frac{\partial^2 w(x, t)}{\partial x^2}\right] - \frac{\partial}{\partial x}\left[\beta(x)\frac{\partial^3 w(x, t)}{\partial x^2 \partial t}\right]. \tag{2.4}$$

If we denote partial differentiation with respect to x by a prime (and partial differentiation with respect to t by an 'overdot'), we may now write the equation of flexural motion as

$$\mu(x)\ddot{w}(x, t) + [EI(x)w''(x, t)]'' + [\beta(x)\dot{w}''(x, t)]'' = Z(x, t). \tag{2.5}$$

This is the equation of vertical symmetric bending of the dry hull, expressed in its most rudimentary form.

In the preceding section we noted that the total deflection is composed of a static ('still water') component and a dynamic component. Let the static deflection in still water be $\bar{w}(x)$, produced by a net loading $\bar{Z}(x)$, comprising distributed weight and buoyancy; evidently

$$[EI(x)\bar{w}''(x)]'' = \bar{Z}(x).$$

Clearly we may replace w and Z in the equation of motion by $(\bar{w} + w)$ and $(\bar{Z} + Z)$ without error; then $w(x, t)$ and $Z(x, t)$ must be reinterpreted as perturbations from the still water values.

In free vibration of the undamped dry beam, $Z(x, t) = 0 = \beta(x)$ for all positions x on the beam and at all times t so that the trial solution

$$w(x, t) = f(x) \sin \omega t$$

requires that

$$-\mu(x)\omega^2 f(x) + [EI(x)f''(x)]'' = 0,$$

where the prime now represents a total derivative with respect to x.

Fig. 2.2. Element of a beam representing the hull of a ship. The thickness of the slice is Δx and the conventions used for shearing force V, bending moment M and vertical load per unit length $Z(x, t)$ are indicated.

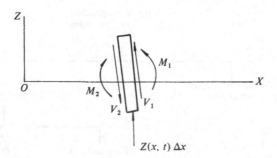

The function $f(x)$ has also to satisfy the boundary conditions

$$EI(x)f''(x) = 0 = [EI(x)f''(x)]',$$

when $x = 0$ or $x = l$.

Provided that $EI(x)$ and $\mu(x)$ are sufficiently well-behaved functions, the general solution of the ordinary differential equation is (see e.g. Ince, 1956)

$$f(x) = AF(x) + BG(x) + CH(x) + DJ(x),$$

where A, B, C and D are constants. If we now write

$$f^{II}(x) \equiv EI(x)\phi''(x),$$
$$f^{III}(x) \equiv [EI(x)\phi''(x)]',$$

the boundary conditions require that

$$AF^{II}(0) + BG^{II}(0) + CH^{II}(0) + DJ^{II}(0) = 0,$$
$$AF^{III}(0) + BG^{III}(0) + CH^{III}(0) + DJ^{III}(0) = 0,$$
$$AF^{II}(l) + BG^{II}(l) + CH^{II}(l) + DJ^{II}(l) = 0,$$
$$AF^{III}(l) + BG^{III}(l) + CH^{III}(l) + DJ^{III}(l) = 0.$$

A non-trivial solution for the constants A, B, C, D exists provided that the determinant

$$\begin{vmatrix} F^{II}(0) & G^{II}(0) & H^{II}(0) & J^{II}(0) \\ F^{III}(0) & G^{III}(0) & H^{III}(0) & J^{III}(0) \\ F^{II}(l) & G^{II}(l) & H^{II}(l) & J^{II}(l) \\ F^{III}(l) & G^{III}(l) & H^{III}(l) & J^{III}(l) \end{vmatrix} = 0.$$

This is an equation in ω and its roots are the natural frequencies $\omega_1, \omega_2, \ldots$ of the beam. Corresponding to the rth root ω_r, values can be found for A, B, C and D which are not all zero. Hence an infinite set of characteristic functions $f_r(x)$ or, as we can more conveniently write, $w_r(x)$, may be found.

The characteristic functions, or 'principal modes', evidently obey the relation

$$\frac{1}{\mu(x)}[EI(x)w_r''(x)]'' = \omega_r^2 w_r(x). \tag{2.6}$$

This is a feature of every function of the set. But each of the functions is essentially different from all the rest; they form an 'orthogonal set'.

Let $w_r(x)$ and $w_s(x)$ be two of the characteristic functions. We see that

$$\omega_r^2 \int_0^l \mu(x)w_r(x)w_s(x)\,dx = \int_0^l [EI(x)w_r''(x)]'' w_s(x)\,dx.$$

If the expression on the right-hand side is integrated twice by parts, it is found that

$$\int_0^l [EI(x)w_r''(x)]''w_s(x)\,\mathrm{d}x$$

$$= [\{EI(x)w_r''(x)\}'w_s(x)]_0^l - \int_0^l [EI(x)w_r''(x)]'w_s'(x)\,\mathrm{d}x$$

$$= [\{EI(x)w_r''(x)\}'w_s(x)]_0^l - [EI(x)w_r''(x)w_s'(x)]_0^l$$

$$+ \int_0^l EI(x)w_r''(x)w_s''(x)\,\mathrm{d}x.$$

But the two integrated terms vanish, as may be seen from the boundary conditions, and so

$$\omega_r^2 \int_0^l \mu(x)w_r(x)w_s(x)\,\mathrm{d}x = \int_0^l EI(x)w_r''(x)w_s''(x)\,\mathrm{d}x.$$

This result must remain valid if the subscripts r and s are interchanged, so

$$\omega_s^2 \int_0^l \mu(x)w_r(x)w_s(x)\,\mathrm{d}x = \int_0^l EI(x)w_r''(x)w_s''(x)\,\mathrm{d}x,$$

and if we now subtract this result from the last we see that, provided $\omega_r \neq \omega_s$,

$$\int_0^l \mu(x)w_r(x)w_s(x)\,\mathrm{d}x = 0 = \int_0^l EI(x)w_r''(x)w_s''(x)\,\mathrm{d}x$$

for $r \neq s$. These are the orthogonality relations.

It follows that the principal modes satisfy the equations

$$\int_0^l \mu(x)w_r(x)w_s(x)\,\mathrm{d}x = a_{rs}\delta_{rs}, \tag{2.7}$$

$$\int_0^l EI(x)w_r''(x)w_s''(x)\,\mathrm{d}x = \omega_r^2 a_{rs}\delta_{rs}, \tag{2.8}$$

where δ_{rs} is the Kronecker delta function defined by

$$\delta_{rs} = \begin{cases} 0 & (\text{for } r \neq s), \\ 1 & (\text{for } r = s). \end{cases}$$

Notice that the values for a_{rs} merely determine the scales of the characteristic functions.

These equations provide for the existence of 'rigid body modes'. Corresponding to $\omega_r = 0$, we find

$$w_r''(x) = 0.$$

That is to say the rigid mode shape can be described by a translation

$$w_0(x) = A$$

and a displacement

$$w_1(x) = B(C - x),$$

where A, B, C are constants. Furthermore, if orthogonality is to be preserved between $w_0(x)$ and $w_1(x)$ it is necessary that

$$\int_0^l \mu(x) AB(C-x)\,\mathrm{d}x = 0$$

and so either A or B (or both) vanish – which is not helpful – or

$$C = \frac{\int_0^l x\mu(x)\,\mathrm{d}x}{\int_0^l \mu(x)\,\mathrm{d}x}.$$

That is to say $C = \bar{x}$, the abscissa of the centre of mass of the beam.

The sequence of principal modes for the free–free non-uniform beam will have shapes like those sketched in fig. 2.3. The first two are associated with natural frequencies of zero value and are distortion-free. The exact forms of the remainder and their natural frequencies depend, of course, on the forms of $\mu(x)$ and $EI(x)$.

It will become clear that the characteristic functions of a ship are of great importance. They may be calculated, along with their corresponding natural frequencies, in several ways (e.g. see Bishop, Gladwell & Michaelson, 1965) using standard procedures. The first few functions of a destroyer are shown in fig. 2.4 for a particular distribution of loading.

2.2.2 Principal coordinates

According to a theorem due to Rayleigh (1894) *any* distortion of the beam may be expressed as an aggregate of distortions in its principal modes. That is to say, for a symmetric deflection,

$$w(x, t) = \sum_{r=0}^{\infty} p_r(t)w_r(x), \qquad (2.9)$$

where $p_r(t)$ is the rth 'principal coordinate'.

Fig. 2.3. A possible sequence of principal modes of symmetric distortion of a ship hull. In identifying the modes the convention adopted is to refer to the number of nodes in the modes. Notice that $\omega_0 = 0 = \omega_1$, while the remaining natural frequencies are non-zero and form a sequence of increasing values.

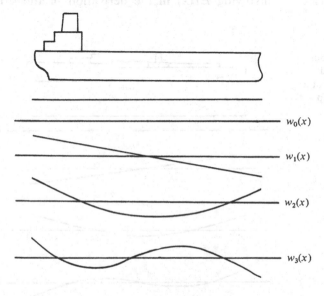

$w_0(x)$

$w_1(x)$

$w_2(x)$

$w_3(x)$

As a matter of approximation, this series may be curtailed. Thus we might admit p_0, p_1, p_2 and p_3 only in a practical analysis. It is a matter for consideration how many modes need to be contemplated in practice, but it is quite possible that just p_0, p_1 and p_2 will do for some purposes, if only in view of the uncertainties that beset the analyst (notably where fluid forces are concerned).

When the solution is substituted into the equation of motion (2.5) the series gives

$$\mu(x) \sum_{r=0}^{\infty} \ddot{p}_r(t) w_r(x) + \sum_{r=0}^{\infty} p_r(t)[EI(x)w_r''(x)]''$$
$$+ \sum_{r=0}^{\infty} \dot{p}_r(t)[\beta(x)w_r''(x)]'' = Z(x, t).$$

If we now multiply this equation by $w_s(x)$ and integrate with respect to x over the range $0 \leq x \leq l$ (i.e. over the length of the ship) the orthogonality conditions show that

$$\sum_{r=0}^{\infty} a_{rs}\delta_{rs}\ddot{p}_r + \sum_{r=0}^{\infty} a_{rs}\delta_{rs}\omega_r^2 p_r + \sum_{r=0}^{\infty} \dot{p}_r \int_0^l [\beta(x)w_r''(x)]'' w_s(x) \, \mathrm{d}x$$

$$= \int_0^l Z(x, t) w_s(x) \, \mathrm{d}x \qquad (s = 0, 1, 2, \ldots).$$

As we shall discover, all too little is known about hull damping, i.e. about the function $\beta(x)$. It is not unreasonable to assume, however, that its distribution is more or less the same as that of the flexural rigidity $EI(x)$. If this assumption is made, the integral on the left-hand side of the equation may be dealt with in the same manner as that involving $EI(x)$ in the derivation of the orthogonality condition.

Fig. 2.4. The first four principal modes and natural frequencies of a typical small warship – a destroyer.

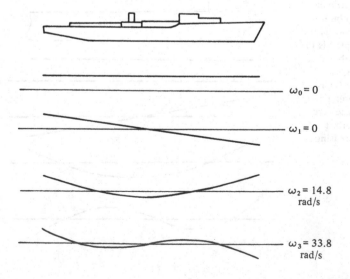

$\omega_0 = 0$

$\omega_1 = 0$

$\omega_2 = 14.8$ rad/s

$\omega_3 = 33.8$ rad/s

That is

$$\int_0^l [\beta(x)w_r''(x)]'' w_s(x)\, dx = \int_0^l \beta(x)w_r''(x)w_s''(x)\, dx.$$

We therefore conclude that

$$\sum_{r=0}^{\infty} a_{rs}\delta_{rs}\ddot{p}_r + \sum_{r=0}^{\infty} \dot{p}_r \int_0^l \beta(x)w_r''(x)w_s''(x)\, dx + \sum_{r=0}^{\infty} a_{rs}\delta_{rs}\omega_r^2 p_r$$

$$= \int_0^l Z(x,t)w_s(x)\, dx \qquad (s = 0, 1, 2, \ldots). \tag{2.10}$$

This is an infinite set of ordinary differential equations governing the principal coordinates $p_0, p_1, p_2, \ldots$.

Suppose, for the sake of argument, that we admit only the four principal modes of lowest order so that $r, s = 0, 1, 2, 3$. The equations of motion then reduce to a set of four simultaneous equations which may be expressed in the form

$$a_{00}\ddot{p}_0 = \int_0^l Z(x,t)w_0(x)\, dx \qquad \text{(corresponding to } s = 0),$$

$$a_{11}\ddot{p}_1 = \int_0^l Z(x,t)w_1(x)\, dx \qquad \text{(for } s = 1),$$

$$a_{22}\ddot{p}_2 + (b_{22}\dot{p}_2 + b_{23}\dot{p}_3) + c_{22}p_2$$

$$= \int_0^l Z(x,t)w_2(x)\, dx \qquad \text{(for } s = 2),$$

$$a_{33}\ddot{p}_3 + (b_{32}\dot{p}_2 + b_{33}\dot{p}_3) + c_{33}p_3$$

$$= \int_0^l Z(x,t)w_3(x)\, dx \qquad \text{(for } s = 3),$$

where we have employed the notation

$$b_{rs} = b_{sr} = \int_0^l \beta(x)w_r''(x)w_s''(x)\, dx. \tag{2.11}$$

It is convenient to write this set of equations in matrix form:

$$\mathbf{A}\ddot{\mathbf{p}} + \mathbf{B}\dot{\mathbf{p}} + \mathbf{C}\mathbf{p} = \mathbf{Z}(t). \tag{2.12}$$

Here, $\mathbf{A}$, $\mathbf{B}$ and $\mathbf{C}$ are the inertia, damping and stiffness matrices respectively, while $\mathbf{p}$ and $\mathbf{Z}$ are column vectors representing response and input loading. That is

$$\mathbf{A} = \begin{bmatrix} a_{00} & 0 & 0 & 0 \\ 0 & a_{11} & 0 & 0 \\ 0 & 0 & a_{22} & 0 \\ 0 & 0 & 0 & a_{33} \end{bmatrix}; \quad \mathbf{B} = \begin{bmatrix} 0 & 0 & 0 & 0 \\ 0 & 0 & 0 & 0 \\ 0 & 0 & b_{22} & b_{23} \\ 0 & 0 & b_{32} & b_{33} \end{bmatrix};$$

$$\mathbf{C} = \begin{bmatrix} 0 & 0 & 0 & 0 \\ 0 & 0 & 0 & 0 \\ 0 & 0 & c_{22} & 0 \\ 0 & 0 & 0 & c_{33} \end{bmatrix}; \quad \mathbf{p}(t) = \begin{bmatrix} p_0 \\ p_1 \\ p_2 \\ p_3 \end{bmatrix};$$

and

$$\mathbf{Z}(t) = \begin{bmatrix} Z_0(t) \\ Z_1(t) \\ Z_2(t) \\ Z_3(t) \end{bmatrix},$$

where

$$Z_s(t) \equiv \int_0^l Z(x, t) w_s(x) \, dx \qquad (s = 0, 1, 2, 3).$$

The principal coordinates fall naturally into two groups:

(a) p_0, p_1 refer to 'rigid body' modes, and

(b) p_2, p_3 relate to distortion modes.

It is convenient to partition the matrix $\mathbf{p}$ therefore into the portions

$$\mathbf{p}(t) = \begin{bmatrix} p_0 \\ p_1 \\ \hline p_2 \\ p_3 \end{bmatrix} = \begin{bmatrix} \mathbf{p}_R \\ \mathbf{p}_D \end{bmatrix}.$$

If we assign the subscripts R (for 'rigid') and D (for 'distortion') in this manner, we have

$$\begin{bmatrix} \mathbf{A}_{RR} & 0 \\ 0 & \mathbf{A}_{DD} \end{bmatrix} \begin{bmatrix} \ddot{\mathbf{p}}_R \\ \ddot{\mathbf{p}}_D \end{bmatrix} + \begin{bmatrix} 0 & 0 \\ 0 & \mathbf{B}_{DD} \end{bmatrix} \begin{bmatrix} \dot{\mathbf{p}}_R \\ \dot{\mathbf{p}}_D \end{bmatrix} + \begin{bmatrix} 0 & 0 \\ 0 & \mathbf{C}_{DD} \end{bmatrix} \begin{bmatrix} \mathbf{p}_R \\ \mathbf{p}_D \end{bmatrix} = \begin{bmatrix} \mathbf{Z}_R \\ \mathbf{Z}_D \end{bmatrix}.$$

Here $\mathbf{A}_{RR}, \mathbf{A}_{DD}$ are diagonal matrices,

$\quad\quad\mathbf{B}_{DD}$ is a square symmetric matrix,

$\quad\quad\mathbf{C}_{DD}$ is a diagonal matrix.

It will be seen that the elements in the diagonal matrices are all positive, while $\mathbf{B}_{DD}$ is a positive definite matrix (this latter feature being explained e.g. by Bishop *et al.* 1965).

Had there been more distortion modes than two admitted into the analysis, the only difference that would be made thereby is that $\mathbf{p}_D$ would be of order greater than two. If a total of n principal coordinates are employed,

$\mathbf{A}_{RR}$ is of order 2×2, while

$\mathbf{A}_{DD}, \mathbf{B}_{DD}$ and $\mathbf{C}_{DD}$ are of order $(n-2) \times (n-2)$.

Indeed if interest centres only on motion in the rigid body modes and it is felt that this will be virtually unaffected by distortions, then it is advantageous to use the equation in the form

$$\mathbf{A}_{RR}\ddot{\mathbf{p}}_R = \mathbf{Z}_R, \tag{2.13}$$

with no distortion modes whatever admitted.

Use of Lagrange's equation

Had we started by noting that, whatever the forms of the functions $w_r(x)$, there must be a set of coordinates $p_r(t)$ which may be used as generalised coordinates, then we could have employed Lagrange's equation to set up the equations of motion. Although we shall not adopt this more abstract approach subsequently, this is an important matter and we shall therefore examine it briefly. (Such an approach appears desirable when discussing complex structures such as oil rigs.) Lagrange's equation will be used in the form

$$\frac{\mathrm{d}}{\mathrm{d}t}\left(\frac{\partial T}{\partial \dot{p}_s}\right) - \frac{\partial T}{\partial p_s} + \frac{\partial U}{\partial p_s} = P_s \qquad (s = 0, 1, 2, \ldots),$$

where the kinetic energy T and potential energy U relate to the ship *in vacuo*. All fluid actions, structural damping and external forces are accounted for in the generalised forces P_s.

The kinetic energy of the hull may be expressed as

$$T = \tfrac{1}{2} \int_0^l \mu(x) \left[\frac{\partial w(x, t)}{\partial t}\right]^2 \mathrm{d}x.$$

If we now introduce the generalised coordinates through the series representation of $w(x, t)$, this becomes

$$T = \tfrac{1}{2} \int_0^l \mu(x) \left[\sum_{r=0}^{\infty} \dot{p}_r(t) w_r(x)\right]^2 \mathrm{d}x.$$

The potential energy is

$$U = \tfrac{1}{2} \int_0^l EI(x) \left[\frac{\partial^2 w(x, t)}{\partial x^2}\right]^2 \mathrm{d}x,$$

which becomes

$$U = \tfrac{1}{2} \int_0^l EI(x) \left[\sum_{r=0}^{\infty} p_r(t) w_r''(x)\right]^2 \mathrm{d}x.$$

The left-hand side of the Lagrange equation is therefore

$$\frac{\mathrm{d}}{\mathrm{d}t}\left(\frac{\partial T}{\partial \dot{p}_s}\right) - \frac{\partial T}{\partial p_s} + \frac{\partial U}{\partial p_s} = \sum_{r=0}^{\infty} \int_0^l (\mu w_r w_s \ddot{p}_r + EI w_r'' w_s'' p_r) \, \mathrm{d}x,$$

for $s = 0, 1, 2, \ldots$, and since we have already established the orthogonality relations we may use them to show that this reduces to

$$a_{ss}\ddot{p}_s + \omega_s^2 a_{ss} p_s \qquad (s = 0, 1, 2, \ldots).$$

Consider now the right-hand side of the Lagrange equation, the quantity P_s. The total work done during a virtual displacement $\delta p_s w_s(x)$ is

$$-\int_0^l \beta(x) \frac{\partial^3 w(x, t)}{\partial t \partial x^2} \delta p_s \frac{\mathrm{d}^2 w_s(x)}{\mathrm{d}x^2} \, \mathrm{d}x \qquad \text{by damping forces,}$$

and

$$\int_0^l Z(x, t) \delta p_s w_s(x) \, \mathrm{d}x \qquad \text{by the external loading.}$$

Denoting the virtual work by δW we therefore find that

$$P_s = \frac{\delta W}{\delta p_s} = -\int_0^l \beta(x)(\sum_{r=0}^{\infty} \dot{p}_r w_r'')w_s'' \, dx$$

$$+ \int_0^l Z(x,t)w_s(x) \, dx \qquad (s = 0, 1, 2, \ldots).$$

That is to say

$$P_s = -\sum_{r=0}^{\infty} b_{rs}\dot{p}_r + Z_s(t) \qquad (s = 0, 1, 2, \ldots),$$

where, it will be remembered, $b_{00} = 0 = b_{11} = b_{01} = b_{10}$. When these results are assembled they give

$$a_{ss}\ddot{p}_s + \omega_s^2 a_{ss} p_s = -\sum_{r=0}^{\infty} b_{rs}\dot{p}_r + Z_s(t) \qquad (s = 0, 1, 2, \ldots),$$

which is what we found previously. If the number of principal modes admitted is limited to some finite number we can adopt the matrix formulation as explained previously.

2.2.3   Still water response in symmetric modes
The loading $Z(x, t)$ is due to hull weight and to fluid actions of all sorts. There will therefore be deflections at $p_0, p_1, p_2, \ldots$ when the ship floats in still water. If the loading is then $\bar{Z}(x)$, the static deflection at the sth principal coordinate is $\bar{p}_s$, where

$$a_{ss}\omega_s^2 \bar{p}_s = \int_0^l \bar{Z}(x)w_s(x) \, dx \qquad (s = 0, 1, 2, \ldots).$$

The rigid body modes are such that $\omega_0 = 0 = \omega_1$, and so

$$\int_0^l \bar{Z}(x) \, dx = 0 = \int_0^l \bar{Z}(x)(x - \bar{x}) \, dx. \qquad (2.14)$$

In other words the hull settles in the water in such a manner that

(a) the weight and buoyancy forces are equal, and
(b) the total moment of the weight and buoyancy forces about the centre of gravity is nil.

Evidently the values of $\bar{p}_0$ and $\bar{p}_1$ are left open, reflecting the fact that zero values of p_0 and p_1 are determined by an arbitrary choice. This is not a simple thing to visualise so it will be of interest to examine the point more closely.

Imagine a ship's hull lying on its side, supported on suitable smooth surfaces with its plane of port and starboard symmetry horizontal. The equilibrium axes $OXYZ$ may then be attached to the hull and this requires no arbitrary choices to be made as regards distortion; for although there would in general be stresses within the hull, they are self-equilibrating. The plane OXZ is to be made to coincide with the plane of symmetry, but then we have two choices open to us:

(a) we might choose to place the origin O at the intended waterline of the ship

(b) the axis OZ might be chosen to point vertically upwards when the ship is in its intended attitude.

The axes are thus attached to the hull as in fig. 2.5(a). When the hull floats in still water (fig. 2.5(b)), it assumes some distorted form, thereby giving non-zero values of $\bar{p}_2, \bar{p}_3, \ldots$. But even if it could be 'pulled straight' by some means, so that the distortion were removed, there is still no assurance that O will lie at the intended waterline or that OZ will be vertical since $\bar{Z}(x)$ may be such as to decree otherwise. Hence there are deflections at $\bar{p}_0$ and $\bar{p}_1$ as well.

To sum up, then, there will be still water deflections at $\bar{p}_0$ and $\bar{p}_1$ whose values are of interest to the marine surveyor and the ship's master. They may affect the safety and the handling characteristics of the vessel, but they are not associated with hull stresses. In addition, there are still water distortions in the distortion modes given by

$$\bar{p}_s = \frac{1}{a_{ss}\omega_s^2} \int_0^l \bar{Z}(x) w_s(x) \, dx \qquad (s = 2, 3, \ldots). \qquad (2.15)$$

These still water responses provide the constant stresses to which stresses caused by rough water and propeller excitation have to be added, and they may be found by considering only the weight and hydrostatic buoyancy forces in the total distribution $Z(x, t)$.

To find what remains, i.e., the time-dependent contributions to the $p_s(t)$, we must first specify what remains of $Z(x, t)$. That is to

Fig. 2.5. The attachment of equilibrium axes to a hull. In (a) we imagine the hull to be dry and smoothly supported on its side so that there is no fluid or weight loading parallel to the fore and aft plane of symmetry. (Strictly speaking distortion in the athwartships direction should also be suppressed by the supports.) Axes $OXYZ$ may then be attached with arbitrary choices to be made as regards the level of O and the orientation of OZ. When the ship floats in still water it distorts as in (b) while the level of O and orientation of OZ may not be as originally intended.

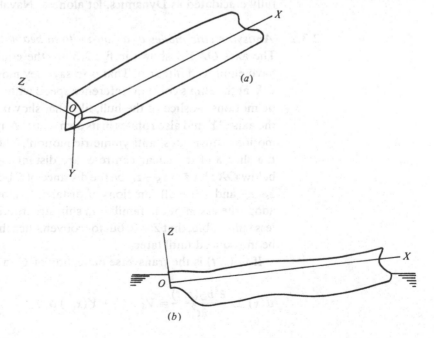

say, we must regard the equations of motion as being those governing the perturbations of the $p_s(t)$ from the still water values, taking the departures of the fluid actions from the hydrostatic distribution as the loading. In particular, hull weight now ceases to enter the calculations, though of course hull mass does not.

The task which now confronts us therefore is the estimation of the time-dependent fluid actions – or, rather, the difference between them and hydrostatic loading.

2.3 Antisymmetric response of a boxlike ship

Whereas longitudinal strength is determined largely by ability to withstand symmetric bending distortions, transverse strength is determined mainly by the combination of horizontal bending and twisting of the hull; consequently this problem is more complex than the former one. Furthermore antisymmetric response in rigid body modes is a matter of concern to those responsible for steering the ship – be they coxwains or the designers of autopilots.

Progress can most easily be made with ships of a more or less conventional 'boxlike' configuration, such as tankers and other vessels with relatively small deck openings. A container ship (whose main deck opening can extend over more than 80% of the beam) presents special difficulties which we shall examine in section 2.4.

It should perhaps be said at the outset that the theory of antisymmetric vibration performed by a non-uniform beam *in vacuo* is not nearly as highly developed as the corresponding theory of symmetric vibration. In attempting to formulate a theory of transverse strength, then, we shall encounter some fresh problems which have not been fully elucidated as Dynamics, let alone as Naval Architecture.

2.3.1 *Antisymmetric modes of a non-uniform beam*

The axes $OXYZ$ shown in fig. 2.6 are the equilibrium axes that we have employed hitherto. That is to say they move parallel to the axis OX at the ship's constant reference speed without performing parasitic motions. A slice of the hull, like that shown, translates parallel to the axis OY and also rotates in its own plane. Any combination of such motions constitutes 'antisymmetric motion'. The centre of mass C of the slice and the shear centre S are distant z_C and z_S respectively below OX; let $\bar{z} = z_S - z_C$ be the distance of S below C. The quantities z_S, z_C and $\bar{z}$ are all functions of distance x along the ship. We shall adopt the assumption, familiar in ship structural theory yet nevertheless vulnerable, that $\bar{z} = 0$; but for convenience this assumption will not be introduced until later.

If $v_C(x, t)$ is the transverse deflection of C in the direction OY,

$$\mu(x)\,\Delta x\,\frac{\partial^2 v_C(x, t)}{\partial t^2} = V_1 - V_2 + Y(x, t)\,\Delta x,$$

where $Y(x, t)$ is the horizontal hydrodynamic force per unit length and we now use V to represent the horizontal shearing force. Reverting to the 'overdot' and 'prime' notation, therefore, we find that

$$\mu(x)\ddot{v}_C(x, t) = V' + Y(x, t). \tag{2.16}$$

If rotatory inertia is ignored, a rotation of the slice of the hull about an axis parallel to OZ requires that

$$M_1 - M_2 + (V_1 + V_2)\frac{\Delta x}{2} = 0$$

so that

$$V = -M', \tag{2.17}$$

M being the bending moment.

Next consider rotation of the slice in its own plane and let $I_C(x)$ be the moment of inertia of the hull per unit length about the axis through C parallel to OX. If $\phi(x, t)$ is the angular rotation of the slice about an axis parallel to OX, then

$$I_C(x) \Delta x \frac{\partial^2 \phi(x, t)}{\partial t^2} = T_1 - T_2 + (V_1 - V_2)(z_S - z_C) + K(x, t)\,\Delta x,$$

T being the twisting moment and $K(x, t)$ the hydrodynamic rolling moment per unit length. Therefore

$$I_C(x)\ddot{\phi}(x, t) = T' + \bar{z}V' + K(x, t). \tag{2.18}$$

If the term V' is eliminated by substitution from the first equation of motion and the kinematic relation

$$v(x, t) = v_C(x, t) + \bar{z}(x)\phi(x, t) \tag{2.19}$$

is introduced, $v(x, t)$ being the transverse deflection of S, then it is found that

$$I_S(x)\ddot{\phi}(x, t) - \mu(x)\bar{z}(x)\ddot{v}(x, t) = T' - \bar{z}(x)Y(x, t) + K(x, t). \tag{2.20}$$

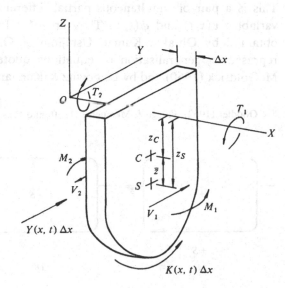

Fig. 2.6. A slice of a hull of thickness Δx. The conventions used for the bending moment M, twisting moment T and shear force V are shown. The quantity $Y(x, t)$ represents fluid loading per unit length, and $K(x, t)$ is the rolling moment per unit length which is due to gravity loading (if any) as well as fluid actions.

In this equation, $I_S(x)$ is the moment of inertia per unit length about an axis that is parallel to OX and which passes through S; it is given by

$$I_S(x) = I_C(x) + \mu(x)[\bar{z}(x)]^2. \tag{2.21}$$

For small transverse bending we may write

$$M = EI(x)v''(x, t) + \beta(x)\dot{v}''(x, t), \tag{2.22}$$

where $EI(x)$ is the flexural rigidity of horizontal bending and $\beta(x)$ now represents viscous damping of structural origin in such motion. Analogously the twisting moment is of the form

$$T = C(x)\phi'(x, t) + \Gamma(x)\dot{\phi}'(x, t), \tag{2.23}$$

where $C(x)$ is the torsional rigidity and $\Gamma(x)$ represents viscous damping of hull twisting, again of structural origin. It is common to relate the torque $T(x)$ to the twist per unit distance along the hull in this way, though this is in fact an approximation because we are here disregarding the warping of cross-sections. The severity of this assumption in ship dynamics is not entirely clear, though it is probably not great.[†] Note, however, that this assumption is quite untenable for a ship of *open* section (such as a container ship). The distinction between the open and closed forms of hull is illustrated in fig. 2.7(a) and (b).

These expressions for M and T allow us to write the equations of motion as

$$\mu(x)\ddot{v}(x, t) - \mu(x)\bar{z}(x)\ddot{\phi}(x, t)$$
$$= Y(x, t) - [EI(x)v''(x, t)]'' - [\beta(x)\dot{v}''(x, t)]'', \tag{2.24}$$

and

$$I_S(x)\ddot{\phi}(x, t) - \mu(x)\bar{z}(x)\ddot{v}(x, t)$$
$$= -\bar{z}(x)Y(x, t) + K(x, t) + [C(x)\phi'(x, t)]' + [\Gamma(x)\dot{\phi}'(x, t)]'. \tag{2.25}$$

This is a pair of simultaneous partial differential equations in the variables $v(x, t)$ and $\phi(x, t)$. They are of a form similar to those obtained by Ohtaka, Kumai, Ushijima & Ohji (1967) and they represent a generalisation of equations quoted, for example, by McGoldrick (1960) and by Leibowitz & Kennard (1961).

† See Goodier (1962), Flügge & Marguerre (1950) and Reissner (1956).

Fig. 2.7. Two distinct types of hull section may be identified. The one shown in (a) is 'open' and is that of a container ship; the 'closed' section (b) is boxlike, and is of the type discussed in this section.

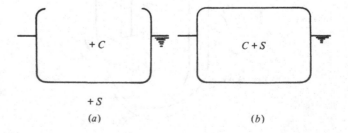

$+C$

$+S$

(a)

$C+S$

(b)

An essential feature of these equations of motion is that they are 'coupled'. That is to say $v(x, t)$ and $\phi(x, t)$ are not independent of each other. The quantities $Y(x, t)$ and $K(x, t)$ represent fluid actions and they alone can be expected to couple the bending and twisting motions. If, however, $Y(x, t) = 0 = K(x, t)$ because the ship is assumed to vibrate freely *in vacuo*, then the coupling depends only on $\bar{z}(x)$ and it will be seen that if S coincides with C, so that $\bar{z}(x) = 0$, the two motion variables become independent of each other. For ships having a 'closed' (or 'boxlike') form, such as tankers, the separation $\bar{z}$ will not be great. Wereldsma (1972), among others, has assumed for these latter types of ship that $\bar{z}(x) = 0$ and, as we shall see, this assumption simplifies matters greatly.

A ship of 'open' section, such as a container ship, will have the shear centre and centre of gravity separated so that $\bar{z}(x) \neq 0$. (Moreover a warping term would have to be introduced in the expression relating $T(x, t)$ to $\phi(x, t)$ for such a vessel.) It is this case which is taken up in section 2.4 of this book.

Suppose that a boxlike hull performs free undamped oscillations *in vacuo*, so that $Y(x, t) = 0 = K(x, t)$ and $\beta(x) = 0 = \Gamma(x)$. The possible motions take place in the principal modes and we may therefore seek solutions of the form

$$v(x, t) = v_r(x) \sin \omega_r t, \qquad \phi(x, t) = \phi_r(x) \sin \Omega_r t,$$

where ω_r and Ω_r are the natural frequencies of antisymmetric motion. The equations of motion then become

$$\omega_r^2 \mu(x) v_r(x) = [EI(x) v_r''(x)]'',$$
$$\Omega_r^2 I_C(x) \phi_r(x) = -[C(x) \phi_r'(x)]'.$$

Consider first the functions $v_r(x)$. They are subject to the boundary conditions

$$EI(x) v_r''(x) = 0 = [EI(x) v_r''(x)]'$$

at $x = 0$ and $x = l$, and by appealing to the theory of section 2.2.1 we see that the modes of a non-uniform beam satisfy orthogonality conditions

$$\left.\begin{aligned}
\int_0^l \mu(x) v_r(x) v_s(x) \, \mathrm{d}x &= a_{rs} \delta_{rs}, \\
\int_0^l EI(x) v_r''(x) v_s''(x) \, \mathrm{d}x &= \omega_r^2 a_{rs} \delta_{rs} = c_{rs} \delta_{rs}.
\end{aligned}\right\} \tag{2.26}$$

The constants a_{rr} are determined partly by the scales chosen for the functions $v_r(x)$.

The modal torsion functions $\phi_r(x)$ have to satisfy the boundary condition

$$C(x) \phi_r'(x) = 0$$

at $x = 0$ and $x = l$. From this requirement we may derive the orthogonality relations as follows. Multiply the equation governing $\phi_r(x)$ by

$\phi_s(x)$ and integrate over the ship's length by the parts method; this gives

$$\Omega_r^2 \int_0^l I_C(x)\phi_r(x)\phi_s(x)\,\mathrm{d}x = \int_0^l C(x)\phi_r'(x)\phi_s'(x)\,\mathrm{d}x.$$

(The boundary conditions ensure that the integrated term vanishes.) Interchanging r and s and subtracting the result from the previous equation we find that

$$(\Omega_r^2 - \Omega_s^2) \int_0^l I_C(x)\phi_r(x)\phi_s(x)\,\mathrm{d}x = 0.$$

The orthogonality conditions now follow from this last result:

$$\left.\begin{aligned}
\int_0^l I_C(x)\phi_r(x)\phi_s(x)\,\mathrm{d}x &= A_{rs}\delta_{rs}, \\
\int_0^l C(x)\phi_r'(x)\phi_s'(x)\,\mathrm{d}x &= \Omega_r^2 A_{rs}\delta_{rs}.
\end{aligned}\right\} \tag{2.27}$$

The value of the constant A_{rs} is determined by the scale of the functions $\phi_r(x)$.

To return to the equations governing the functions $v_r(x)$ and $\phi_r(x)$ we may note that they provide for the existence of 'rigid body modes'. Corresponding to $\omega_r = 0$, we find

$$v_r''(x) = 0.$$

The rigid mode shapes are therefore

$$v_0(x) = A,$$
$$v_1(x) = B(\bar{x} - x),$$

as before.

In the same way, if $\Omega_r = 0$ there is a requirement that

$$\phi_r'(x) = 0,$$

from which we see that the rigid rotation

$$\phi_0(x) = D$$

is a mode of zero natural frequency. Notice that the rigid rotation $\phi_0(x)$ is associated with a distribution $v(x) = z_C(x)\phi_0(x)$.

2.3.2   *Principal coordinates in antisymmetric motion of a boxlike ship*

Any antisymmetric deflection of a ship may be expressed as the sum of deflections in its antisymmetric principal modes. That is to say, for a boxlike ship we may write

$$v(x, t) = \sum_{r=0}^{\infty} p_r(t)v_r(x), \qquad \phi(x, t) = \sum_{i=0}^{\infty} q_i(t)\phi_i(x),$$

where $p_r(t)$ and $q_i(t)$ are the rth and ith principal coordinates for the two independent sets of modes. If these expressions are substituted into the uncoupled equations

$$\mu(x)\ddot{v}(x,t)+[\beta(x)\dot{v}''(x,t)]''+[EI(x)v''(x,t)]''=Y(x,t),$$
$$I_S(x)\ddot{\phi}(x,t)-[\Gamma(x)\dot{\phi}'(x,t)]'-[C(x)\phi'(x,t)]'=K(x,t),$$

then it is found that

$$\mu\sum_{r=0}^{\infty}\ddot{p}_r v_r+\sum_{r=0}^{\infty}\dot{p}_r(\beta v_r'')''+\sum_{r=0}^{\infty}p_r(EIv_r'')''=Y(x,t),$$

$$I_S\sum_{i=0}^{\infty}\ddot{q}_i\phi_i-\sum_{i=0}^{\infty}\dot{q}_i(\Gamma\phi_i')'-\sum_{i=0}^{\infty}q_i(C\phi_i')'=K(x,t).$$

If the first of these equations is multiplied by $v_s(x)$ and the second is multiplied by $\phi_j(x)$ and both are integrated with respect to x over the range $0 \le x \le l$, the orthogonality conditions give

$$\sum_{r=0}^{\infty}a_{rs}\delta_{rs}\ddot{p}_r+\sum_{r=0}^{\infty}\dot{p}_r\int_0^l(\beta v_r'')''v_s\,\mathrm{d}x+\sum_{r=0}^{\infty}a_{rs}\delta_{rs}\omega_r^2 p_r$$

$$=\int_0^l Y(x,t)v_s\,\mathrm{d}x \qquad (s=0,1,2,\dots)$$

and

$$\sum_{i=0}^{\infty}A_{ij}\delta_{ij}\ddot{q}_i-\sum_{i=0}^{\infty}\dot{q}_i\int_0^l(\Gamma\phi_i')'\phi_j\,\mathrm{d}x-\sum_{i=0}^{\infty}q_i\int_0^l(C\phi_i')'\phi_j\,\mathrm{d}x$$

$$=\int_0^l K(x,t)\phi_j\,\mathrm{d}x \qquad (j=0,1,2,\dots).$$

We have already seen, however, that

$$\int_0^l(C\phi_i')'\phi_j\,\mathrm{d}x=-\int_0^l C\phi_i'\phi_j'\,\mathrm{d}x=-\Omega_i^2\int_0^l I_C\phi_i\phi_j\,\mathrm{d}x=-A_{ij}\delta_{ij}\Omega_i^2.$$

If, as with the symmetric modes, we assume that the damping functions $\beta(x)$ and $\Gamma(x)$ are of the same general forms as the appropriate rigidities ($EI(x)$ and $C(x)$ respectively) we may integrate the damping terms by parts to simplify them. That is, we may write

$$\int_0^l(\beta v_r'')''v_s\,\mathrm{d}x=\int_0^l \beta v_r''v_s''\,\mathrm{d}x=b_{rs}$$

and

$$-\int_0^l(\Gamma\phi_i')'\phi_j\,\mathrm{d}x=\int_0^l\Gamma\phi_i'\phi_j'\,\mathrm{d}x=B_{ij},$$

where the $b_{rs}=b_{sr}$ and $B_{ij}=B_{ji}$ are damping coefficients. Thus the equations governing the principal coordinates become

$$a_{ss}\ddot{p}_s+\sum_{r=0}^{\infty}b_{rs}\dot{p}_r+a_{ss}\omega_s^2 p_s$$

$$=\int_0^l Y(x,t)v_s\,\mathrm{d}x \qquad (s=0,1,2,\dots) \tag{2.28}$$

and

$$A_{jj}\ddot{q}_j + \sum_{i=0}^{\infty} B_{ij}\dot{q}_i + A_{jj}\Omega_j^2 q_j$$

$$= \int_0^l K(x,t)\phi_j \, dx \qquad (j = 0, 1, 2, \dots). \qquad (2.29)$$

This is an infinite set of simultaneous ordinary differential equations.

Just as with the deflection in symmetric modes we can limit the number of principal modes admitted in an analysis. If this is done it becomes possible to employ the matrix formulation. And, as before, the matrices may be partitioned so as to separate rigid from distortion modes. It must be remembered, however, that there is but one rigid body mode of rotation.

2.3.3 *Still water response of a boxlike ship in its antisymmetric modes*
The distributed force $Y(x, t)$ is due to fluid actions, since gravity loading does not occur in the horizontal direction. In still water, the symmetry of the ship ensures that there is no net fluid loading at any section and so $Y(x, t) = \bar{Y}(x) = 0$. It follows that there are no still water responses $\bar{p}_s$ at those coordinates p_s which are associated with hull distortion. That is to say, although the relevant equation of motion suggests that

$$a_{ss}\omega_s^2 \bar{p}_s = \int_0^l \bar{Y}(x)v_s \, dx \qquad (s = 0, 1, 2, \dots),$$

the evanescence of the integral term ensures that $\bar{p}_s = 0$ when $s = 2, 3, \dots$. The exceptions occur with the rigid mode deflections $\bar{p}_0$ and $\bar{p}_1$, which are indeterminate because $\omega_0 = 0 = \omega_1$. Again we may conclude that this is due to the element of choice in attaching the axes to the ship.

Turning now to the rotation equation, we see that the still water response $\bar{q}_j$ at the jth principal coordinate is given by

$$A_{jj}\Omega_j^2 \bar{q}_j = \int_0^l \bar{K}(x)\phi_j \, dx \qquad (j = 0, 1, 2, \dots).$$

Now $\Omega_0 = 0$ and, since ϕ_0 is simply a constant,

$$\int_0^l \bar{K}(x) \, dx = 0. \qquad (2.30)$$

The hull will so adjust its attitude as to satisfy this relationship. If the roll angle is zero when the ship is upright a value of $\bar{q}_0$ will thus be defined with its associated athwartships displacement $\bar{q}_0 z_C(x)\phi_0(x)$.

The moment $\bar{K}(x)$ is due to hydrostatic fluid actions and gravity loading. If the latter is symmetrically disposed, port and starboard, along the ship's length then so will be the former and none of the still water responses will be significant. With a boxlike ship, loaded normally, this is likely to be a good approximation to the truth. But if the loading is not well distributed then we see that $\bar{q}_0$ defines a bias, or an angle of 'loll', and $\bar{q}_s$ is given by

$$\bar{q}_s = \frac{1}{A_{ss}\Omega_s^2} \int_0^l \bar{K}(x)\phi_s(x)\,\mathrm{d}x \qquad (s = 1, 2, 3, \ldots). \tag{2.31}$$

2.4   Antisymmetric response of ships with large deck openings

In the treatment of antisymmetric response in section 2.3 attention was drawn to an important special case that raises difficulties. This latter case is exemplified by the container ship whose large deck openings will cause distortions of horizontal bending and twisting to be associated with significant warping. The warping has the effect of greatly reducing the torsional rigidity of the hull† and this results in close coupling of the bending and twisting distortions. An aerial photograph of a small container ship is shown in fig. 2.8; it is worth noting, however, that container ships are by no means the only vessels with large deck openings. Seagoing barges and suction dredgers, for instance, also have this feature.

As in section 2.3 we shall first examine the free vibration that the dry hull could perform *in vacuo*. Unfortunately comparatively little is known about the antisymmetric vibration of a thin-walled non-uniform beam having an open section (cf. fig. 2.7.) Most published work in this field refers to aircraft structures and, for them, it is possible to make practical estimates of modes and natural frequencies (e.g. see Gere, 1954) on the basis of certain major assumptions; thus it is usual to assume that an 'axis of shear centres' can be identified. Unfortunately a vessel like a container ship may not present sufficiently simple problems for the usual assumptions to be tenable since it is, in effect, a beam that is not merely subject to non-uniform torsion but is itself violently non-uniform. (The large deck opening does not extend over the whole ship length as in a 'Red Indian' canoe for instance.)

While examining a strategy for estimating stresses in a container ship, then, we must remember that a need exists here for research of a quite basic nature. The main difficulty is not with the dry hull, however, for the principal modes and natural frequencies do not present difficulties when finite element techniques are used. The problem is how to achieve an adequate representation of the hull which is sufficiently simple to permit estimates to be made of the fluid actions.

2.4.1   *Antisymmetric modes of a non-uniform beam of open section*

It will be helpful to refer to fig. 2.9, rather than to fig. 2.6, though the meanings of the various symbols remain exactly the same. The essential thing is that the vertical distance between C and S is now large and that S may lie below the keel.

---

† This can be readily illustrated by twisting a cardboard shoe box and observing the enormous difference that removing the top makes.

Fig. 2.8. An aerial view of a typical small container ship, the MS *Jersey Fisher*, showing the extensive deck opening. (Photograph by courtesy of the Royal Navy.)

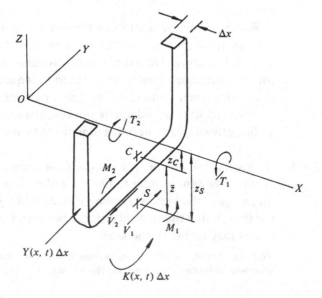

Fig. 2.9. Transverse slice of the hull showing the positions of the centre of gravity C and shear centre S, and indicating the sign conventions for the horizontal shearing force V, horizontal bending moment M and twisting moment T. The fluid sway force is $Y(x, t)$ per unit length and the rolling moment is $K(x, t)$ per unit length.

The position of the shear centre S depends only on the properties of the structure whereas the position of the centre of mass C depends partly on the loading condition of the ship. Fig. 2.10 is taken from a paper by Meek, Adams, Chapman, Reibel & Wieske (1972) and shows how the location of the shear centre varies along the length of a container ship of the *Liverpool Bay* class.

The theory presented in section 2.3.1 remains valid here until we come to specify the twisting moment. As we have already observed, the torsional rigidity is substantially altered by the admission of warping and we now have

$$T = C(x)\phi'(x, t) - [C_1(x)\phi''(x, t)]' + \Gamma(x)\dot{\phi}'(x, t). \qquad (2.32)$$

The new term introduces the warping rigidity $C_1(x)$ that is discussed by Timoshenko (1945) and Vlasov (1961) for a uniform beam. This means that, whereas the bending equation remains the same as before, the equation governing rotation becomes

$$I_S(x)\ddot{\phi}(x, t) - \mu(x)\bar{z}(x)\ddot{v}(x, t) - \{C(x)\phi'(x, t) - [C_1(x)\phi''(x, t)]'\}'$$
$$- [\Gamma(x)\dot{\phi}'(x, t)]' = -\bar{z}(x)Y(x, t) + K(x, t). \qquad (2.33)$$

The coupled equations of motion have associated boundary conditions and in establishing what they are we shall make use of results given by Timoshenko (1945). Warping occurs as the hull twists so that plane cross-sections distort as a result of deflections $u(x, t)$ in the longitudinal direction, i.e. perpendicular to the cross-sections.

The warping displacement from the average at any section is

$$u(x, t; s) = \phi'(x, t)(D - \alpha),$$

where $D(x)$ is a constant for a given cross-section, $\alpha(x, s)$ is a quantity which varies over a given cross-section with s the distance from a traction-free edge measured round the thin wall at the section x. If warping is restrained, the longitudinal stress is

$$\sigma_x(x, t; s) = Eu'(x, t) = E(D - \alpha)\phi''(x, t),$$

where it is assumed that the longitudinal rate of change in the quantity $(D - \alpha)$ is negligible. Thus the warping and longitudinal stress at any

Fig. 2.10. Variation of the location of the shear centre along the length of a *Liverpool Bay* class container ship, measured with respect to the keel (see Meek *et al.*, 1972). The overall length of the ship is 289.55 m.

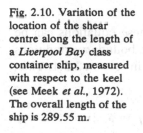
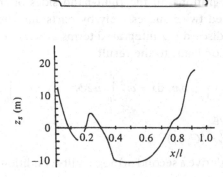

point at any cross-section are directly proportional to $\phi'(x, t)$ and $\phi''(x, t)$ respectively.

If it is assumed that, in antisymmetric distortion, the hull acts as a free–free beam, there must be no torque and no warping restraint at the two ends. The boundary conditions are thus

$$
\left.\begin{aligned}
&T = C(x)\phi'(x, t) - [C_1(x)\phi''(x, t)]' = 0, \\
&\sigma_x = 0 = \phi''(x, t), \\
&\text{at } x = 0 \text{ and } x = l.
\end{aligned}\right\}
\tag{2.34}
$$

During free vibration of the undamped hull *in vacuo*, $Y(x, t) = 0 = K(x, t)$ and $\beta(x) = 0 = \Gamma(x)$ for all x and t. Since there is no longer any question of the equations of motion being uncoupled, the motions of bending and twisting both occur with the same natural frequencies. With the trial solutions

$$
v(x, t) = v_r(x) \sin \omega_r t, \qquad \phi(x, t) = \phi_r(x) \sin \omega_r t,
$$

the coupled equations of bending and rotation become

$$
\mu(x)\omega_r^2[v_r(x) - \bar{z}(x)\phi_r(x)] = [EI(x)v_r''(x)]'',
$$
$$
\omega_r^2[I_S(x)\phi_r(x) - \mu(x)\bar{z}(x)v_r(x)] = -\{C(x)\phi_r'(x) - [C_1(x)\phi_r''(x)]'\}'
$$

respectively.

The characteristic functions $v_r(x)$ and $\phi_r(x)$ must satisfy boundary conditions which we shall assume to be

$$
EI(x)v_r''(x) = 0 = [EI(x)v_r''(x)]',
$$
$$
C(x)\phi_r'(x) - [C_1(x)\phi_r''(x)]' = 0 = \phi_r''(x)
$$

at $x = 0$ and $x = l$.

If the first of the pair of equations of motion is multiplied by $v_s(x)$ and integrated over the beam's length it is found that

$$
\omega_r^2 \int_0^l (\mu v_r v_s - \mu \bar{z}\phi_r v_s) \, \mathrm{d}x = \int_0^l (EIv_r'')'' v_s \, \mathrm{d}x,
$$

and if r and s are interchanged,

$$
\omega_s^2 \int_0^l (\mu v_s v_r - \mu \bar{z}\phi_s v_r) \, \mathrm{d}x = \int_0^l (EIv_s'')'' v_r \, \mathrm{d}x.
$$

The integrations on the right-hand sides of these equations can be performed twice successively by parts and the boundary conditions used to discard the integrated terms as we saw in section 2.3.1. Then subtraction leads to the result

$$
(\omega_r^2 - \omega_s^2) \int_0^l \mu v_r v_s \, \mathrm{d}x - \omega_r^2 \int_0^l \mu \bar{z}\phi_r v_s \, \mathrm{d}x + \omega_s^2 \int_0^l \mu \bar{z}v_r \phi_s \, \mathrm{d}x = 0.
$$

Writing

$$
T_r = C\phi_r' - (C_1\phi_r'')',
$$

we may derive a second orthogonality condition in a like manner, using the equation governing twisting motions. Thus

$$\omega_r^2 \int_0^l (I_S\phi_r\phi_s - \mu\bar{z}v_r\phi_s)\, dx = -\int_0^l T_r'\phi_s\, dx,$$

and

$$\omega_s^2 \int_0^l (I_S\phi_s\phi_r - \mu\bar{z}v_s\phi_r)\, dx = -\int_0^l T_s'\phi_r\, dx.$$

By the previous method, but this time integrating only once by parts on the right-hand sides, we find that

$$(\omega_r^2 - \omega_s^2)\int_0^l I_S\phi_r\phi_s\, dx - \omega_r^2 \int_0^l \mu\bar{z}v_r\phi_s\, dx + \omega_s^2 \int_0^l \mu\bar{z}v_s\phi_r\, dx$$

$$= \int_0^l (T_r\phi_s' - T_s\phi_r')\, dx.$$

But

$$\int_0^l (T_r\phi_s' - T_s\phi_r')\, dx$$

$$= \int_0^l [C\phi_r' - (C_1\phi_r'')']\phi_s'\, dx - \int_0^l [C\phi_s' - (C_1\phi_s'')']\phi_r'\, dx$$

$$= \int_0^l (C_1\phi_s'')'\phi_r'\, dx - \int_0^l (C_1\phi_r'')'\phi_s'\, dx$$

$$= [C_1(\phi_s''\phi_r' - \phi_r''\phi_s')]_0^l = 0.$$

Thus the second orthogonality condition is

$$(\omega_r^2 - \omega_s^2)\int_0^l I_S\phi_r\phi_s\, dx - \omega_r^2 \int_0^l \mu\bar{z}v_r\phi_s\, dx + \omega_s^2 \int_0^l \mu\bar{z}v_s\phi_r\, dx = 0.$$

If the two orthogonality conditions are added together they give

$$(\omega_r^2 - \omega_s^2)\int_0^l [\mu v_r v_s - \mu\bar{z}(\phi_r v_s + v_r\phi_s) + I_S\phi_r\phi_s]\, dx = 0.$$

We thus arrive at the orthogonality relationship

$$\int_0^l [\mu v_r v_s - \mu\bar{z}(\phi_r v_s + v_r\phi_s) + I_S\phi_r\phi_s]\, dx = a_{rs}\delta_{rs}, \qquad (2.35)$$

where a_{rs} is a constant whose value is determined by the scale chosen for the coupled modes and δ_{rs} is the Kronecker delta function. We merely note here that a special case arises if $\omega_r = 0 = \omega_s$ and will return to this shortly.

This orthogonality condition can be expressed in a simpler form if the notation

$$y = v - \bar{z}\phi$$

is adopted, with suitable subscripts on y, v and ϕ. (Notice that y represents the sideways deflection of the centre of mass in the direction OY, as indicated in fig. 2.9).

The condition then becomes

$$\int_0^l (\mu y_r y_s + I_C \phi_r \phi_s)\, dx = a_{rs}\delta_{rs}, \tag{2.36}$$

where

$$I_S = I_C + \mu\bar{z}^2.$$

A second orthogonality relationship can be found by adding the bending and twisting equations involving only ω_r^2. It is found that

$$\int_0^l [(EIv_r'')''v_s - T_r'\phi_s]\, dx = \omega_r^2 a_{rs}\delta_{rs} = c_{rs}\delta_{rs}. \tag{2.37}$$

If we now integrate by parts and impose the boundary conditions we find that

$$\int_0^l (EIv_r''v_s'' + C\phi_r'\phi_s' + C_1\phi_r''\phi_s'')\, dx = \omega_r^2 a_{rs}\delta_{rs}.$$

It may be seen by inspection that the equations of motion have non-trivial solutions for which $\omega_r^2 = 0$. Those solutions are the rigid body modes and they will be referred to as the '0', '1' and '2' modes:

$$v_0(x) = A, \qquad \phi_0(x) = 0 \qquad \text{(for } r = 0\text{)};$$

$$v_1(x) = B(C - x), \qquad \phi_1(x) = 0 \qquad \text{(for } r = 1\text{)};$$

$$v_2(x) = z_S(x)D, \qquad \phi_2(x) = D \qquad \text{(for } r = 2\text{)},$$

where A, B, C and D are arbitrary constants. Notice that $v_2(x)$ does not satisfy the bending equation, being merely a concomitant of the rotation $\phi_2(x)$.

The standing of the rigid body modes as regards orthogonality is of particular interest. From the derivation of the orthogonality conditions we see that the Kronecker delta δ_{rs} may have to be omitted if both r and s are of the group 0, 1, 2. Thus if mode 1 is given the particular form

$$v_1(x) = B\left(1 - \frac{2x}{l}\right), \qquad \phi_1(x) = 0,$$

being a yawing (or, perhaps more accurately, a 'slewing') displacement about the amidships section, we have no assurance that a_{01}, a_{02}, a_{12} will all vanish. In fact

$$a_{01} = \int_0^l \mu v_0 v_1\, dx = a_{10},$$

$$a_{02} = \int_0^l \mu v_0 (v_2 - \bar{z}\phi_2)\, dx = a_{20},$$

$$a_{12} = \int_0^l \mu v_1 (v_2 - \bar{z}\phi_2)\, dx = a_{21}.$$

It will evidently be necessary, henceforth, to remember that these quantities are in general non-zero.

It is instructive, before proceeding, to notice how a slight simplification can be made. This simplification is in fact implicit in much of the literature on seakeeping theory. It is a well-known result in vibration analysis that when a system has repeated natural frequencies it possesses indeterminate principal modes in the sense that any linear combination of the corresponding principal modes is itself a principal mode. If one of these modes is selected arbitrarily, the other (or others) may be determined in such a way as to preserve orthogonality (e.g. see Mahalingam & Bishop, 1974). Here, though, we have a degenerate case with three repeated natural frequencies of zero value in which this possibility no longer exists.† The constant a_{02} cannot now vanish, though either a_{01} or a_{12} can be made to do so.

Let the centre of mass of the ship lie at the section $\bar{x}$ and select $v_1(x)$ in such a way as to vanish at $x = \bar{x}$; i.e. let the arbitrary constant $C = \bar{x}$. Then,

$$a_{01} = AB \int_0^l \mu(x)(\bar{x} - x)\, \mathrm{d}x = a_{10}$$

and, since

$$\bar{x} = \frac{\int_0^l \mu x\, \mathrm{d}x}{\int_0^l \mu\, \mathrm{d}x},$$

it follows that $a_{01} = 0 = a_{10}$. In other words if the $r = 0$ mode is combined with a suitable multiple of the previous mode $r = 1$, a fresh mode $r = 1$ is defined:

$$v_1(x) = B\left(1 - \frac{x}{\bar{x}}\right), \qquad \phi_1(x) = 0.$$

This is the form that we employed in sections 2.2.1 and 2.3.1 and it preserves orthogonality with the $r = 0$ mode.

It is a common simplifying assumption, either made implicitly or explicitly in the seakeeping literature, that $\bar{x} = l/2$. We now see what its effect is in terms of orthogonality.

2.4.2   *Principal coordinates in antisymmetric motion of a hull of open section*

The antisymmetric distortion of the open-section hull may be expressed as the sum of distortions in the antisymmetric principal modes. That is, we may write

$$v(x, t) = \sum_{r=0}^{\infty} p_r(t)v_r(x), \qquad \phi(x, t) = \sum_{r=0}^{\infty} p_r(t)\phi_r(x),$$

where the quantities $p_r(t)$ are a set of principal coordinates. As we have noted, $\omega_r = 0$ for $r = 0, 1, 2$, and free motion in these modes raises some special considerations. For $r \geqslant 3$ the modal shapes are of a

† This apparent paradox may be resolved by observing that for $r = 0, 1, 2$ the equations are uncoupled.

compound nature, being composed of functions $v_r(x)$ and $\phi_r(x)$ that are interrelated by the coupling of the relevant equations.

With these expressions for $v(x, t)$ and $\phi(x, t)$, the coupled differential equations of motion become

$$\sum_{r=0}^{\infty} [\mu(v_r - \bar{z}\phi_r)\ddot{p}_r + (\beta v_r'')''\dot{p}_r + (EIv_r'')''p_r] = Y(x, t),$$

$$\sum_{r=0}^{\infty} \{(I_s\phi_r - \mu\bar{z}v_r)\ddot{p}_r - (\Gamma\phi_r')'\dot{p}_r - [C\phi_r' - (C_1\phi_r'')']'p_r\}$$

$$= -\bar{z}Y(x, t) + K(x, t).$$

These may be simplified in the usual way by means of the orthogonality relations. Multiply the first by $v_s(x)$ and then integrate over the range $0 \le x \le l$; multiply the second by $\phi_s(x)$ and integrate similarly and then add the two resulting equations together. This gives

$$\sum_{r=0}^{\infty} \left\{ a_{rs}\delta_{rs}(\ddot{p}_r + \omega_r^2 p_r) + \left[\int_0^l (\beta v_r'')'' v_s \, dx - \int_0^l (\Gamma\phi_r')' \phi_s \, dx \right] \dot{p}_r \right\}$$

$$= \int_0^l [(v_s - \bar{z}\phi_s) Y + \phi_s K] \, dx,$$

for $s = 0, 1, 2, 3, \ldots$.

The damping term containing β may be simplified by integrating twice by parts so that

$$\int_0^l (\beta v_r'')'' v_s \, dx = [(\beta v_r'')' v_s]_0^l - [\beta v_r'' v_s']_0^l + \int_0^l \beta v_r'' v_s'' \, dx.$$

If, as we have done hitherto, we assume that the damping in flexure follows the same general laws as the flexural rigidity, the integrated terms vanish. It follows that the damping associated with bending is again specified by the constant

$$b_{rs} = \int_0^l \beta v_r'' v_s'' \, dx. \tag{2.38}$$

The damping term containing Γ needs a single integration, i.e.

$$-\int_0^l (\Gamma\phi_r')' \phi_s \, dx = -[\Gamma\phi_r' \phi_s]_0^l + \int_0^l \Gamma\phi_r' \phi_s' \, dx.$$

Now, by the same reasoning, we may define

$$B_{rs} = \int_0^l \Gamma\phi_r' \phi_s' \, dx \tag{2.39}$$

as the measure of damping associated with twisting.

The single equation of motion governing the principal coordinates $p_0, p_1, p_2, p_3, \ldots$ is

$$\sum_{r=0}^{\infty} [a_{rs}\delta_{rs}(\ddot{p}_r + \omega_r^2 p_r) + (b_{rs} + B_{rs})\dot{p}_r] = \int_0^l [(v_s - \bar{z}\phi_s) Y + \phi_s K] \, dx,$$

$$\tag{2.40}$$

$s = 0, 1, 2, 3, \ldots$. The integral on the right-hand side is, of course, the generalised force at the coordinate p_s. We shall examine the forms that it might take later on.

Particular significance is attached to the rigid body modes for $r = 0, 1, 2$, since they are the ones to which 'seakeeping' theory refers. It is seen that, for these modes, $b_{rs} = 0 = B_{rs}$ no matter to what mode s refers. Furthermore, $\omega_r = 0$ so that the equation may be written in the modified form

$$a_{0s}\delta_{0s}\ddot{p}_0 + a_{1s}\delta_{1s}\ddot{p}_1 + a_{2s}\delta_{2s}\ddot{p}_2 + \sum_{r=3}^{\infty} [a_{rs}\delta_{rs}(\ddot{p}_r + \omega_r^2 p_r) + (b_{rs} + B_{rs})\dot{p}_r]$$

$$= \int_0^l [(v_s - \bar{z}\phi_s)Y + \phi_s K]\,\mathrm{d}x, \tag{2.41}$$

for $s = 0, 1, 2, 3, \ldots$. We may therefore separate out the equations for $s = 0, 1$ and 2, whence

$$\left.\begin{aligned}
\ddot{p}_0 \int_0^l \mu v_0^2\,\mathrm{d}x + \ddot{p}_1 \int_0^l \mu v_0 v_1\,\mathrm{d}x + \ddot{p}_2 \int_0^l \mu z_C v_0 \phi_2\,\mathrm{d}x &= \int_0^l v_0 Y\,\mathrm{d}x, \\
\ddot{p}_0 \int_0^l \mu v_0 v_1\,\mathrm{d}x + \ddot{p}_1 \int_0^l \mu v_1^2\,\mathrm{d}x + \ddot{p}_2 \int_0^l \mu z_C v_1 \phi_2\,\mathrm{d}x &= \int_0^l v_1 Y\,\mathrm{d}x, \\
\ddot{p}_0 \int_0^l \mu z_C v_0 \phi_2\,\mathrm{d}x + \ddot{p}_1 \int_0^l \mu z_C v_1 \phi_2\,\mathrm{d}x + \ddot{p}_2 \int_0^l (I_C + \mu z_C^2)\phi_2^2\,\mathrm{d}x \\
= \int_0^l (z_C Y + K)\phi_2\,\mathrm{d}x, \\
\sum_{r=3}^{\infty} [a_{rs}\delta_{rs}(\ddot{p}_r + \omega_r^2 p_r) + (b_{rs} + B_{rs})\dot{p}_r] = \int_0^l [(v_s - \bar{z}\phi_s)Y + \phi_s K]\,\mathrm{d}x,
\end{aligned}\right\} \tag{2.42}$$

where $s = 3, 4, 5, \ldots$. The quantities on the right-hand sides of these equations are, respectively the generalised forces at p_0, p_1, p_2 and p_s.

If the number of distortion modes is now limited for the purposes of calculation, a matrix formulation again becomes available. Its form is quite straightforward and there is no necessity to develop the theory of section 2.2.2 again in the present context.

2.4.3 *Still water antisymmetric response of a ship having an open section*
The distributed force $Y(x, t) - \bar{Y}(x)$ is due to hydrostatic action alone when the ship lies in still water, since gravity loading does not act in the horizontal direction. Symmetry of the hull ensures that, in equation (2.40), $\bar{Y}(x) = 0$ at all x and so

$$\sum_{r=0}^{\infty} a_{rs}\delta_{rs}\omega_r^2 \bar{p}_r = \int_0^l \bar{K}(x)\phi_s\,\mathrm{d}x,$$

$\bar{K}(x)$ being the still water value of $K(x, t)$. If $s = 0$, $\phi_s = 0$ so that the integral vanishes; it follows that

$$a_{00}\omega_0^2 \bar{p}_0 + a_{10}\omega_1^2 \bar{p}_1 + a_{20}\omega_2^2 \bar{p}_2 = 0,$$

and since $\omega_0 = 0 = \omega_1 = \omega_2$ the values $\bar{p}_0, \bar{p}_1, \bar{p}_2$ are unspecified. The same conclusion is reached by taking $s = 1$.

Suppose $s = 2$; then

$$a_{02}\omega_0^2\bar{p}_0 + a_{12}\omega_1^2\bar{p}_1 + a_{22}\omega_2^2\bar{p}_2 = \int_0^l \bar{K}(x)\phi_2 \, dx.$$

Now since $\omega_0 = 0 = \omega_1 = \omega_2$ and $\phi_2(x) = D$, a constant, this equation becomes simply

$$\int_0^l \bar{K}(x) \, dx = 0, \tag{2.43}$$

and once again the values of $\bar{p}_0, \bar{p}_1$ and $\bar{p}_2$ are undetermined.

Yet again, then, we discover that the rigid body deflections in still water reflect the element of choice that is involved when the axes are attached to the hull. We also find that the ship so adjusts its attitude in the water as to give no net rolling moment due to gravity loading plus hydrostatic loading.

Turning now to the distortion modes $s = 3, 4, 5, \ldots$ we first note that the Kronecker delta function must no longer be ignored. Accordingly, the still water response at p_s is

$$\bar{p}_s = \frac{1}{a_{ss}\omega_s^2} \int_0^l \bar{K}(x)\phi_s(x) \, dx \qquad (s = 3, 4, 5, \ldots). \tag{2.44}$$

This is not necessarily zero and can in fact be large. Thus, if during the loading of a container ship all the heavy containers are inadvertently loaded forward on the starboard side and aft on the port side, the hull may go to sea with a dangerous combination of horizontal bending and twisting; the corresponding excessive static stresses will be augmented by the fluctuating stresses caused by waves and the action of the propeller.

2.5 The principle of orthogonality

It is already apparent that the orthogonality of principal modes plays a part of the utmost importance in modal analysis. In this chapter, and henceforth, we present orthogonality conditions as mathematical relationships to which the various modes conform, for we shall have repeatedly to use those relationships. But it must not be thought that the analytical results are without any physical significance. Indeed the reader may well have recognised that the forms of the various mathematical results strongly suggest that physical interpretations can be found.

There are several ways of interpreting the orthogonality relationships. For our present purposes, however, we need merely note that they specify conditions that must be met by any conservative mechanical system, i.e. by any system which is subject neither to damping nor to external excitation. Suppose that such a system were to vibrate freely and simultaneously in two of its principal modes, with two different natural frequencies. The sum of the kinetic and potential energies must be constant since there is no means of supplying or removing energy. This can only be the case if the total energy is the sum of two contributions, one from the motion in each principal mode. The modes cannot jointly contribute to the total energy since their contribution would be time-dependent, and the orthogonality conditions give expression to this fact. This matter is dealt with in general terms by Bishop & Johnson (1960) and the reader may wish to establish the result for the particular cases we have dealt with in this chapter and for those we shall introduce in chapter 3.

3 More accurate analysis of hull dynamics

That's, if anything, *too* unbending –
Too aggressively stiff and grand;
The Gondoliers

3.1 Allowances for shear and rotatory inertia

Practical methods of estimating hydrodynamic actions on a hull depend on the assumption that the hull is 'beamlike'. Unfortunately the hull is not always slender and so it may be desirable to perform a more accurate analysis than that of chapter 2 even for motion in modes of low order. The distance between adjacent nodes in modes of higher order is certainly not 'much greater than' the hull depth. Retention of the assumption that the hull is beamlike therefore suggests that some refinement of the Bernoulli–Euler theory is needed. We shall consider what this implies in this chapter.

There are two ways in which the theory of chapter 2 can be made more general. In the first place we may note that both the bending and the shearing force in the beam are associated with distortion, whereas hitherto we have only admitted distortion in bending. Secondly, slices of the hull rotate about axes that are perpendicular to the planes in which they oscillate; these 'rotatory motions' are associated with non-negligible inertia effects which we have previously ignored.

It is to these matters that we now turn. We shall deal first with symmetric responses and then with antisymmetric.

3.2 Equations of symmetric motion

Fig. 3.1 shows a slice of the hull which moves relative to the equilibrium axes $OXYZ$, the plane OXZ coinciding with the plane of port and starboard symmetry. Vertical motion of the slice is governed by the equation

$$\mu(x)\ddot{w}(x, t) = V'(x, t) + Z(x, t), \tag{3.1}$$

where $\mu(x)$ is the mass per unit length, $w(x, t)$ is the upward deflection, $V(x, t)$ is the shear force whose positive sense is indicated in the figure and $Z(x, t)$ is the upward applied force per unit length. Rotatory motion of the slice is such that

$$I_y(x)\ddot{\theta}(x, t) = M'(x, t) + V(x, t), \tag{3.2}$$

$I_y(x)$ being the moment of inertia per unit length, $\theta(x, t)$ the slope attributable to bending and $M(x, t)$ the bending moment. Upward deflection of the hull is given by

$$w'(x, t) = \theta(x, t) + \gamma(x, t), \tag{3.3}$$

where $\gamma(x, t)$ represents the shear strain.

These three equations are familiar in the theory of a 'Timoshenko beam'.[†] To them we must add expressions for the shearing force and bending moment. The equations that will be adopted embody damping terms and are akin to those employed by Kumai (1958), namely

$$V(x, t) = kAG(x)[\gamma(x, t) + \alpha(x)\dot{\gamma}(x, t)], \tag{3.4}$$

$$M(x, t) = EI(x)[\theta'(x, t) + \beta(x)\dot{\theta}'(x, t)]. \tag{3.5}$$

Here $\alpha(x)$ represents distributed shear damping and $\beta(x)$ represents the damping of bending. The quantity $EI(x)$ is the flexural rigidity and $kAG(x)$ is the shear rigidity, where k is a constant which depends on the shape of the hull cross-section.[‡]

The five equations may be combined to give a pair of simultaneous partial differential equations from which $V(x, t)$ and $M(x, t)$ are excluded. They are

$$\mu\ddot{w} - [kAG(\gamma + \alpha\dot{\gamma})]' = Z, \tag{3.6}$$

$$I_y\ddot{\theta} - [EI(\theta' + \beta\dot{\theta}')]' - kAG(\gamma + \alpha\dot{\gamma}) = 0. \tag{3.7}$$

It will be seen that these equations relate the distortion parameters $w(x, t)$, $\theta(x, t)$ and $\gamma(x, t)$ and that these three quantities also satisfy the relationship

$$w' = \theta + \gamma. \tag{3.8}$$

[†] e.g. see Timoshenko, Young & Weaver (1974).
[‡] e.g. see Cowper (1966).

Fig. 3.1. Slice of a hull showing the conventions used for shear force V and bending moment M. The axes $OXYZ$ are 'equilibrium axes' in which the frame moves in the OX direction with the constant reference speed of the ship.

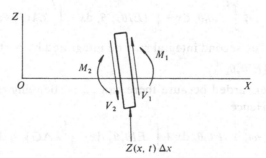

3.2.1   *Principal modes of symmetric motion*
In order to extract the principal modes of the dry hull, let $Z = 0 = \alpha = \beta$ and suppose that

$$
\left.\begin{aligned}
w(x, t) &= w_r(x) \sin \omega_r t, \\
\theta(x, t) &= \theta_r(x) \sin \omega_r t, \\
\gamma(x, t) &= \gamma_r(x) \sin \omega_r t.
\end{aligned}\right\}
\tag{3.9}
$$

The equations of motion now show that

$$
\left.\begin{aligned}
&-\omega_r^2 \mu w_r - (kAG\gamma_r)' = 0, \\
&-\omega_r^2 I_y \theta_r - (EI\theta_r')' - kAG\gamma_r = 0, \\
&w_r' = \theta_r + \gamma_r.
\end{aligned}\right\}
\tag{3.10}
$$

By manipulating these equations we can find simple orthogonality relations between the functions $w_r(x)$, $\theta_r(x)$ and $\gamma_r(x)$.

Multiply the first of the equations by w_s and integrate over the length of the ship, so that

$$
-\omega_r^2 \int_0^l \mu w_r w_s \, dx - \int_0^l (kAG\gamma_r)' w_s \, dx = 0.
$$

If the second integral is evaluated by parts, the integrated term

$$
[kAG\gamma_r w_s]_0^l
$$

vanishes because the free–free hull has no shearing force at the boundaries $x = 0$ and $x = l$. We thus discover that

$$
-\omega_r^2 \int_0^l \mu w_r w_s \, dx + \int_0^l kAG\gamma_r w_s' \, dx = 0.
\tag{3.11}
$$

If the subscripts r and s are interchanged, this becomes

$$
-\omega_s^2 \int_0^l \mu w_r w_s \, dx + \int_0^l kAG\gamma_s w_r' \, dx = 0.
$$

By subtracting these last two equations we now find that

$$
(\omega_r^2 - \omega_s^2) \int_0^l \mu w_r w_s \, dx - \int_0^l kAG(\gamma_r w_s' - \gamma_s w_r') \, dx = 0.
\tag{3.12}
$$

Return now to the second of the equations (3.10), multiply by θ_s and integrate. This gives the equation

$$
-\omega_r^2 \int_0^l I_y \theta_r \theta_s \, dx - \int_0^l (EI\theta_r')' \theta_s \, dx - \int_0^l kAG\gamma_r \theta_s \, dx = 0.
$$

The second integral may be integrated by parts and the term

$$
[EI\theta_r' \theta_s]_0^l
$$

discarded because there can be no bending moment at $x = 0$ and l. Hence

$$
-\omega_r^2 \int_0^l I_y \theta_r \theta_s \, dx + \int_0^l EI\theta_r' \theta_s' \, dx - \int_0^l kAG\gamma_r \theta_s \, dx = 0,
\tag{3.13}
$$

and, on reversing the subscripts, we also find that

$$-\omega_s^2 \int_0^l I_y\theta_r\theta_s \, dx + \int_0^l EI\theta_r'\theta_s' \, dx - \int_0^l kAG\gamma_s\theta_r \, dx = 0.$$

Subtraction of these last two equations now gives

$$(\omega_r^2 - \omega_s^2) \int_0^l I_y\theta_r\theta_s \, dx + \int_0^l kAG(\gamma_r\theta_s - \gamma_s\theta_r) \, dx = 0. \qquad (3.14)$$

The orthogonality conditions are found by adding equations (3.12) and (3.14) and, after employing the last of equations (3.10), we find that

$$(\omega_r^2 - \omega_s^2) \int_0^l (\mu w_r w_s + I_y\theta_r\theta_s) \, dx = 0.$$

It follows that

$$\int_0^l (\mu w_r w_s + I_y\theta_r\theta_s) \, dx = a_{rs}\delta_{rs}, \qquad (3.15)$$

where δ_{rs} is the Kronecker delta function such that

$$\delta_{rs} = \begin{cases} 0 & \text{(for } r \neq s), \\ 1 & \text{(for } r = s). \end{cases}$$

A second orthogonality condition can readily be found from the first, i.e. from equation (3.13). By adding equations (3.11) and (3.13) and modifying the sum in the light of the last of equations (3.10), it is found that

$$\int_0^l (EI\theta_r'\theta_s' + kAG\gamma_r\gamma_s) \, dx = \omega_r^2 a_{rs}\delta_{rs} = c_{rs}. \qquad (3.16)$$

By letting $r = s$ in the two orthogonality equations we arrive at expressions which, we shall find, give the generalised mass and generalised stiffness corresponding to p_s. They are, respectively,

$$a_{ss} = \int_0^l (\mu w_s^2 + I_y\theta_s^2) \, dx,$$

$$c_{ss} = \omega_s^2 a_{ss} = \int_0^l (EI\theta_s'^2 + kAG\gamma_s^2) \, dx.$$

The rigid body modes require separate attention. Such is the way we have written the third of equations (3.10), the second of the two zero-frequency modes

$$w_0(x) = A, \qquad \theta_0(x) = 0 = \gamma_0(x) \qquad \text{(for } r = 0),$$

$$w_1(x) = B(C - x), \qquad \theta_1(x) = 0 = \gamma_1(x) \qquad \text{(for } r = 1),$$

is apparently ruled out. But that equation can be ignored in the discussion of these modes. If orthogonality is to be preserved between these two modes it is necessary that

$$\int_0^l \mu AB(C - x) \, dx = 0$$

or

$$C = \frac{\int_0^l \mu x \, dx}{\int_0^l \mu \, dx} = \bar{x},$$

where $\bar{x}$ is the abscissa of the centre of mass. That is to say, if we require that $w_0(0) = 1 = w_1(0)$ to comply with the rule we shall adopt in scaling the modes, it is necessary to write

$$w_0(x) = 1, \qquad \theta_0(x) = 0 = \gamma_0(x) \qquad \text{(for } r = 0\text{)},$$

$$w_1(x) = 1 - \frac{x}{\bar{x}}, \qquad \theta_1(x) = 0 = \gamma_1(x) \qquad \text{(for } r = 1\text{)}.$$

Then

$$a_{00} = \int_0^l \mu \, dx,$$

$$a_{11} = \int_0^l \mu \left(1 - \frac{x}{\bar{x}}\right)^2 dx,$$

$$c_{00} = 0 = c_{11}.$$

3.2.2   *Modal analysis of forced symmetric oscillation*

The symmetric distortion of the hull may be expressed as the sum of distortions in the symmetric principal modes; i.e. we may write

$$\left.\begin{aligned}
w(x, t) &= \sum_{r=0}^{\infty} p_r(t) w_r(x), \\
\theta(x, t) &= \sum_{r=0}^{\infty} p_r(t) \theta_r(x), \\
\gamma(x, t) &= \sum_{r=0}^{\infty} p_r(t) \gamma_r(x),
\end{aligned}\right\} \tag{3.17}$$

where $p_r(t)$ is the rth principal coordinate. The mode number r can no longer be taken reliably as the number of nodal points in the mode to which it relates; it must now be assigned by arranging the natural frequencies of the distortion modes in ascending order ($\omega_2, \omega_3, \dots$) and using the appropriate subscript of ω. With these expressions for $w(x, t)$ and $\gamma(x, t)$, equation (3.6) becomes

$$\sum_{r=0}^{\infty} \mu w_r \ddot{p}_r - \sum_{r=0}^{\infty} [kAG(\gamma_r p_r + \alpha \gamma_r \dot{p}_r)]' = Z(x, t). \tag{3.18}$$

Multiplying by w_s and integrating over the length of the hull we find that

$$\sum_{r=0}^{\infty} \ddot{p}_r \int_0^l \mu w_r w_s \, dx - \sum_{r=0}^{\infty} \int_0^l w_s [kAG(\gamma_r p_r + \alpha \gamma_r \dot{p}_r)]' \, dx = \int_0^l w_s Z \, dx.$$

If this second integral is evaluated by parts, the term

$$[w_s \{kAG(\gamma_r p_r + \alpha \gamma_r \dot{p}_r)\}]_0^l = 0$$

because the shearing force (as given by the term in curly brackets) vanishes at the ends of the beam. Hence,

$$\sum_{r=0}^{\infty} \ddot{p}_r \int_0^l \mu w_r w_s \, dx + \sum_{r=0}^{\infty} p_r \int_0^l kAG\gamma_r w_s' \, dx$$

$$+ \sum_{r=0}^{\infty} \dot{p}_r \int_0^l \alpha kAG\gamma_r w_s' \, dx = \int_0^l w_s Z \, dx. \tag{3.19}$$

We next treat equation (3.7) in a similar manner. Substituting for $\theta(x, t)$ and $\gamma(x, t)$, multiplying by θ_s and integrating, we find that

$$\sum_{r=0}^{\infty} \ddot{p}_r \int_0^l I_y \theta_r \theta_s \, dx - \sum_{r=0}^{\infty} \int_0^l \theta_s [EI(\theta_r' p_r + \beta \theta_r' \dot{p}_r)]' \, dx$$

$$- \sum_{r=0}^{\infty} \int_0^l \theta_s kAG(\gamma_r p_r + \gamma_r \dot{p}_r) \, dx = 0.$$

Again the integrated term obtained from the second integral when it is integrated by parts is zero because, this time, the bending moment (as given by the term in square brackets) vanishes at $x = 0$ and $x = l$. Thus the previous equation becomes

$$\sum_{r=0}^{\infty} \ddot{p}_r \int_0^l I_y \theta_r \theta_s \, dx + \sum_{r=0}^{\infty} p_r \int_0^l EI\theta_r' \theta_s' \, dx + \sum_{r=0}^{\infty} \dot{p}_r \int_0^l \beta EI\theta_r' \theta_s' \, dx$$

$$- \sum_{r=0}^{\infty} p_r \int_0^l kAG\gamma_r \theta_s \, dx - \sum_{r=0}^{\infty} \dot{p}_r \int_0^l \alpha kAG\gamma_r \theta_s \, dx = 0. \tag{3.20}$$

If equations (3.19) and (3.20) are added, they give

$$\sum_{r=0}^{\infty} \ddot{p}_r \int_0^l (\mu w_r w_s + I_y \theta_r \theta_s) \, dx + \sum_{r=0}^{\infty} p_r \int_0^l [EI\theta_r' \theta_s' + kAG\gamma_r(w_s' - \theta_s)] \, dx$$

$$+ \sum_{r=0}^{\infty} \dot{p}_r \int_0^l [\alpha kAG\gamma_r(w_s' - \theta_s) + \beta EI\theta_r' \theta_s'] \, dx = \int_0^l Z w_s \, dx.$$

The orthogonality conditions (3.15) and (3.16), together with the last of equations (3.10), now show that

$$a_{ss}\ddot{p}_s + \omega_s^2 a_{ss} p_s + \sum_{r=0}^{\infty} \dot{p}_r (\alpha_{rs} + \beta_{rs})$$

$$= \int_0^l Z w_s \, dx \qquad (s = 0, 1, 2, \dots), \tag{3.21}$$

where α_{rs}, β_{rs} are damping coefficients given by

$$\alpha_{rs} = \int_0^l \alpha kAG\gamma_r \gamma_s \, dx, \tag{3.22}$$

$$\beta_{rs} = \int_0^l \beta EI\theta_r' \theta_s' \, dx. \tag{3.23}$$

3.2.3 Matrix formulation

The equations (3.21) have precisely the same form as those arrived at in section 2.2.2, save only that the matrix **B** has now to be written

$$\mathbf{B} = \begin{bmatrix} \mathbf{0} & \vdots & \mathbf{0} \\ \cdots & \cdots & \cdots \\ \mathbf{0} & \vdots & \mathbf{B}_{DD} \end{bmatrix}, \tag{3.24}$$

where

$$\mathbf{B}_{DD} = \begin{bmatrix} \alpha_{22}+\beta_{22} & \alpha_{23}+\beta_{23} \dots \alpha_{2n}+\beta_{2n} \\ \alpha_{32}+\beta_{32} & \alpha_{33}+\beta_{33} \dots \alpha_{3n}+\beta_{3n} \\ \dotfill \\ \alpha_{n2}+\beta_{n2} & \alpha_{n3}+\beta_{n3} \dots \alpha_{nn}+\beta_{nn} \end{bmatrix}. \tag{3.25}$$

This alteration merely reflects the fact that we have now to distinguish between the damping associated with bending and that related to shearing.

3.3 Equations of antisymmetric motion

In chapter 2 we introduced antisymmetric deflection in two stages, referring first to the 'boxlike' ship and then to a hull with large deck openings. This approach was adopted only for ease of presentation and it will be found that, in fact, the former case is merely a special form of the latter. Had we elected to present the theory for the ship with large deck openings first, we could have arrived at that for the boxlike hull by investigating the effects of letting $\bar{z}(x) = 0 = C_1(x)$. This is the standpoint we shall now adopt in introducing allowances for shear deflection and rotatory inertia in transverse bending.

Fig. 3.2 shows a slice of the hull in which C is the centre of mass of the slice and S is the shear centre. The shear force is represented by V, while M is the bending moment. The notation and conventions are in fact the same as those employed in section 2.4.1.

The equation governing motion parallel to OY is

$$\mu(x)\ddot{v}_C(x, t) = V'(x, t) + Y(x, t)$$

where $v_C(x, t)$ is the deflection of C and $Y(x, t)$ is the applied force per unit length. Since the deflection of S is

$$v(x, t) = v_C(x, t) + \bar{z}(x)\phi(x, t),$$

Fig. 3.2. Slice of hull showing the notation and conventions employed. The shear centre is S, and C is the centre of mass of the slice. V represents shearing force and the bending moment is M.

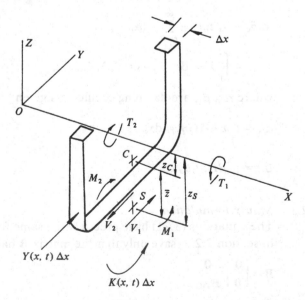
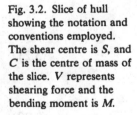

where $\phi(x, t)$ is the angle through which the slice rotates, this equation may be written

$$\mu(x)[\ddot{v}(x, t) - \bar{z}(x)\ddot{\phi}(x, t)] = V'(x, t) + Y(x, t). \tag{3.26}$$

Rotation of the slice about the vertical axis through C is such that

$$I_z(x)\ddot{\theta}(x, t) = M'(x, t) + V(x, t). \tag{3.27}$$

Here $I_z(x)$ is the moment of inertia per unit length and $\theta(x, t)$ is that contribution to the slope which is due to bending. Deflection of the hull in the OY direction is due to bending and to shear deformation. It is such that

$$v'(x, t) = \theta(x, t) + \gamma(x, t), \tag{3.28}$$

where $\gamma(x, t)$ is now the transverse shear strain.

These three equations govern transverse bending of the 'Timoshenko beam'. As with the symmetric motion we have to supplement them with suitable expressions for V and M. As before, we take

$$V(x, t) = kAG(x)[\gamma(x, t) + \alpha(x)\dot{\gamma}(x, t)], \tag{3.29}$$

$$M(x, t) = EI(x)[\theta'(x, t) + \beta(x)\dot{\theta}'(x, t)]. \tag{3.30}$$

In general, antisymmetric bending is associated with twisting of the hull. And the equations governing the twisting motion remain those of sections 2.3.1 and 2.4.1. That is to say,

$$I_S(x)\ddot{\phi}(x, t) - \mu(x)\bar{z}(x)\ddot{v}(x, t) = T' - \bar{z}(x)Y(x, t) + K(x, t), \tag{3.31}$$

where

$$T = C(x)\phi'(x, t) - [C_1(x)\phi''(x, t)]' + \Gamma(x)\dot{\phi}'(x, t). \tag{3.32}$$

3.3.1 Principal modes of antisymmetric motion

When the hull oscillates *in vacuo* with no applied actions and with no damping admitted, the equations governing its bending motion reduce to

$$\mu(\ddot{v} - \bar{z}\ddot{\phi}) = V',$$

$$v' = \theta + \gamma = \theta + \frac{V}{kAG},$$

$$I_z\ddot{\theta} = M' + V,$$

$$M = EI\theta'.$$

(We shall take up the rotation equations separately, later on.) Assume that motion in the rth principal mode is such that

$$v(x, t) = v_r(x) \sin \omega_r t,$$

$$\theta(x, t) = \theta_r(x) \sin \omega_r t,$$

$$\phi(x, t) = \phi_r(x) \sin \omega_r t,$$

$$M(x, t) = M_r(x) \sin \omega_r t,$$

$$V(x, t) = V_r(x) \sin \omega_r t.$$

Then the equations of motion give

$$-\mu\omega_r^2 v_r + \mu\bar{z}\omega_r^2\phi_r = V_r', \tag{3.33}$$

$$v_r' = \theta_r + \frac{V_r}{kAG}, \tag{3.34}$$

$$V_r = -M_r' - I_z\omega_r^2\theta_r, \tag{3.35}$$

$$M_r = EI\theta_r'. \tag{3.36}$$

By eliminating M_r from the last two equations we find that

$$V_r = -(EI\theta_r')' - I_z\omega_r^2\theta_r,$$

so

$$-\mu\omega_r^2 v_r + \mu\bar{z}\omega_r^2\phi_r = -[(EI\theta_r')' + I_z\omega_r^2\theta_r]'.$$

Multiplying this result by $v_s(x)$ and integrating with respect to x, we find that

$$-\omega_r^2\int_0^l \mu v_r v_s\,\mathrm{d}x + \omega_r^2\int_0^l \bar{z}\mu\phi_r v_s\,\mathrm{d}x = -\int_0^l [(EI\theta_r')' + I_z\omega_r^2\theta_r]'v_s\,\mathrm{d}x$$

and, when the term on the right-hand side is evaluated by parts, it gives

$$-[\{(EI\theta_r')' + I_z\omega_r^2\theta_r\}v_s]_0^l + \int_0^l [(EI\theta_r')' + I_z\omega_r^2\theta_r]v_s'\,\mathrm{d}x.$$

The integrated term is nil because the contents of the curly brackets are $-V_r$, which vanishes at the boundaries $x = 0$ and $x = l$. The integral may again be evaluated by parts to give

$$[(EI\theta_r')v_s']_0^l - \int_0^l EI\theta_r'v_s''\,\mathrm{d}x + \omega_r^2\int_0^l I_z\theta_r v_s'\,\mathrm{d}x,$$

and again the integrated term vanishes since the bracketed term is M_r which is zero at the two extremities. In other words,

$$-\omega_r^2\int_0^l \mu v_r v_s\,\mathrm{d}x + \omega_r^2\int_0^l \mu\bar{z}\phi_r v_s\,\mathrm{d}x = -\int_0^l EI\theta_r'v_s''\,\mathrm{d}x + \omega_r^2\int_0^l I_z\theta_r v_s'\,\mathrm{d}x. \tag{3.37}$$

This result must still hold if the suffices r and s are interchanged so that

$$-\omega_s^2\int_0^l \mu v_r v_s\,\mathrm{d}x + \omega_s^2\int_0^l \mu\bar{z}\phi_s v_r\,\mathrm{d}x = -\int_0^l EI\theta_s'v_r''\,\mathrm{d}x + \omega_s^2\int_0^l I_z\theta_s v_r'\,\mathrm{d}x. \tag{3.38}$$

If the last two equations are subtracted, it is now found that

$$(\omega_r^2 - \omega_s^2)\int_0^l \mu v_r v_s\,\mathrm{d}x - \omega_r^2\int_0^l \mu\bar{z}\phi_r v_s\,\mathrm{d}x + \omega_s^2\int_0^l \mu\bar{z}\phi_s v_r\,\mathrm{d}x$$

$$= \int_0^l EI(\theta_r'v_s'' - \theta_s'v_r'')\,\mathrm{d}x - \omega_r^2\int_0^l I_z\theta_r v_s'\,\mathrm{d}x + \omega_s^2\int_0^l I_z\theta_s v_r'\,\mathrm{d}x. \tag{3.39}$$

If equations (3.34), (3.35) and (3.36) are combined it is found that

$$v'_r = \theta_r - \left[\frac{(EI\theta'_r)' + I_z\omega_r^2\theta_r}{kAG}\right]. \tag{3.40}$$

When this result is differentiated with respect to x, multiplied by $EI\theta'_s$ and then integrated with respect to x, it gives

$$\int_0^l EI\theta'_s v''_r \, dx = \int_0^l EI\theta'_s \theta'_r \, dx - \int_0^l \left[\frac{(EI\theta'_r)' + I_z\omega_r^2\theta_r}{kAG}\right]' EI\theta'_s \, dx.$$

If the last integration in this equation is performed by parts, it is found to give

$$\left[\left\{\frac{(EI\theta'_r)' + I_z\omega_r^2\theta_r}{kAG}\right\}EI\theta'_s\right]_0^l - \int_0^l \left[\frac{(EI\theta'_r)' + I_z\omega_r^2\theta_r}{kAG}\right](EI\theta'_s)' \, dx,$$

of which the integrated term vanishes since $M_s = 0$ when $x = 0$ and $x = l$. Therefore

$$\int_0^l EI\theta'_s v''_r \, dx = \int_0^l EI\theta'_s \theta'_r \, dx + \int_0^l \frac{(EI\theta'_r)'(EI\theta'_s)'}{kAG} \, dx$$
$$+ \omega_r^2 \int_0^l \frac{I_z\theta_r(EI\theta'_s)'}{kAG} \, dx. \tag{3.41}$$

When the suffices r and s in this last equation are interchanged it is found that

$$\int_0^l EI\theta'_r v''_s \, dx = \int_0^l EI\theta'_r\theta'_s \, dx + \int_0^l \frac{(EI\theta'_r)'(EI\theta'_s)'}{kAG} \, dx$$
$$+ \omega_s^2 \int_0^l \frac{I_z\theta_s(EI\theta'_r)'}{kAG} \, dx. \tag{3.42}$$

Subtraction now reveals that

$$\int_0^l EI(\theta'_r v''_s - \theta'_s v''_r) \, dx = \omega_s^2 \int_0^l \frac{I_z\theta_s(EI\theta'_r)'}{kAG} \, dx - \omega_r^2 \int_0^l \frac{I_z\theta_r(EI\theta'_s)'}{kAG} \, dx. \tag{3.43}$$

Return now to equation (3.40), multiply it throughout by $\omega_s^2 I_z\theta_s$ and integrate with respect to x. This gives

$$\omega_s^2 \int_0^l I_z\theta_s v'_r \, dx = \omega_s^2 \int_0^l I_z\theta_r\theta_s \, dx - \omega_s^2 \int_0^l \frac{(EI\theta'_r)'I_z\theta_s}{kAG} \, dx$$
$$- \omega_r^2\omega_s^2 \int_0^l \frac{I_z^2\theta_r\theta_s}{kAG} \, dx. \tag{3.44}$$

Interchanging r and s, we find that

$$\omega_r^2 \int_0^l I_z\theta_r v'_s \, dx = \omega_r^2 \int_0^l I_z\theta_r\theta_s \, dx - \omega_r^2 \int_0^l \frac{(EI\theta'_s)'I_z\theta_r}{kAG} \, dx$$
$$- \omega_r^2\omega_s^2 \int_0^l \frac{I_z^2\theta_r\theta_s}{kAG} \, dx. \tag{3.45}$$

When these last two equations are subtracted they give

$$\omega_s^2 \int_0^l I_z \theta_s v_r' \, dx - \omega_r^2 \int_0^l I_z \theta_r v_s' \, dx$$

$$= (\omega_s^2 - \omega_r^2) \int_0^l I_z \theta_r \theta_s \, dx - \omega_s^2 \int_0^l \frac{(EI\theta_r')' I_z \theta_s}{kAG} \, dx$$

$$+ \omega_r^2 \int_0^l \frac{(EI\theta_s')' I_z \theta_r}{kAG} \, dx. \tag{3.46}$$

If equations (3.43) and (3.46) are now added together, they give

$$\int_0^l EI(\theta_r' v_s'' - \theta_s' v_r'') \, dx + \omega_s^2 \int_0^l I_z \theta_s v_r' \, dx - \omega_r^2 \int_0^l I_z \theta_r v_s' \, dx$$

$$= (\omega_s^2 - \omega_r^2) \int_0^l I_z \theta_r \theta_s \, dx. \tag{3.47}$$

Comparison of this result with that of equation (3.39) reveals that

$$(\omega_r^2 - \omega_s^2) \int_0^l \mu v_r v_s \, dx - \omega_r^2 \int_0^l \mu \bar{z} \phi_r v_s \, dx + \omega_s^2 \int_0^l \mu \bar{z} \phi_s v_r \, dx$$

$$= (\omega_s^2 - \omega_r^2) \int_0^l I_z \theta_r \theta_s \, dx. \tag{3.48}$$

We next go back to equations (3.31) and (3.32) governing rotation. For undamped free vibration in the rth principal mode,

$$\omega_r^2 (I_S \phi_r - \mu \bar{z} v_r) = -[C\phi_r' - (C_1 \phi_r'')']' = -T_r' \text{ (say)}. \tag{3.49}$$

If this equation is multiplied throughout by ϕ_s and integrated it becomes

$$\omega_r^2 \int_0^l (I_S \phi_r \phi_s - \mu \bar{z} v_r \phi_s) \, dx = -\int_0^l T_r' \phi_s \, dx \tag{3.50}$$

and so, by interchanging r and s, we find that

$$\omega_s^2 \int_0^l (I_S \phi_r \phi_s - \mu \bar{z} v_s \phi_r) \, dx = -\int_0^l T_s' \phi_r \, dx.$$

Subtraction now shows that

$$(\omega_r^2 - \omega_s^2) \int_0^l I_S \phi_r \phi_s \, dx - \omega_r^2 \int_0^l \mu \bar{z} v_r \phi_s \, dx + \omega_s^2 \int_0^l \mu \bar{z} v_s \phi_r \, dx$$

$$= -\int_0^l T_r' \phi_s \, dx + \int_0^l T_s' \phi_r \, dx.$$

When they are evaluated by parts the integrals on the right-hand side of this last equation become

$$-[T_r \phi_s]_0^l + \int_0^l T_r \phi_s' \, dx + [T_s \phi_r]_0^l - \int_0^l T_s \phi_r' \, dx.$$

The integrated terms vanish because T_r and T_s are both zero when $x = 0$ and $x = l$. Furthermore it was shown in section 2.4.1 that the remaining integrals also sum to zero. It follows, therefore, that

$$(\omega_r^2 - \omega_s^2) \int_0^l I_S \phi_r \phi_s \, dx - \omega_r^2 \int_0^l \mu \bar{z} v_r \phi_s \, dx + \omega_s^2 \int_0^l \mu \bar{z} v_s \phi_r \, dx = 0.$$

(3.51)

Equations (3.48) and (3.51) may be added together to give

$$(\omega_r^2 - \omega_s^2) \int_0^l [\mu v_r v_s + I_S \phi_r \phi_s + I_z \theta_r \theta_s - \mu \bar{z}(v_r \phi_s + v_s \phi_r)] \, dx = 0.$$

We therefore reach the orthogonality condition

$$\int_0^l [\mu v_r v_s + I_S \phi_r \phi_s + I_z \theta_r \theta_s - \mu \bar{z}(v_r \phi_s + v_s \phi_r)] \, dx = a_{rs} \delta_{rs},$$

(3.52)

where δ_{rs} is the Kronecker delta function. Notice, however, that (as we showed in section 2.4.1) a special case may arise when $\omega_r = 0 = \omega_s$.

A second orthogonality condition can now be found. Let

$$\gamma_r = \frac{V_r}{kAG} = v_r' - \theta_r.$$

It can now be shown by substituting back into equation (3.52) from equations (3.37) and (3.41), and also from equation (3.50) noting that

$$-\int_0^l T_r' \phi_s \, dx = \int_0^l T_r \phi_s' \, dx,$$

that

$$\int_0^l [EI\theta_r' \theta_s' + kAG\gamma_r \gamma_s + T_r \phi_s'] \, dx = \omega_r^2 a_{rs} \delta_{rs} = c_{rs} \delta_{rs}.$$

(3.53)

Explicit expressions for the generalised mass and generalised stiffness corresponding to p_s can now be written down. If $r = s$ in the orthogonality equations, these quantities are

$$a_{ss} = \int_0^l (\mu v_s^2 + I_S \phi_s^2 + I_z \theta_s^2 - 2\mu \bar{z} v_s \phi_s) \, dx$$

(3.54)

and

$$c_{ss} = \omega_s^2 a_{ss} = \int_0^l (EI\theta_s'^2 + kAG\gamma_s^2 + T_s \phi_s') \, dx$$

(3.55)

respectively.

We next consider the rigid body modes. These are

$$\left.\begin{aligned}
v_0(x) &= 1, & \theta_0(x) &= 0 = \gamma_0(x) = \phi_0(x) & &\text{(for } r = 0\text{)}, \\
v_1(x) &= 1 - \frac{x}{\bar{x}}, & \theta_1(x) &= 0 = \gamma_1(x) = \phi_1(x) & &\text{(for } r = 1\text{)}, \\
v_2(x) &= z_S(x), & \theta_2(x) &= 0 = \gamma_2(x), & \phi_2(x) &= 1 & &\text{(for } r = 2\text{)}.
\end{aligned}\right\}$$

(3.56)

The modal shape $v_1(x)$ does not satisfy equation (3.34) and $v_2(x)$ does not satisfy the bending equations, but the reasons are obvious. The modes corresponding to $r = 0$ and $r = 1$ satisfy the orthogonality condition (3.52) and are therefore mutually orthogonal. But

$$
\left.\begin{aligned}
a_{02} &= \int_0^l \mu z_C \, \mathrm{d}x = a_{20} \neq 0 \qquad \text{(for } r, s = 0, 2\text{)}, \\
a_{12} &= \int_0^l \mu z_C \left(1 - \frac{x}{\bar{x}}\right) \mathrm{d}x = a_{21} \neq 0 \qquad \text{(for } r, s = 1, 2\text{)}.
\end{aligned}\right\} \tag{3.57}
$$

In other words the Kronecker delta function in equation (3.52) must be ignored for these combinations of the modal indices.

3.3.2   *Modal analysis of forced antisymmetric oscillation*
Antisymmetric distortion of the dry hull may be expressed as the sum of distortions in the antisymmetric principal modes. That is, we may write

$$
\left.\begin{aligned}
v(x, t) &= \sum_{r=0}^{\infty} p_r(t) v_r(x), \\
\theta(x, t) &= \sum_{r=0}^{\infty} p_r(t) \theta_r(x), \\
\gamma(x, t) &= \sum_{r=0}^{\infty} p_r(t) \gamma_r(x), \\
\phi(x, t) &= \sum_{r=0}^{\infty} p_r(t) \phi_r(x),
\end{aligned}\right\} \tag{3.58}
$$

where $p_r(t)$ is the rth principal coordinate.

Equation (3.26) now becomes

$$
\sum_{r=0}^{\infty} \mu(v_r - \bar{z}\phi_r)\ddot{p}_r - \sum_{r=0}^{\infty} [kAG(\gamma_r p_r + \alpha \gamma_r \dot{p}_r)]' = Y(x, t).
$$

If this equation is multiplied throughout by v_s and then integrated over the length of the ship it is found that

$$
\sum_{r=0}^{\infty} \ddot{p}_r \int_0^l \mu(v_r - \bar{z}\phi_r) v_s \, \mathrm{d}x - \sum_{r=0}^{\infty} \int_0^l [kAG(\gamma_r p_r + \alpha \gamma_r \dot{p}_r)]' v_s \, \mathrm{d}x
$$
$$
= \int_0^l Y v_s \, \mathrm{d}x.
$$

Integrated by parts, the second integral gives the term

$$
[\{kAG(\gamma_r p_r + \alpha \gamma_r \dot{p}_r)\} v_s]_0^l
$$

and this vanishes because the quantity in curly brackets is the shearing force which vanishes at the ends of the beam. That is to say

$$
\sum_{r=0}^{\infty} \ddot{p}_r \int_0^l \mu(v_r - \bar{z}\phi_r) v_s \, \mathrm{d}x + \sum_{r=0}^{\infty} p_r \int_0^l kAG\gamma_r v_s' \, \mathrm{d}x
$$
$$
+ \sum_{r=0}^{\infty} \dot{p}_r \int_0^l \alpha kAG\gamma_r v_s' \, \mathrm{d}x = \int_0^l Y v_s \, \mathrm{d}x. \tag{3.59}
$$

We now transfer our attention to equation (3.27) and treat it in a similar manner, this time multiplying by θ_s. Thus

$$\sum_{r=0}^{\infty} \ddot{p}_r \int_0^l I_z \theta_r \theta_s \, dx - \sum_{r=0}^{\infty} \int_0^l [EI(\theta_r' p_r + \beta \theta_r' \dot{p}_r)]' \theta_s \, dx$$

$$- \sum_{r=0}^{\infty} \int_0^l kAG(\gamma_r p_r + \alpha \gamma_r \dot{p}_r) \theta_s \, dx = 0.$$

By the same reasoning as before we can reject the integrated term when the second integral is evaluated by parts – the bending moment is nil at the boundaries – and we are left with

$$\sum_{r=0}^{\infty} \ddot{p}_r \int_0^l I_z \theta_r \theta_s \, dx + \sum_{r=0}^{\infty} p_r \int_0^l EI \theta_r' \theta_s' \, dx + \sum_{r=0}^{\infty} \dot{p}_r \int_0^l \beta EI \theta_r' \theta_s' \, dx$$

$$- \sum_{r=0}^{\infty} p_r \int_0^l kAG \gamma_r \theta_s \, dx - \sum_{r=0}^{\infty} \dot{p}_r \int_0^l \alpha kAG \gamma_r \theta_s \, dx = 0. \qquad (3.60)$$

Next we turn to equation (3.31), multiplying by ϕ_s and integrating:

$$\sum_{r=0}^{\infty} \ddot{p}_r \int_0^l (I_s \phi_r - \mu \bar{z} v_r) \phi_s \, dx - \sum_{r=0}^{\infty} \int_0^l [C \phi_r' p_r - (C_1 \phi_r'')' p_r + \Gamma \phi_r' \dot{p}_r]' \phi_s \, dx$$

$$= - \int_0^l \bar{z} \phi_s Y \, dx + \int_0^l \phi_s K \, dx.$$

If the second integral is evaluated by parts, the integrated term vanishes since the term $[C \phi_r' p_r - (C_1 \phi_r'')' p_r + \Gamma \phi_r' \dot{p}_r]$ is equal to zero at both boundaries. We are thus left with the equation

$$\sum_{r=0}^{\infty} \ddot{p}_r \int_0^l (I_s \phi_r - \mu \bar{z} v_r) \phi_s \, dx + \sum_{r=0}^{\infty} p_r \int_0^l T_r \phi_s' \, dx + \sum_{r=0}^{\infty} \dot{p}_r \int_0^l \Gamma \phi_r' \phi_s' \, dx$$

$$= - \int_0^l \bar{z} \phi_s Y \, dx + \int_0^l \phi_s K \, dx. \qquad (3.61)$$

We now add equations (3.59), (3.60), (3.61) together. This gives

$$\sum_{r=0}^{\infty} \ddot{p}_r \int_0^l [\mu v_r v_s + I_s \phi_r \phi_s + I_z \theta_r \theta_s - \mu \bar{z}(\phi_r v_s + \phi_s v_r)] \, dx$$

$$+ \sum_{r=0}^{\infty} \dot{p}_r \int_0^l [\alpha kAG \gamma_r (v_s' - \theta_s) + \beta EI \theta_r' \theta_s' + \Gamma \phi_r' \phi_s'] \, dx$$

$$+ \sum_{r=0}^{\infty} p_r \int_0^l [EI \theta_r' \theta_s' + kAG \gamma_r (v_s' - \theta_s) + T_r \phi_s'] \, dx$$

$$= \int_0^l [Y(v_s - \bar{z} \phi_s) + K \phi_s] \, dx.$$

and this is true for $s = 0, 1, 2, \ldots$. That is to say,

$$\sum_{r=0}^{\infty} [a_{rs} \delta_{rs} \ddot{p}_r + \omega_r^2 a_{rs} \delta_{rs} p_r + (\alpha_{rs} + \beta_{rs} + \Gamma_{rs}) \dot{p}_r]$$

$$= \int_0^l [Y(v_s - \bar{z} \phi_s) + K \phi_s] \, dx, \qquad (3.62)$$

for $s = 0, 1, 2, \ldots$ and where

$$
\left.\begin{array}{l}
\alpha_{rs} = \displaystyle\int_0^l \alpha kAG\gamma_r\gamma_s \, dx, \\[12pt]
\beta_{rs} = \displaystyle\int_0^l \beta EI\theta_r'\theta_s' \, dx, \\[12pt]
\Gamma_{rs} = \displaystyle\int_0^l \Gamma\phi_r'\phi_s' \, dx.
\end{array}\right\}
\qquad (3.63)
$$

If the rigid body modes of equations (3.56) are adopted this gives

$$
\left.\begin{array}{ll}
a_{00}\ddot{p}_0 + a_{20}\ddot{p}_2 = \displaystyle\int_0^l Y \, dx & \text{(for } s = 0), \\[12pt]
a_{11}\ddot{p}_1 + a_{12}\ddot{p}_2 = \displaystyle\int_0^l Y\left(1 - \frac{x}{\bar{x}}\right) dx & \text{(for } s = 1), \\[12pt]
a_{02}\ddot{p}_0 + a_{12}\ddot{p}_1 + a_{22}\ddot{p}_2 = \displaystyle\int_0^l (K + z_C Y) \, dx & \text{(for } s = 2), \\[12pt]
a_{ss}\ddot{p}_s + \displaystyle\sum_{r=3}^{\infty} (\alpha_{rs} + \beta_{rs} + \Gamma_{rs})\dot{p}_r + c_{ss}p_s \\[12pt]
\quad = \displaystyle\int_0^l [Y(v_s - \bar{z}\phi_s) + K\phi_s] \, dx & \text{(for } s > 2).
\end{array}\right\}
\qquad (3.64)
$$

Seakeeping analysis of antisymmetric motion

The first three of equations (3.64) govern motions in the antisymmetric rigid body modes. They are of interest in seakeeping analysis and are therefore of importance. Now, it is an essential feature of modal analysis that equations governing motions in rigid body modes not only fit in with those governing the distortion modes in an analytical sense but also are what one would derive on the basis of elementary rigid dynamics. It is worthwhile, then, to show that the first three of equations (3.64) can be derived in a less roundabout way.

Consider first translation of the rigid ship in the OY direction. The net external force in that direction is equal to the mass of the ship multiplied by the acceleration of its centre of mass in that direction. That is to say,

$$
\int_0^l Y \, dx = \int_0^l \mu\ddot{v}_C \, dx = \int_0^l \mu(\ddot{v} - \bar{z}\ddot{\phi}) \, dx.
$$

Now

$$
\ddot{v}(x, t) = \sum_{r=0}^{2} \ddot{p}_r(t)v_r(x), \qquad \ddot{\phi}(x, t) = \sum_{r=0}^{2} \ddot{p}_r(t)\phi_r(x)
$$

and so

$$
\int_0^l Y \, dx = \ddot{p}_0 \int_0^l \mu \, dx + \ddot{p}_1 \int_0^l \mu\left(1 - \frac{x}{\bar{x}}\right) dx + \ddot{p}_2 \int_0^l \mu z_C \, dx.
$$

Thus

$$\int_0^l Y \, dx = a_{00}\ddot{p}_0 + a_{02}\ddot{p}_2$$

since $a_{01} = 0 = a_{10}$.

It is convenient now to turn to the theory of moving axes. Let $Gx'y'z'$ be a frame of *body* axes fixed to the ship with its origin at the ship's centre of mass and distant $\bar{x}$ from O. The axis Gx' points forward, Gy' to port and Gz' upwards in the plane of hull symmetry. For small angular displacements we may take the angular velocities about Gx' and Gz' as

$$p = \dot{p}_2(t) \quad \text{and} \quad r = \frac{\dot{p}_1(t)v_1(x)}{x - \bar{x}} = -\frac{\dot{p}_1(t)}{\bar{x}}$$

respectively. The moments of inertia about Gx' and Gz' are

$$I_{x'} = \int_0^l [I_C + \mu(z_C - z_G)^2] \, dx,$$

$$I_{z'} = \int_0^l \mu(x - \bar{x})^2 \, dx$$

respectively, where z_G is the distance of the origin G below the axis OX. The product of inertia

$$I_{x'z'} = -\int_0^l \mu(x - \bar{x})(z_C - z_G) \, dx.$$

Consider first the yaw equation. The theory of rigid dynamics states that, for the ship, the net applied yawing moment about Gz' is equal to $I_{z'}\dot{r} - I_{x'z'}\dot{p}$. That is to say

$$\int_0^l Y(x - \bar{x}) \, dx = -\frac{\ddot{p}_1}{\bar{x}} \int_0^l \mu(x - \bar{x})^2 \, dx + \ddot{p}_2 \int_0^l \mu(x - \bar{x})(z_C - z_G) \, dx.$$

If this equation is divided throughout by $-\bar{x}$, it gives

$$\int_0^l Yv_1 \, dx = \ddot{p}_1 \int_0^l \mu v_1^2 \, dx + \ddot{p}_2 \int_0^l \mu v_1 z_C \, dx,$$

since

$$z_G \int_0^l \mu v_1 \, dx = 0.$$

We have thus found the second of the equations of rigid body motion in (3.64).

The applied rolling moment, according to theory of rigid dynamics, is equal to $I_{x'}\dot{p} - I_{x'z'}\dot{r}$. Here, then

$$\int_0^l [K + Y(z_C - z_G)] \, dx = \ddot{p}_2 \int_0^l [I_C + \mu(z_C - z_G)^2] \, dx$$

$$-\frac{\ddot{p}_1}{\bar{x}} \int_0^l \mu(x - \bar{x})(z_C - z_G) \, dx.$$

It follows that

$$\int_0^l (K + Yz_C)\, dx = z_G \int_0^l Y\, dx + \ddot{p}_2 \int_0^l (I_C + \mu z_C^2)\, dx - \ddot{p}_2 z_G \int_0^l \mu z_C\, dx$$

$$+ \ddot{p}_1 \int_0^l \mu v_1 z_C\, dx.$$

If we now substitute for

$$\int_0^l Y\, dx$$

using the sway equation, we find that this equation reduces to

$$\int_0^l (K + Yz_C)\, dx = \ddot{p}_0 \int_0^l \mu z_C\, dx + \ddot{p}_1 \int_0^l \mu v_1 z_C\, dx$$

$$+ \ddot{p}_2 \int_0^l (I_C + \mu z_C^2)\, dx.$$

This is as stated in our theory previously.

We see, then, that the first three of equations (3.64) are simply those which may be found by assuming the ship to be rigid from the outset. The merit of deriving them as we did previously is that they are then seen to conform to a much more general pattern – the hull can, in fact, distort.

3.4 Antisymmetric motion of a boxlike ship

As we said in section 3.3 the theory for a boxlike hull is a special case of that already presented. But it became clear in chapter 2 that, while this is true, the relationship requires special consideration. It will be remembered that we defined a boxlike ship as one in which the centre of mass C of each slice of the hull coincides with the shear centre S. That is to say $\bar{z} = 0$. Now if $\bar{z} = 0$, equations (3.26) and (3.31) are entirely independent of each other. Variations of $v(x, t)$, $\theta(x, t)$ and $\gamma(x, t)$ are independent of those of $\phi(x, t)$. Thus the last of equations (3.58) should now be replaced by some such expression as

$$\phi(x, t) = \sum_{i=0}^{\infty} q_i(t)\phi_i(x). \tag{3.65}$$

Now equations (3.62) and (3.64) fall into two distinct sets of equations save only for the rigid body component $v(x, t) = q_0(t)z_C(x)\phi_0(x)$ which accompanies a rigid rotation of roll.

The transverse motion of horizontal bending is now governed by the equations

$$\left.\begin{aligned}
a_{00}\ddot{p}_0 &= \int_0^l Y\, dx, \\[2mm]
a_{11}\ddot{p}_1 &= \int_0^l Y\left(1 - \frac{x}{\bar{x}}\right) dx, \\[2mm]
a_{ss}\ddot{p}_s + \sum_{r=2}^{\infty} (\alpha_{rs} + \beta_{rs})\dot{p}_r + c_{ss}p_s &= \int_0^l Yv_s\, dx \qquad (s = 2, 3, \ldots),
\end{aligned}\right\} \tag{3.66}$$

where, now,

$$a_{ss} = \int_0^l (\mu v_s^2 + I_z \theta_s^2)\, dx,$$

$$c_{ss} = \int_0^l (EI\theta_s'^2 + kAG\gamma_s^2)\, dx. \tag{3.67}$$

The rigid body modes are

$$v_0(x) = 1, \qquad \theta_0(x) = 0 = \gamma_0(x),$$

$$v_1(x) = 1 - \frac{x}{\bar{x}}, \qquad \theta_1(x) = 0 = \gamma_1(x). \tag{3.68}$$

Turning next to the twisting deflections we extract the set

$$a_{00}\ddot{q}_0 = \int_0^l K\, dx,$$

$$a_{jj}\ddot{q}_j + \sum_{i=1}^\infty \Gamma_{ij}\dot{q}_j + c_{jj}q_j = \int_0^l K\phi_j\, dx \qquad (j = 1, 2, \ldots) \tag{3.69}$$

from equations (3.64). Now,

$$a_{jj} = \int_0^l I_S \phi_j^2\, dx, \qquad c_{jj} = \int_0^l T_j \phi_j'\, dx, \tag{3.70}$$

and the one rigid body mode is

$$\phi_0(x) = 1 \qquad \text{with} \qquad v(x) = z_S(x) = z_C(x). \tag{3.71}$$

Since there is now no warping of the hull cross-sections, the modal twisting moment takes the simplified form

$$T_j = C(x)\phi_j'(x, t). \tag{3.72}$$

3.5   General conclusions

In this chapter we have improved the simple Bernoulli–Euler theory of chapter 2. Bending distortions are those admitted in the theory of a Timoshenko beam. As we shall see, the effect of making this change is usually fairly minor so far as the shapes of those principal modes of lowest order are concerned. But the effect on the corresponding natural frequencies is often substantial. (Notice that the Timoshenko beam necessarily has lower natural frequencies than the corresponding Bernoulli–Euler beam since its stiffness is reduced by the allowance for shear deflection and its inertia is increased by the rotatory inertia.)

At first sight it may seem surprising that, despite the added complications, the theory in this chapter produces exactly the same equations as those arrived at in the last. Thus generalised masses, damping coefficients, stiffnesses, displacements and forces at the principal coordinates all perform the same functions as before, though of course the expressions for them are different. In fact, however, it would be surprising if this were not so; for *both* theories conform to the general pattern of linear vibration theory which may be developed by the use of Lagrange's equation (e.g. see Bishop & Johnson, 1960). As we shall discover, the practical effect of this is of the greatest importance.

4 The characteristics of practical hulls

Haughty, humble, coy, or free,
   Little care I what maid may be.
So that a maid is fair to see,
   Every maid is the maid for me!
Princess Ida

4.1   Hull data

In chapters 2 and 3 we have discussed the dynamical characteristics of hulls in abstract terms. Interest has centred on principal modes and on the equations governing modal responses. We shall now turn our attention to practical hulls, to try to show how theory accords with reality.

In chapter 2, bending distortions were analysed on the basis of the simple Bernoulli–Euler theory of beams. In chapter 3 we introduced the more complicated theory of the Timoshenko beam. The more elementary theory can be regarded as a special form of the more complex one. Although this observation is hardly a surprising one, it is of considerable importance. It means that if we can draw up a convenient scheme of presenting the necessary data and deriving the requisite results for a Timoshenko beam representation, it will also cater for the Bernoulli–Euler representation.

Such is the present difficulty of obtaining data of the sort required, it is hardly surprising that those which have been found and used here are apparently open to question. The estimation and specification of structural properties are by no means without their difficulties. The deductions from the given data, on the other hand, are relatively straightforward; thus the modes and frequencies should correspond accurately to the given data, whether those data be dependable or not.

4.2   Data relevant to symmetric response

Such have been the methods used hitherto in estimating ship response, there is sufficient relevant information available on very few practical hulls, relating to symmetric response. There is almost none readily available on antisymmetric response. We shall first discuss hull data relevant to symmetric response. In doing so we shall refer to three hulls:

(a) The 'uniform ship' which is not a ship at all, but merely a uniform beam whose behaviour can be discussed in simple terms by reference to 'exact' theory;

(b) A hull of 'fine form' which, to be a little more precise, is that of a destroyer (that has, alas, long since been scrapped);

(c) A large tanker of 250 000 DWT.

The uniform ship will serve two purposes. First of all the results obtained for it will allow us to check the accuracy of the necessarily approximate techniques of calculation that will have to be used for ship hulls. Secondly, by giving the uniform ship fairly representative dimensions we can find fairly representative values of natural frequencies, and so obtain a better grasp of the order of things.

4.2.1 *Inertia and stiffness data for symmetric response*

There are, of course, many ways of extracting the principal modes and natural frequencies of a non-uniform beam vibrating in flexure (e.g. see Bishop, Gladwell & Michaelson, 1965). Perhaps the most useful in ship dynamics is the so-called Prohl–Myklestad method (e.g. see Myklestad, 1944; Bishop, 1956) which is familiar in vibration theory and which can be made to cater for a Timoshenko beam. We have drawn up and used variants of that technique in deriving the results that are quoted in this book.

The hull is imagined to be cut into slices. It is convenient to make all the slices of the same thickness. The number of slices to be employed depends on the degree of accuracy that is required and on what use is subsequently to be made of the derived modes and frequencies. If interest will centre on excitation of high frequency of a mode of fairly high order, then more slices will be needed than if only the lowest distortion modes matter. In practice, twenty slices at least should be used for most calculations. Experience suggests that fifty slices are likely to be adequate for most practical purposes. Generally speaking, the greater the number of slices used the greater will be the accuracy of subsequent calculations. (When frequencies and modes are high, and

Fig. 4.1. The dry hull of an old destroyer. For the purposes of calculation the hull is thought of as indicated in the diagram drawn beneath. The length is divided into slices of equal thickness (in this case 20 of them) and the mass is concentrated at the centre of each slice.

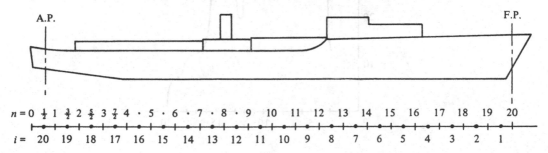

we are concerned with a mode that is of higher order than the 4th or 5th, then it is necessary to refine the equations of motion anyway.)

Fig. 4.1 shows the hull of a destroyer divided into twenty slices. The mass of each slice is assumed to be concentrated halfway across the thickness of each slice.† The slices are denoted by $i = 1, 2, \ldots, 20$ starting from the bow. Fig. 4.2 shows the distribution of mass per unit length, $\mu(x)$; it is this curve which, mathematically speaking, is first approximated by a stepped curve when the various concentrated masses are assigned to the slices.

In the theory we have referred to the distance x as measured forward from the stern. It is convenient to employ a compatible system of identification for the slices. Noting that 21 stations are identified along the hull when 20 slices are formed by 19 imaginary 'cuts', we let $n = 0, 1, 2, \ldots, 20$ denote those stations. This is made clear in fig. 4.1, where it will be seen that the masses are concentrated at the stations $n = 1/2, 3/2, \ldots, 39/2$.

When rotatory inertia is introduced its distribution along the hull has to be specified in some way. The full-line curve in fig. 4.3 shows the rotatory inertia per unit length along the hull. To all intents and purposes the curve represents data that should be specified by the designer. This curve, too, has to be approximated in a stepped form before values of rotatory inertia can be allotted to the individual slices.

It happens, as we shall discover, that the inclusion of rotatory inertia does not make much difference to the modes and frequencies. It therefore may not matter much if the curve of $I_y(x)$ is only a rough one – and that of fig. 4.3 certainly does appear to raise some questions.

† This is not the only possibility. It may be equally divided between the two surfaces of the slice. We have merely quoted one method for the sake of definiteness.

Fig. 4.2. The variation of mass/unit length of the destroyer shown in fig. 4.1.

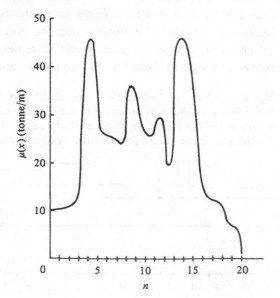

Since $I_y(x)$ is a somewhat complicated quantity to calculate accurately, it is of interest to note an approximation, even though it is extremely crude. The second moment of area about the neutral axis is

$$I(x) = \int_A y^2 \, dA,$$

where A is the area of cross-section of the hull and y is distance from the neutral axis of bending. Moreover the rotatory moment of inertia of a short length Δx of the hull is

$$I_y(x) \, \Delta x = \Delta x \int_A \rho y^2 \, dA,$$

where ρ is the density of the hull material. If, then, it is assumed that rotatory inertia is accounted for entirely by the structurally active material of the hull, then it follows that

$$I_y(x) \doteqdot \rho I(x). \tag{4.1}$$

The broken line in fig. 4.3 gives this approximation (the density of steel being taken as $7850 \ \mathrm{kg/m^3}$). The agreement with the data that were actually specified for the ship is very poor indeed; but, as we shall see, the difference that the error would make to the natural frequencies of low order is most unlikely to be large.

In order to calculate the principal modes of the dry hull we need information, not only on the inertial properties but also on stiffness. That is to say it is necessary to specify the flexural rigidity $EI(x)$ and, if possible, the shearing rigidity $kAG(x)$. It will be convenient to refer, here, to the second moment of area $I(x)$ and to the shear area $kA(x)$.

Fig. 4.3. The distribution of rotatory inertia per unit length $I_y(x)$, of the destroyer. The broken curve shows the approximation to $I_y(x)$ that is found when the rule of thumb of equation (4.1) is used.

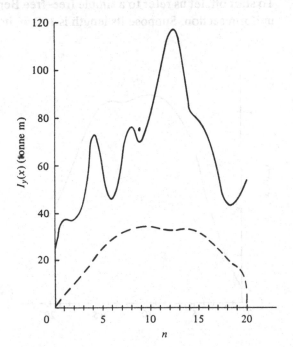

Estimation of the effective values of $I(x)$ at various sections of the hull is one of the established tasks of the naval architect and little need be said on the subject here. Fig. 4.4 shows the variation of this quantity with position for the destroyer of fig. 4.1, though a stepped approximation to this curve would have to be used for the Prohl–Myklestad calculation.

The estimation of shear area is less familiar. Elementary theory suggests that, generally speaking, $kA(x)$ should be taken as the cross-sectional area of those plates in the hull structure whose planes are vertical or nearly so. But it is unlikely that the matter is really straightforward and, since shear deflection does modify natural frequencies significantly, this is a matter that we shall leave open for the present. Fig. 4.5 shows the distribution of $kA(x)$ that was thought appropriate for the destroyer of fig. 4.1, though here again a stepped approximation would normally be employed in the computation of modes and frequencies.

4.2.2 Symmetric principal modes and natural frequencies

As we have already suggested, much of the task that is immediately before us is of a familiar sort. There is therefore no need to rehearse the details of the calculations in this book. What is, perhaps, unfamiliar, is the application of the standard techniques to ship hulls. Our concern, then, will chiefly be to present results so that the reader can acquire a 'feel' for the orders of magnitude involved.

The 'uniform ship'

To start off, let us refer to a simple free–free Bernoulli–Euler beam of uniform section. Suppose its length is 100 m, its total mass 2×10^6 kg

Fig. 4.4. The variation of second moment of area $I(x)$ for the destroyer.

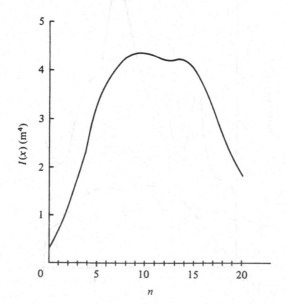

and its flexural rigidity $0.64 \times 10^{12} \, \text{N m}^2$. For this imaginary 'hull', therefore,

$$\mu(x) = \frac{2 \times 10^6}{100} = 20 \times 10^3 \, \text{kg/m},$$

$$I(x) = \frac{0.64 \times 10^{12}}{0.207 \times 10^{12}} = 3.092 \, \text{m}^4,$$

Young's modulus for steel being 0.207×10^{12} Pascals. A glance at figs. 4.2 and 4.4 reveals that our 'uniform ship' has values of $\mu(x)$ and $I(x)$ which are comparable with those of the destroyer of fig. 4.1. (In fact the destroyer was about 107 m long instead of the 100 m we have selected.)

The principal modes and natural frequencies of a uniform beam are derived and tabulated elsewhere (see Bishop & Johnson, 1960). The rth non-zero frequency ω_r is given by

$$\omega_r = \frac{\alpha_r^2}{l^2}\left(\frac{EI}{\mu}\right)^{\frac{1}{2}} = \frac{\alpha_r^2}{10^4}\left(\frac{0.64 \times 10^{12}}{20 \times 10^3}\right)^{\frac{1}{2}} = 0.5657\alpha_r^2,$$

where α_r is the rth root of the frequency equation

$$\cos \alpha_r \cosh \alpha_r - 1 = 0 \qquad (r = 2, 3, \ldots).$$

The corresponding characteristic functions $w_r(x)$, scaled so as to give $w_r(0) = 1$, are

$$w_r(x) = \tfrac{1}{2}\{\cosh(\alpha_r x/l) + \cos(\alpha_r x/l) - \sigma_r[\sinh(\alpha_r x/l) + \sin(\alpha_r x/l)]\},$$

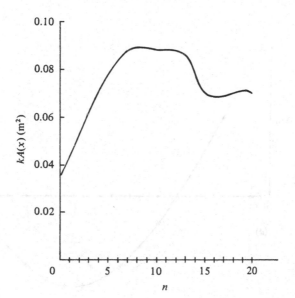

Fig. 4.5. The variation of shear area $kA(x)$ for the destroyer.

where

$$\sigma_r = \frac{\cosh \alpha_r - \cos \alpha_r}{\sinh \alpha_r - \sin \alpha_r} \qquad (r = 2, 3, \dots).$$

The modal bending moment is

$$M_r = EI \frac{\mathrm{d}^2 w_r(x)}{\mathrm{d}x^2}$$

and the modal shearing force is

$$V_r = -EI \frac{\mathrm{d}^3 w_r(x)}{\mathrm{d}x^3}.$$

Since the characteristic function and its derivatives are tabulated, it is a straightforward matter to plot curves of $w_r(x)$, $M_r(x)$, $V_r(x)$ for $r = 2, 3, \dots$. The curves so obtained are given in

fig. 4.6 (a), (b), (c) for $r = 2$,

fig. 4.7 (a), (b), (c) for $r = 3$,

fig. 4.8 (a), (b), (c) for $r = 4$.

Suppose now that the uniform ship is divided into 20 sections (or 'slices') and that the frequencies and modes are calculated instead by the Prohl–Myklestad method. The results obtained in this way for $r = 2, 3, 4$ are those marked by crosses in figs. 4.6, 4.7 and 4.8. When we employ 50 sections instead of 20 we find even closer agreement. Moreover the exact and computed natural frequencies are as shown in Table 4.1, from which it will be seen that agreement is excellent even up to the 7-node mode – i.e. up to an order of mode at which the validity of the assumptions is decidedly strained.

Fig. 4.6. The curves refer to the $r = 2$ mode of a uniform beam (i.e. the first principal mode that involves distortion). The curves are calculated from 'exact' theory for a Bernoulli–Euler beam and the crosses show the predictions of a Prohl–Myklestad approximation using a 20-slice idealisation of the beam. The mode is shown, with its scale, in (a). The corresponding bending moment $M_2(x)$ and shearing force $V_2(x)$ are shown in (b) and (c) respectively.

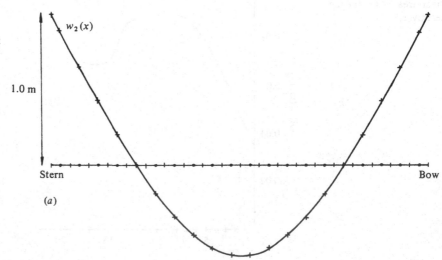

Fig. 4.6. (Continued)

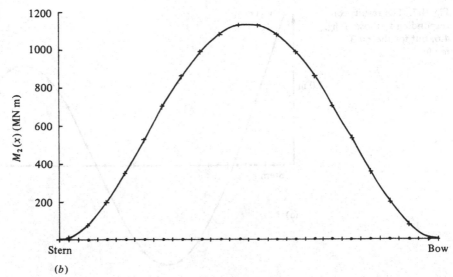

(b)

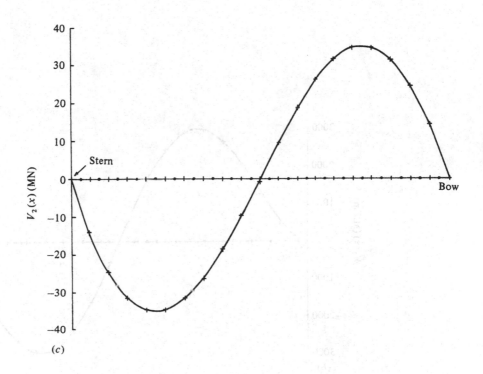

(c)

Fig. 4.7. The results corresponding to those of fig. 4.6, but for the $r = 3$ mode.

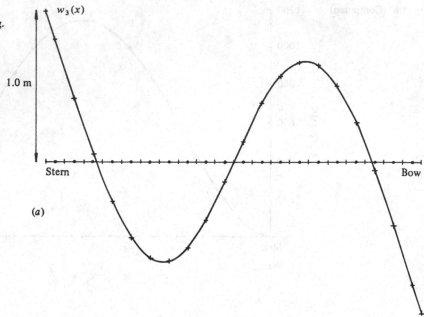

(a)

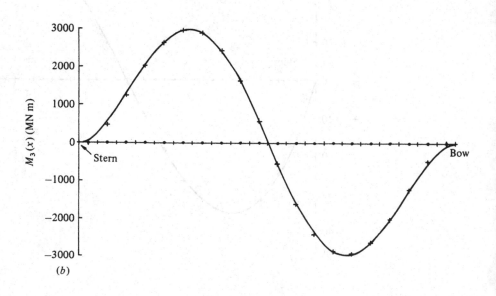

(b)

Fig. 4.7. (Continued)

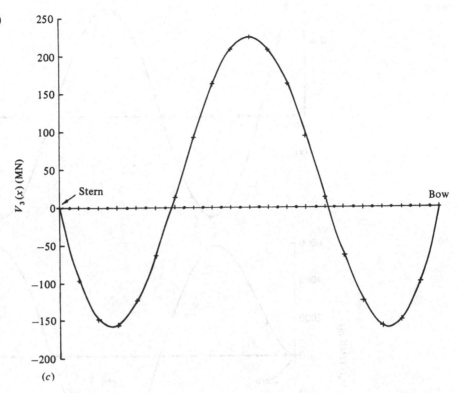

(c)

Table 4.1. *Exact and computed natural frequencies*
of the uniform ship

Modal index r	Exact value of natural frequency ω_r (rad/s)	Computed value using 20 sections ω_r (rad/s)	Computed value using 50 sections ω_r (rad/s)
2	12.656	12.675	12.661
3	34.888	34.781	34.877
4	68.393	67.587	68.281
5	113.06	110.25	112.64
6	168.9	161.8	167.8
7	235.9	220.9	233.6

Fig. 4.8. The results corresponding to those of fig. 4.6, but for the $r = 4$ mode.

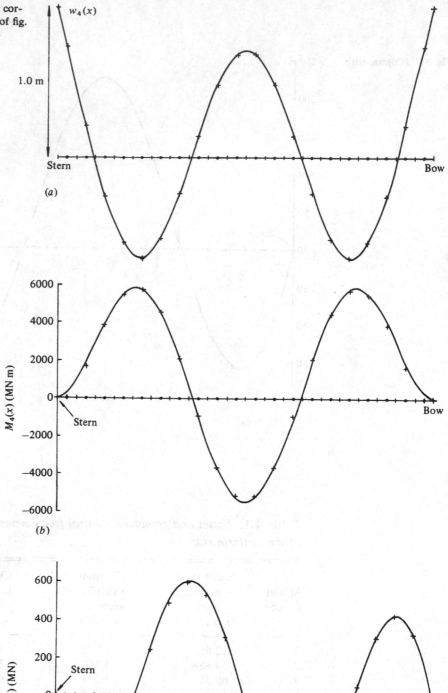

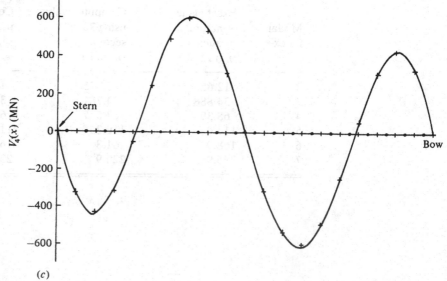

This example clearly shows the striking accuracy that is obtained with the Prohl–Myklestad method. It also gives us a first glimpse of the natural frequencies of a dry hull. For a ship like the old destroyer of fig. 4.1, it would appear that the lowest non-zero natural frequencies are, very roughly,

2 Hz, $5\frac{1}{2}$ Hz, 11 Hz, 18 Hz, ...

A destroyer hull

Let us now turn to the destroyer hull of fig. 4.1, to which the curves of figs. 4.2, 4.3, 4.4 and 4.5 relate. There is now no question of an exact solution and it is convenient to employ the Prohl–Myklestad technique from the outset. Four separate sets of results may be obtained by that means, corresponding to the assumptions set out in Table 4.2. By finding the modes and frequencies in this way we can obtain some idea of the relative importance of the two corrections, and see how big an effect they have on the predictions of the Bernoulli–Euler theory. In all the computations for this destroyer hull, 20 sections are employed.

To start with the natural frequencies, results are as shown in Table 4.3. All values quoted are in rad/s. In a preliminary way we see that the 'best' estimates (i.e. those of Case IV) of the six lowest non-zero natural frequencies are[†]

2.16 Hz, 4.33 Hz, 6.37 Hz, 9.24 Hz, 11.80 Hz, 13.76 Hz.

It will be seen that the addition of each correction lowers all the natural frequencies. This is in line with the general theory of linear vibration since the correction for shear effectively reduces the stiffness of the beam while the correction for rotatory inertia effectively increases its inertia (e.g. see Bishop & Johnson, 1960). But the table

---

[†] The rough and ready predictions we made when discussing the 'uniform ship' correspond to the Bernoulli–Euler beam and are comparable with those of Case I, viz.

2.35 Hz, 5.37 Hz, 9.11 Hz, 15.41 Hz.

The agreement is quite good, considering how crude the underlying assumptions were.

Table 4.2. *Possible assumptions in the idealisation of a hull as a beam*

Case	Correction made for shear deflections	Correction made for rotatory inertia
I	No	No
II	Yes	No
III	No	Yes
IV	Yes	Yes

shows that, of the two corrections, that for shear is by far the more important where these low-order modes are concerned.

Each natural frequency is, of course, associated with a principal mode. It is found that, while the modes too are hardly altered by the rotatory inertia correction so long as the mode is of low order, the inclusion of an allowance for shear may be important. That the differences found between Cases I and II may be significant can be readily demonstrated. The curves of fig. 4.9(a), (b) and (c) show the variations of $w_2(x)$, $M_2(x)$, $V_2(x)$, the crosses relating to Case I and the dots to Case II. Similarly figs. 4.10(a), (b) and (c) relate to the third mode and figs. 4.11(a), (b) and (c) to the fourth.

It is clear from figs. 4.9–4.11 that the simple theory of chapter 2 (which refers to Case I in Table 4.2) is not really satisfactory. We shall therefore use either Case II or Case IV henceforth and take advantage of the extra accuracy so obtained by referring to modes up to and including that of 7th order. (Generally speaking, it would not be

Table 4.3. *Calculated natural frequencies (in rad/s) for symmetric principal modes of the dry destroyer hull*

Modal index r	ω_r (rad/s) in case number:			
	I	II	III	IV
2	14.78	13.65	14.66	13.57
3	33.76	27.43	33.25	27.22
4	57.25	40.32	56.05	40.05
5	96.82	58.38	93.55	58.05
6	140.62	74.63	133.63	74.17
7	184.99	86.99	173.58	86.45

Fig. 4.9. The curves (a), (b) and (c) refer to the scaled mode $w_r(x)$, the modal bending moment $M_r(x)$ and the modal shearing force $V_r(x)$ for the $r = 2$ mode of the destroyer. In each case, the crosses refer to Case I (see Table 4.2) and the dots to Case II.

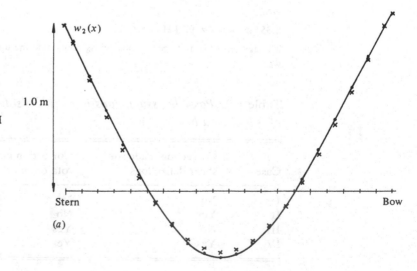

Fig. 4.9. (Continued)

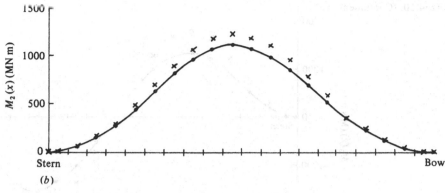

(b)

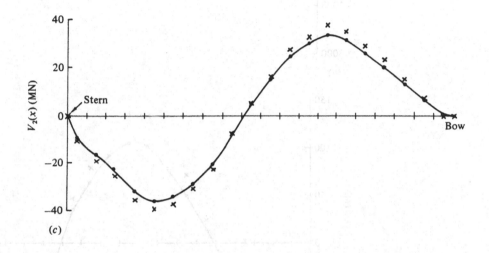

(c)

Fig. 4.10. Results as in fig. 4.9, but for the $r = 3$ mode of the destroyer.

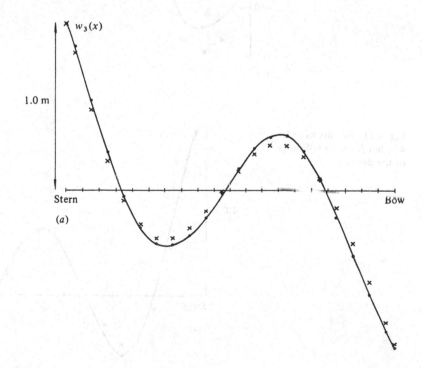

(a)

Fig. 4.10. (Continued)

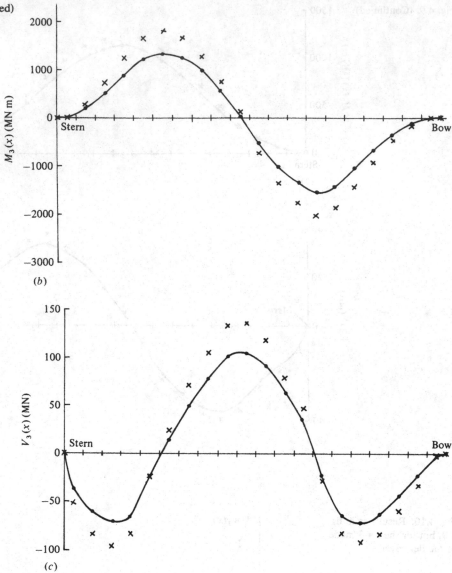

(b)

(c)

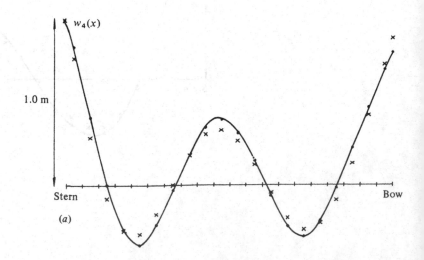

Fig. 4.11. Results as in fig. 4.9, but for the $r = 4$ mode of the destroyer.

$w_4(x)$

1.0 m

Stern

Bow

(a)

Fig. 4.11. (Continued)

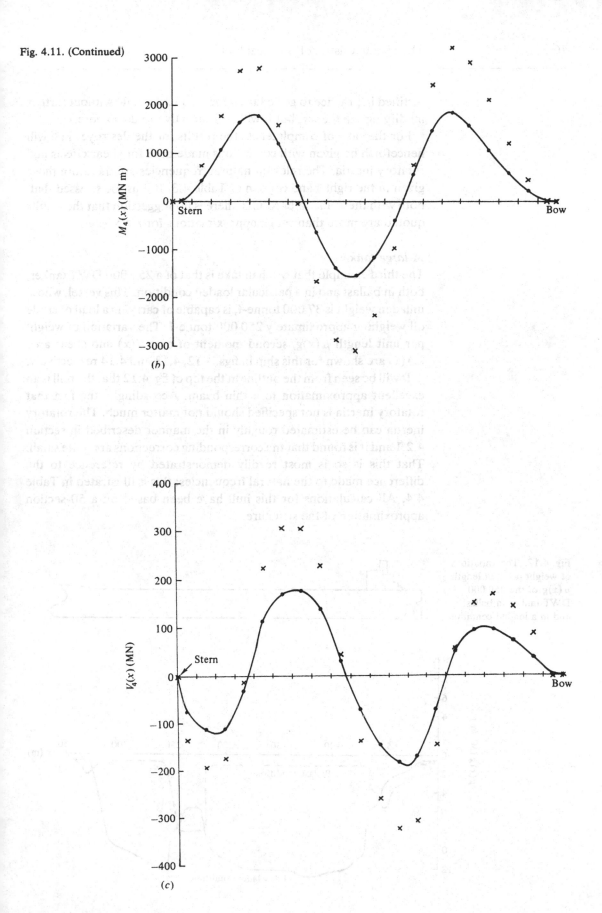

justified in practice to go so far up the frequency scale without further modifying the theory, but it will be instructive to do so here.)

For the sake of completeness the results for the destroyer hull will henceforth be given with corrections made both for shear effects and rotatory inertia. The relevant natural frequencies are therefore those given in the right-hand column of Table 4.3. It must be stressed that, both with these and later results, there is no suggestion that the results quoted are more than rough approximations for $r \geqslant 5$, say.

A large tanker

The third example that we shall take is that of a 250 000 DWT tanker, both in ballast and in a particular loaded condition. This vessel, whose unladen weight is 37 000 tonne-f, is capable of carrying a load of crude oil weighing approximately 250 000 tonne-f. The variation of weight per unit length $\mu(x)g$, second moment of area $I(x)$ and shear area $kA(x)$ are shown for this ship in figs. 4.12, 4.13 and 4.14 respectively.

It will be seen from the outline at the top of fig. 4.12 that the hull is an excellent approximation to a thin beam. Accordingly, the fact that rotatory inertia is not specified should not matter much. The rotatory inertia can be estimated roughly in the manner described in section 4.2.1 and it is found that the corresponding corrections are quite small. That this is so is most readily demonstrated by reference to the difference made to the natural frequencies; this is illustrated in Table 4.4. All calculations for this hull have been based on a 50-section approximation of the structure.

Fig. 4.12. The variations of weight per unit length $\mu(x)g$ of the 250 000 DWT tanker in ballast and in a loaded condition.

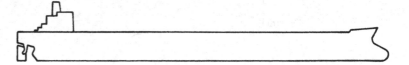

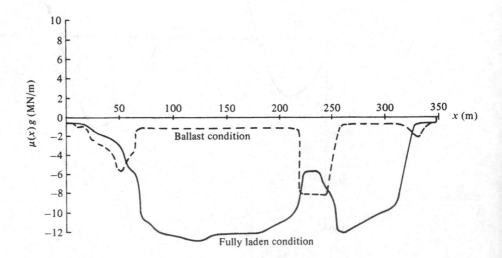

The principal modes $w_r(x)$ of the dry hull are heavily dependent on whether the tanker is in ballast or loaded, as one would expect. So, too, are the associated curves of modal shearing force $V_r(x)$ and modal bending moment $M_r(x)$. But none of these curves is sensitively dependent on whether or not an allowance is made for the effects of rotatory inertia.

Fig. 4.13. The curve of second moment of area $I(x)$ for the tanker.

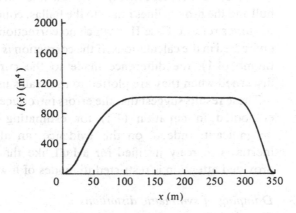

Fig. 4.14. The curve of shear area $kA(x)$ for the tanker.

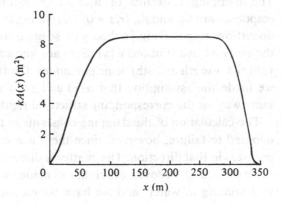

Table 4.4. *Calculated natural frequencies (in rad/s) for symmetric principal modes of a dry tanker hull*

Modal index r	Ballast condition		Loaded condition	
	ω_r (rad/s) in case no:		ω_r (rad/s) in case no:	
	II ($I_y = 0$)	IV (I_y estimated)	II ($I_y = 0$)	IV (I_y estimated)
2	6.73	6.68	4.01	4.00
3	15.83	15.69	10.05	10.01
4	29.90	29.55	16.71	16.63
5	44.42	43.76	24.10	23.98
6	57.82	57.02	33.11	32.95
7	69.54	68.84	42.02	41.78

Reading downwards in columns, the curves of fig. 4.15(a), (b), (c), (d) show:

$w_r(x)$ which must be interpreted as having an ordinate of 1 m at
             $x = 0$,
$M_r(x)$ corresponding to the scaled $w_r(x)$,
$V_r(x)$ corresponding to the scaled $w_r(x)$,

for $r = 2, 3, 4, 5$ respectively. The full-line curves are for the loaded hull and the broken lines relate to the ballast condition. Both of the sets of curves refer to Case II in which no correction for rotatory inertia is embodied in the calculations. If the correction *is* made using the rule of thumb (4.1), the difference made to the curves cannot easily be discerned when they are plotted to the scales used in fig. 4.15.

These results suggest that the errors introduced by the rule of thumb embodied in equation (4.1) for estimating $I_y(x)$ may be quite insignificant. Indeed, on this evidence, an allowance for rotatory inertia is scarcely justified for a hull like that of the large tanker, provided attention is restricted to modes of low order.

4.2.3   Damping of symmetric distortions

The damping constants of interest in estimations of symmetric responses are α_{rs} and β_{rs} $(r, s \neq 0, 1)$, relating to shearing and bending distortions respectively. In theory these quantities can be calculated if the relevant distributions $\alpha(x)$, $\beta(x)$ are known, along with the relevant (i.e. the rth and sth) principal modes. (Notice that, in the theory, we made the assumption that $\alpha(x)$ and $\beta(x)$ are distributed 'in the same way' as the corresponding structural rigidities.)

The calculation of the damping constants in this way appears to be doomed to failure, however, since there are some large obstacles to progress in that direction. This matter is discussed by Betts, Bishop & Price (1977a). Moreover, observations made on a ship are made while it is floating in water, and we have no means of deciding how the observed damping of a ship should be apportioned between that of structural and that of hydrodynamic origin.

The rate of energy loss by hysteresis is probably much greater than could be accounted for by integration throughout the whole hull under the assumption that field stresses are related to the bending moment and shearing force in any simple fashion. It seems probable – and the question is by no means closed – that the disparity exists for three main reasons. The first of these arises from the method used to join the plates of which the hull is constructed. Welding greatly increases local mean stresses, and does so over a significant volume of hull material (see Faulkner, 1973); this in turn significantly increases the rate of energy dissipation by hysteresis. If, on the other hand, the plates are riveted, slight slip can occur in the vicinity of the rivets so that energy losses may occur due to the friction of rubbing. Generally speaking, a riveted hull has greater damping than a welded one.

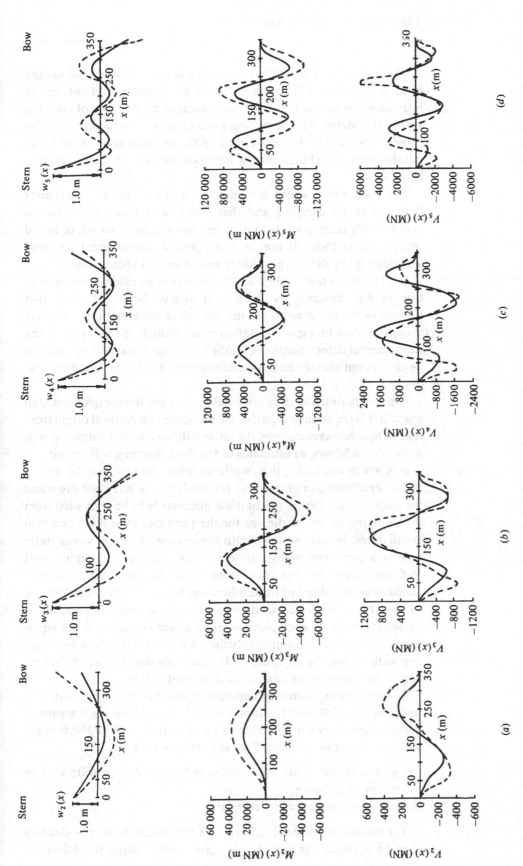

Fig. 4.15. Curves of $w_r(x)$, $M_r(x)$ and $V_r(x)$ for the tanker. The full-line curves are for the loaded hull and the broken lines refer to the hull in ballast and they correspond to Case II of Table 4.2. (The results for Case IV differ only very slightly from these.) The sets of curves are arranged so that (a) refers to the $r = 2$ mode, (b) refers to the $r = 3$ mode, (c) refers to the $r = 4$ mode and (d) refers to the $r = 5$ mode.

The second hidden source of damping is again to be found in high local mean stresses. These are due to stress concentrations, either introduced by design (as at deck openings) or by the state of cracking within the metal. All the plating used in ship construction contains small cracks; the problem confronting the designer is not whether or not the cracks exist but, rather, whether or not they will grow – and if so how rapidly.

The third way in which one might seek to explain the discrepancy between actual damping and that which would be found from a rudimentary calculation is to examine the effects of cargo, of liquid stores and so forth. It may well be possible to augment damping significantly, by the use of suitable baffles in tanks for instance.

It is fair to say that negligible progress has been made in the estimation of hull damping by theoretical means. Nor is it likely that progress will ever be rapid, for the difficulties are enormous. We must therefore turn to experimentation even though it suffers from the fundamental defect mentioned earlier, viz. that tests have perforce to be carried out with the hull in water and not while it is unsupported and dry.

Some experimental data are available on the damping of wet ship hulls. If it is required to separate the damping of structural origin from that of hydrodynamic origin, the latter will have to be subtracted with as much confidence as estimates of the fluid damping will permit.

It is worth remarking that, while we have already seen the way in which structural damping enters the analysis, we have not discussed the question of whether or not it has necessarily to be separated from the damping effects of the sea for the purposes of making practical calculations. Indeed, we might also remind ourselves that a ship under way is a non-conservative system whose 'inertia', 'damping' and 'stiffness' matrices are not symmetric, as we shall discover. Consequently the use of these adjectives to describe the matrices is a loose one and, in particular, there is no assurance that all 'damping' is associated with the coefficients of generalised velocities in the equations of motion, as is commonly believed. This may well be a fine point in practice since the damping of distortion modes is so small, but we appear to have no reassurance on that point either.

To return to experimental determinations of damping, we may note that they have followed traditional lines. Viscous damping is assumed, and it is hard to see why anything more sophisticated should be thought worth pursuing in this field. Two approaches are available:

(a) measurement of vibration decay in free oscillation following an initial disturbance,
(b) resonance testing using an exciter.

The initial disturbance used to start a free vibration may be due to a slam or a sudden anchor drop. Again, even though the effects of restricted water are almost inescapable in such tests, a small ship may be lifted at a bollard using a crane pulling a bar whose breaking strength is known.

The exciters used in forced vibration tests are usually of the inertial type employing contra-rotating unbalanced wheels. The response is measured and damping factors are obtained either from amplitude plots or, alternatively and better, from polar plots (e.g. see Kennedy & Pancu, 1947; or Bishop & Gladwell, 1963).

Provided it is assumed that hydrodynamic damping is negligible, these experimental methods afford practical means of determining the direct coefficients $b_{rr} = \alpha_{rr} + \beta_{rr}$, though not of separating the damping of shearing from the damping of bending. The cross-coefficients α_{rs}, β_{rs} ($r \neq s$), by contrast, would be very hard to extract from these results, though it is common to assume that the cross-terms may be neglected. This latter assumption is almost certainly justified for the cross damping coupling of the principal modes of a distorting dry hull.

All in all, the damping of the distortion modes of a dry hull poses some very difficult questions. It remains a research field of great importance. In these circumstances, the assumption of linear viscous damping is fully justified and we are forced to rely on rule-of-thumb

Table 4.5. *Magnitude of logarithmic decrement δ_r in rth symmetric wet mode*

Reference	Suggested relation	Limitations	Comments
Kumai (1958)	$\delta_2 = 3.5/l$ $\delta_r = \delta_2(\omega_r/\omega_2)^{\frac{3}{4}}$	$r = 2$ only $r > 2$	$80\text{ m} \leqslant l \leqslant 100\text{ m}$ For cargo ships and tankers in ballast condition
Tomita (1960)	$\delta_2 = 2.4 \times 10^{-3} \omega_2$ $\delta_r = 6.8 \times 10^{-3} \omega_r^{0.7}$	$r = 2$ only $r > 2$	Dry cargo ships in ballast. δ_r increased by 0.02 for general cargo
Hirowatari (1963)	$\delta_r = 1.065 \times 10^{-2} \omega_r^{1/2}$ $\delta_r = 6.1 \times 10^{-5} \omega_r^2$	$\omega_r \leqslant 31.5\text{ rad/s}$ $\omega_r > 31.5\text{ rad/s}$	Index increased at higher modes 'due to effects of local vibration'
Johnson, Ayling & Couchman (1962)	$\delta_r = C_r \omega_r$	$C_2 = 2 \times 10^{-3}$ $C_3 = 1.9 \times 10^{-3}$ $C_4 = 1.8 \times 10^{-3}$ $C_5 = 1.75 \times 10^{-3}$	Derived from semi-empirical analysis
Aertssen & de Lembre (1971)	$\delta_r = 7.3 \times 10^{-3} \omega_r$	$\omega_r < 20\text{ rad/s}$	Based mainly on results for $r = 2$ mode

estimates which are based more or less logically on experimental evidence. These estimates are of

logarithmic decrement, δ_r, or
damping factor, $\nu_r = \delta_r/2\pi$,

where $r = 2, 3, \ldots$ but it has to be said that the estimates given by various authorities differ widely. The reliability of the estimates is a matter for debate.

Table 4.5 is taken from the survey by Betts *et al.* (1977a) and it sets out the more important empirical formulae that have been quoted in the literature for welded hulls. The quantity ω_r is in rad/s. Notice that the formulae relate to ships in water and not to dry hulls so that predictions of the formulae can only be used for a dry hull if hydrodynamic damping can validly be assumed negligible.

Using the data already quoted for the destroyer and the tanker we may estimate the hull damping. The values of the damping factors ν_r that are found by the various methods given in Table 4.5 are set out in Table 4.6 under the assumption that hydrodynamic damping is negligible. They are based on calculations in which allowance is made for shear deflection.

Table 4.6. *Values of hull damping factors ν_r*

r	Kumai	Tomita	Hirowatari	Johnson *et al.*	Aertssen & de Lembre
(a) *The destroyer* (Case IV of Table 4.2)					
2	0.005	0.005	0.006	0.004	0.016
3	0.009	0.011	0.009	0.008	—
4	0.012	0.014	0.016	0.012	—
5	0.015	0.019	0.033	0.016	—
6	0.019	0.022	0.053	—	—
7	0.021	0.025	0.073	—	—
(b) *The large tanker in ballast* (Case II of Table 4.2)					
2	—	0.003	0.004	0.002	0.008
3	—	0.007	0.007	0.005	0.018
4	—	0.011	0.009	0.009	—
5	—	0.015	0.019	0.012	—
6	—	0.019	0.032	—	—
7	—	0.021	0.047	—	—
(c) *The large tanker in the loaded condition* (Case II of Table 4.2)					
2	—	0.002	0.003	0.001	0.005
3	—	0.005	0.005	0.003	0.012
4	—	0.008	0.007	0.005	0.019
5	—	0.010	0.008	0.007	—
6	—	0.013	0.011	—	—
7	—	0.015	0.017	—	—

The predictions contained in Table 4.6 do little to foster confidence in our knowledge of hull damping. The simple truth is that that knowledge is abysmal. We note, for instance, that the loaded tanker is much more lightly damped than the same ship in ballast because frequencies are much lower, but it is not at all clear what allowance to make for the cargo damping; see Betts *et al.* (1977a). It is not our purpose to dwell on these matters here, but we may confidently predict that much research will be done on the subject of hull damping in the coming years.

At the present stage we do not know how sensitive our predictions of hull response will be to the estimates of modal damping factors. This is something that we shall examine later, in chapter 8. At this stage we merely suppose that some sort of 'democratic' selection will be adequate. On this basis we might assume in the light of Table 4.6 that the values given in Table 4.7 will be satisfactory.†

It would be unwise to leave this matter of damping factors without pointing out that, on one crucial matter, there is really no disagreement. It is that the structural damping is very small. We are certainly dealing with 'high Q' systems. In fact the extremes given in Table 4.7 correspond to

$$Q = 250 \text{ for } \nu_r = 0.002,$$
$$Q = 17 \text{ for } \nu_r = 0.030,$$

where Q, it will be recalled, is the resonant magnification factor for the mode concerned – see Bishop & Johnson (1960). One other comment is in order: it is that, size for size, one would expect the damping of a warship hull to be smaller than that of a merchant hull, by reason of higher standards of design detail and construction work.

† The figures for the destroyer are certainly too low, because the ship in question was partially of riveted construction. We are here pretending that it was allwelded.

Table 4.7. *Representative selection of damping factors*

	Value of ν_r for:		
r	Destroyer	Tanker in ballast	Tanker loaded
2	0.006	0.004	0.002
3	0.009	0.007	0.005
4	0.014	0.010	0.008
5	0.020	0.015	0.010
6	0.025	0.024	0.012
7	0.030	0.030	0.016

4.2.4   *Generalised characteristics relating to symmetric motion*
The equations of motion contain constant coefficients which specify the hull's dynamical characteristics. These quantities are

the generalised masses a_{rs},
the generalised damping coefficients b_{rs},
the generalised stiffnesses c_{rs},

and they are the elements of the matrices **A**, **B** and **C**. It is evidently necessary to assign values to these parameters before numerical results can be deduced for a given hull.

Our purpose now is to show how these values may be arrived at and to give an idea of what their orders of magnitude are. It is essential to note, however, that in general the values of these coefficients depend on the scaling of the principal modes. We have adopted the convention that unit value of a principal coordinate corresponds to a positive deflection at the stern of one metre. If unit deflection were taken, instead, to entail a 2-metre deflection at the stern in all the principal modes, then all the generalised masses, stiffnesses and damping coefficients would be increased fourfold. The values of the coefficients depend, that is, on the modal scaling.

Generalised mass and stiffness
According to the theory given in section 3.2.1, the inertia coefficient a_{rs} is given by

$$\int_0^l (\mu w_r w_s + I_y \theta_r \theta_s)\, dx = a_{rs} \delta_{rs} \tag{4.2}$$

(see equation (3.15)). If no allowance is to be made for rotatory inertia, I_y must be taken as zero in accordance with equation (2.7) – and the functions $w_r(x)$, $w_s(x)$ will be somewhat different.

The orthogonality of the principal modes ensures that the cross-terms

$$\int_0^l (\mu w_r w_s + I_y \theta_r \theta_s)\, dx \qquad (r \neq s)$$

vanish and any reputable program for computing the modes will check that this is so. (These integrals may not be *exactly* zero by reason of the approximations introduced in the process of integration.) By way of example, the computations for the destroyer of fig. 4.1 give the results contained in Table 4.8.

Having checked that all the cross-coefficients a_{rs} ($r \neq s$) are sensibly zero, we may list the inertia constants a_{rr} that make up the leading diagonal of the inertia matrix **A** of the dry hull. For the hulls we have discussed, the results are given in Table 4.9; those for the destroyer contain an allowance for shear and rotatory inertia, all those for the tanker embody an allowance for shear effects.

The generalised stiffnesses c_{rs} are related to the generalised masses as in equation (3.16). That is to say

$$c_{rs} = \omega_r^2 a_{rs}.$$

The cross-coefficients, for $r \neq s$, again vanish and we may therefore use the foregoing results to find the constants c_{rr} from the equation

$$c_{rr} = \omega_r^2 a_{rr}. \tag{4.3}$$

This is obviously a more economical approach than computing the quantities

$$c_{rr} = \int_0^l (EI\theta_r'^2 + kAG\gamma_r^2)\, dx \tag{4.4}$$

(see section 3.2.1). Typical results found from equation (4.3) are given in Table 4.10, corresponding to those appearing in Tables 4.3, 4.4 and 4.9.

Table 4.8. *Values of a_{rs} for the destroyer (in tonne m^2) – Case IV of Table 4.2*

Mode	0	1	2	3	4	
0	2559.7	−0.5	1.0	0.1	1.4	.
1	−0.5	629.7	0.3	0.0	0.3	.
2	1.0	0.3	338.2	−0.2	2.0	.
3	0.1	0.0	−0.2	247.6	0.3	.
4	1.4	0.3	2.0	0.3	248.7	.
.	.	.	.	.	.	.

Table 4.9. *Values of generalised mass*

Modal index r	Value of a_{rr} in tonne m^2 for:				
	Destroyer ($I_y \neq 0$)	Tanker in ballast		Tanker loaded	
		$I_y = 0$	I_y estimated	$I_y = 0$	I_y estimated
0	2560	77 888	77 888	284 699	284 699
1	630	23 621	23 621	53 174	53 174
2	338	22 010	22 251	20 091	20 170
3	248	7 868	8 034	11 101	11 191
4	249	3 433	3 541	8 776	8 873
5	518	4 834	4 931	7 163	7 254
6	592	2 683	2 792	6 580	6 675
7	713	2 616	2 566	6 889	7 003

Generalised damping

The generalised damping is represented by the coefficients

$$b_{rs} = \int_0^l (\alpha k AG\gamma_r\gamma_s + \beta EI\theta_r'\theta_s') \, dx, \tag{4.5}$$

as shown in section 3.2.2. We have seen that direct estimation of these quantities is a virtually hopeless task. This being so, we have assumed that the hull damping does not couple the principal modes (so that the $b_{rs} = 0$ if $r \neq s$) and have obtained rough and ready values of the modal damping factors v_r.

According to elementary vibration theory (see Bishop & Johnson, 1960),

$$b_{rr} = 2a_{rr}\omega_r v_r. \tag{4.6}$$

Table 4.10. *Values of generalised stiffness*

Modal index r	Value of c_{rr} in MN m for:				
	Destroyer $(I_y \neq 0)$	Tanker in ballast		Tanker loaded	
		$I_y = 0$	I_y estimated	$I_y = 0$	I_y estimated
0	0	0	0	0	0
1	0	0	0	0	0
2	62.2	997	993	323	323
3	184	1 972	1 978	1 121	1 121
4	399	3 069	3 092	2 450	2 454
5	1746	9 536	9 443	4 160	4 171
6	3257	8 970	9 078	7 213	7 247
7	5329	12 650	12 160	12 164	12 224

Table 4.11. *Values of generalised damping*

Modal index r	Value of b_{rr} in tonne m^2/s for:		
	Destroyer $(I_y \neq 0)$	Tanker in ballast $(I_y = 0)$	Tanker loaded $(I_y = 0)$
0	0	0	0
1	0	0	0
2	55	1 185	322
3	122	1 744	1116
4	279	2 053	2346
5	1203	6 442	3453
6	2195	7 446	5229
7	3698	10 915	9263

Thus we are in a position to calculate the generalised damping coefficients using the data contained in Tables 4.3, 4.4, 4.7 and 4.9. They are given in Table 4.11.

4.3 Data relevant to antisymmetric response

It is not easy to find adequate data on actual ships for the calculation of structural characteristics in symmetric motion. The information that has been used for the destroyer and the large tanker was not found in the open literature and was specially assembled for the purposes of research. The position is much worse where *anti*symmetric response is concerned and, in particular, relevant information on structural damping is almost non-existent.

Antisymmetric hull vibration of low frequency is taken up by Ohtaka *et al.* (1967) in an important and interesting paper. They not only discuss the acquisition of most of the required data but also present modes and frequencies of certain representative vessels. Unfortunately their results are not strictly relevant to our present needs, however, because they are for 'wet' hulls – i.e. for ships floating in the sea rather than free–free *in vacuo*.

In order to extract results of the type we need it is necessary to devise a technique like that of the Prohl–Myklestad method. This has been done for antisymmetric motions by Bishop, Price & Temarel (1979). Those writers used the paper by Ohtaka *et al.* (1967) for the purposes of checking results as if the somewhat artificial assumption were made that their data referred to dry hulls rather than to hulls afloat in the sea.

It is not yet possible to present results for antisymmetric oscillations of typical dry hulls, simply because the necessary data are lacking. The characteristics of the one ship for which adequate data have been compiled are quoted later in chapter 10 where we take up the wider matter of antisymmetric responses.

5 Ship distortion in still water

When at anchor here I ride,
My bosom swells with pride...
HMS Pinafore

5.1 Hydrostatic fluid actions

In still water, the fluid actions on a ship are simply those of hydrostatics and, as we have already seen in chapters 2 and 3, theory readily lends itself to modal analysis of the distortions. In chapter 4 we pointed out, however, that the necessary data for actual ships are not readily available for the analysis of antisymmetric distortions. In this chapter, then, we shall examine specific cases of symmetric distortion in still water.

5.1.1 Bending moment and shearing force

Fig. 5.1(a) shows the curves of buoyancy force $Z_0(x)$ per unit length and weight $\mu(x)g$ per unit length for the destroyer. The two curves are in the 'stepped' form appropriate to a 20-slice representation of the hull. The net upward force per unit length is

$$\bar{Z}(x) = Z_0(x) - \mu(x)g \tag{5.1}$$

and it is shown in fig. 5.1 (b).

The distribution of static shearing force, $\bar{V}(x)$, is given by

$$\bar{V}(x) = -\int_0^x \bar{Z}(x)\,\mathrm{d}x. \tag{5.2}$$

This quantity is represented by the curve of fig. 5.2. The static bending moment, $\bar{M}(x)$, is found by integrating $\bar{V}(x)$:

$$\bar{M}(x) = -\int_0^x \bar{V}(x)\,\mathrm{d}x. \tag{5.3}$$

The variation of bending moment is shown by the curve of fig. 5.3. From this latter curve the bending stresses σ at any section can be estimated using the familiar expression

$$\sigma = \frac{\bar{M}y}{I},$$

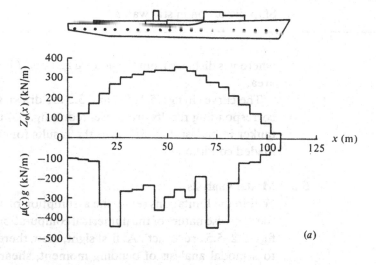

Fig. 5.1. A destroyer hull divided into 20 discrete masses. Curves (a) represent the buoyancy force/unit length, $Z_0(x)$ and the weight/unit length $\mu(x)g$. The net upward force/unit length is $\bar{Z}(x)$ as shown in curve (b).

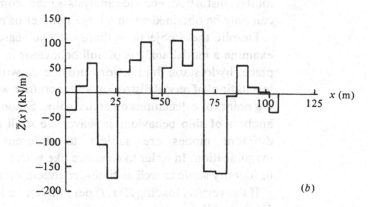

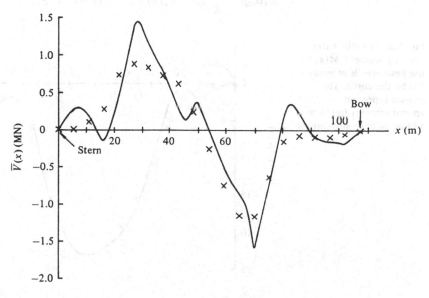

Fig. 5.2. The still water shearing force $\bar{V}(x)$ for the destroyer is represented by the curve. The crosses represent approximations found by adding modal contributions.

where y is distance from the neutral axis and I is the second moment of area.

The curves in figs. 5.1, 5.2 and 5.3 are drawn for the destroyer. The corresponding results are presented in fig. 5.4 for the 250 000 DWT tanker in ballast. Fig. 5.5 gives the results for the same tanker in the loaded condition.

5.2   Modal analysis

Within the limitations set by the assumption of a beamlike hull and by the stepwise nature of the numerical computations, results like those of figs. 5.2–5.5 are 'exact'. At first sight, then, there is no point in turning to a modal analysis of bending moment, shearing force or (for that matter) distortion. For such analysis is more complicated and accuracy can only be obtained when a large number of modes are used.

Despite these objections there are two reasons why we shall now examine a modal analysis of hull behaviour in still water. In the first place, hydrostatic fluid actions provide us with the simplest sort of 'excitation' of modal distortion and therefore with a convenient starting point for calculations on actual ships. Secondly, in our subsequent analysis of ship behaviour in waves we shall find that responses in different modes are subject to different levels of dynamic magnification. In order to compare like with like, then, it is necessary to identify static as well as time-dependent modal responses.

If the vertical loading $Z(x, t)$ per unit length has the particular form $\bar{Z}(x)$, and if the static response is expressed in the modal form

$$w(x) = \sum_{s=0}^{\infty} \bar{p}_s w_s(x),$$

equation (3.21) shows that

$$\omega_s^2 a_{ss} \bar{p}_s = \int_0^l \bar{Z}(x) w_s(x)\, dx \qquad (s = 0, 1, 2, \ldots). \tag{5.4}$$

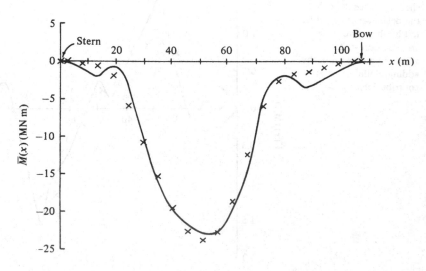

Fig. 5.3. The still water bending moment $\bar{M}(x)$ for the destroyer is represented by the curve. The crosses represent approximations found by adding modal contributions.

Exactly the same result was found in section 2.2.3 where the more rudimentary theory of the Bernoulli–Euler beam is referred to.

It was explained in section 2.2.3 that, since $\omega_0 = 0 = \omega_1$, the values of $\bar{p}_0$ and $\bar{p}_1$ are left undetermined. This remains true when the corrections for shear and rotatory inertia are included, as also is the result

$$\bar{p}_s = \frac{1}{\omega_s^2 a_{ss}} \int_0^l \bar{Z}(x) w_s(x) \, dx \qquad (s = 2, 3, \ldots). \tag{5.5}$$

It follows that if we know $Z_0(x)$ and $\mu(x)g$ (and hence $\bar{Z}(x)$), and if we know the dynamical characteristics of the hull as explained in chapter 4, we are in a position to calculate the distortion at the principal coordinates, $\bar{p}_s$, for $s = 2, 3, \ldots$.

Having found the values of $\bar{p}_2, \bar{p}_3, \ldots, \bar{p}_n$ we can find approximations to the distortion $\bar{w}_{\text{dist}}(x)$ (i.e. the total deflection $\bar{w}(x)$ minus deflections in the rigid modes), the shearing force and the bending

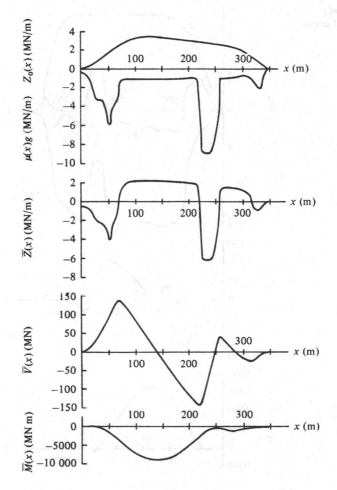

Fig. 5.4. Curves corresponding to those of figs. 5.1–5.3 but relating to the 250 000 DWT tanker in ballast.

moment. That is,

$$\bar{w}_{dist}(x) \doteq \sum_{s=2}^{n} \bar{p}_s w_s(x),$$ (5.6)

$$\bar{V}(x) \doteq \sum_{s=2}^{n} \bar{p}_s V_s(x),$$ (5.7)

$$\bar{M}(x) \doteq \sum_{s=2}^{n} \bar{p}_s M_s(x).$$ (5.8)

The more values $\bar{p}_s$ that are used in these series (i.e. the greater the number n) the better will be the approximation.

5.3 Results for particular ships

When the lines of a hull have been adequately specified, the immersed area of cross-section S is known for any 'slice' of thickness Δx. The appropriate buoyancy force on the slice is $S\rho g \Delta x$ where ρ is the density of water, and so $Z_0(x) = S\rho g$. By subtracting the weight per

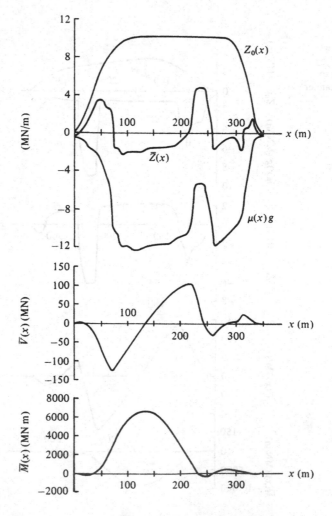

Fig. 5.5. Still water curves for the 250 000 DWT tanker in its loaded condition.

unit length from this quantity we arrive at the net upward force per unit length, $\bar{Z}(x)$.

The generalised applied force at p_s is the quantity

$$\bar{P}_s = \int_0^l \bar{Z}(x) w_s(x) \, dx. \tag{5.9}$$

This is readily computed, using the known sth principal mode $w_s(x)$ scaled in the manner we have selected (i.e. such that $w_s(0) = 1$ metre). Hence $\bar{p}_s$ may be calculated from equation (5.5). Notice that we are free to select any of the four levels of accuracy listed in Table 4.2.

5.3.1   *The destroyer*

For the destroyer we shall refer only to Case IV. That is to say, we shall quote results in which allowance has been made for shear deflection and for rotatory inertia. They are given in Table 5.1.

Admittedly with insufficient justification, we have obtained results up to and including that for the 7-node mode. Now, by forming the sums (5.7) and (5.8), we may see how accurately the 'exact results' are matched by the modal approximation, while the sum (5.6) will show how the hull can be expected to distort in still water. The results are best illustrated graphically.

Fig. 5.6 shows the estimated distortion of the destroyer hull in still water. It will be remembered that the values of $\bar{p}_0$ and $\bar{p}_1$ are not determined when the hull assumes its bodily position as regards depth of immersion and trim. Now we see that the superimposed distortion is such that the maximum displacement is about 17 mm downwards (at the stern).

Figs. 5.2 and 5.3, it will be recalled, show curves of the 'exact' shearing force and bending moment respectively. Adjacent to them are the approximate values obtained by using the seven calculated values of the principal coordinates as given by equations (5.7) and (5.8).

Table 5.1. *Still water distortion of the destroyer (Case IV)*

Modal index s	Natural frequency ω_s (rad/s)	Generalised mass a_{ss} (tonne m²)	Generalised force $\bar{P}_s$ (KN m)	Generalised distortion $\bar{p}_s$
2	13.57	338	−1139	−0.0183
3	27.22	248	−255	−0.0014
4	40.05	249	+937	+0.0023
5	58.05	518	−156	−0.0001
6	74.17	592	−189	−0.0001
7	86.45	713	+1190	+0.0002

Convergence of modal series

The accuracy with which the series in equations (5.6)–(5.8) converge to the 'exact' values is of interest. First of all it must be remembered that the functions $w_s(x)$ are assumed to be accurate, and this implies that 'a large number' of slices is used in modelling the hull. In our numerical work we have used only 20 slices for the destroyer (whereas 50 are employed for the tanker). Accordingly the calculated modal shapes with a greater number of nodes than, say, 4 or 5 cannot be particularly reliable. Working with $w_7(x)$, we have purposely placed a considerable strain on the assumptions – a strain that cannot be eased by taking still more modal functions. In other words, the positioning of the crosses in figs. 5.2 and 5.3 is probably considerably worse on this account than would ever be the case in a routine analysis in which more slices are used.

There remains the question of how quickly the series converge even when the forms of the modal functions are dependable. While this matter is of some mathematical interest, however, its practical significance is seldom of great significance; for, as we shall see when we come to study dynamic magnification in the modes, it is the distortion in an individual mode rather than the aggregate of all such distortions that matters.

Referring back to section 3.2 we see that

$$\bar{M}(x) = EI(x)[\bar{w}'_{\text{dist}}(x) - \bar{\gamma}(x)]', \tag{5.10}$$

$$\bar{V}(x) = -\bar{M}'(x). \tag{5.11}$$

Thus if the distortion is specified in the series form

$$\left.\begin{array}{l} \bar{w}_{\text{dist}}(x) = \sum_{s=2}^{\infty} \bar{p}_s w_s(x), \\[2mm] \bar{\gamma}(x) = \sum_{s=2}^{\infty} \bar{p}_s \gamma_s(x), \end{array}\right\} \tag{5.12}$$

the bending moment and shearing force are found by successive differentiation of series. It is to be expected that this differentiation can

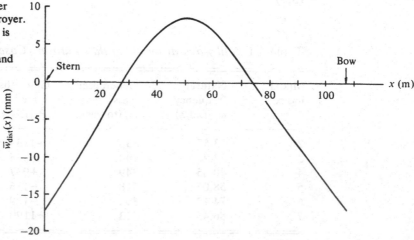

Fig. 5.6. The still water distortion of the destroyer. Note that no account is taken of the rigid deflections of heave and pitch.

be performed only at the expense of slower convergence of the series. This can readily be demonstrated analytically for a uniform beam. In other words, whereas the modal convergence is likely to be very rapid indeed for $\bar{w}_{\text{dist}}(x)$, it will be somewhat less rapid for $\bar{M}(x)$ and less rapid still for $\bar{V}(x)$.

5.3.2   *The tanker*

It has already been mentioned that a 50-slice idealisation has been used for the tanker hull. This degree of sophistication should be ample to permit our use of the seven distortion modes of lowest order. The results, comparable with those of Table 5.1 are given in Table 5.2 for the tanker in ballast and in Table 5.3 for the tanker in the loaded condition. In both cases we have used the best available information in that a correction has been made for shear deformation and for estimated rotatory inertia.

Figs. 5.7, 5.8 and 5.9 show the distortion $\bar{w}_{\text{dist}}$, the shearing force $\bar{V}(x)$ and the bending moment $\bar{M}(x)$ curves for the tanker in its ballast condition. Figs. 5.10, 5.11 and 5.12 show the corresponding curves for the loaded condition. Again, the marks near the curves show the approximations obtained by adding the modal contributions referred to in Tables 5.2 and 5.3.

Table 5.2. *Still water distortion of the tanker in ballast (Case IV)*

Modal index s	Natural frequency ω_s (rad/s)	Generalised mass a_{ss} (tonne m^2)	Generalised force $\bar{P}_s$ (KN m)	Generalised distortion $\bar{p}_s$
2	6.68	22 251	−169 080	−0.1702
3	15.69	8 034	−153 260	−0.0775
4	29.55	3 541	+38 046	+0.0123
5	43.76	4 931	+73 758	+0.0078
6	57.03	2 792	−34 374	−0.0038
7	68.84	2 566	+20 721	+0.0017

Table 5.3. *Still water distortion of the loaded tanker (Case IV)*

Modal index s	Natural frequency ω_s (rad/s)	Generalised mass a_{ss} (tonne m^2)	Generalised force $\bar{P}_s$ (KN m)	Generalised distortion $\bar{p}_s$
2	4.00	20 170	+71 603	+0.2219
3	10.01	11 191	+93 841	−0.0837
4	16.63	8 873	+19 739	+0.0080
5	23.98	7 254	−9 298	−0.0022
6	32.95	6 675	+21 873	+0.0030
7	41.78	7 003	−66 364	−0.0054

Fig. 5.7. The still water distortion of the tanker in ballast.

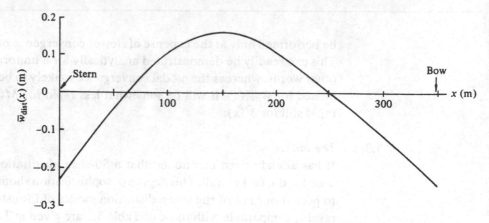

Fig. 5.8. The still water shearing force $\bar{V}(x)$ for the tanker in ballast is represented by the curve. The crosses represent approximations found by adding together the contributions due to deflections in the lowest six distortion modes (i.e. modes 2 to 7 inclusive).

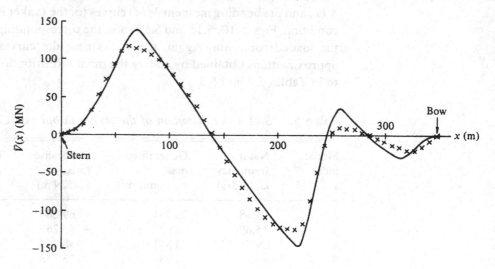

Fig. 5.9. The bending moment curve corresponding to that of shearing force shown in fig. 5.8.

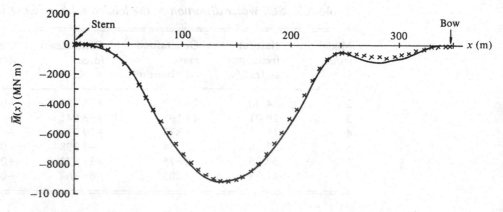

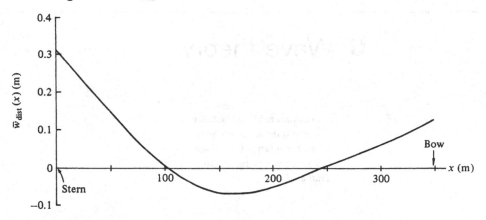

Fig. 5.10. The still water distortion of the tanker in its loaded condition.

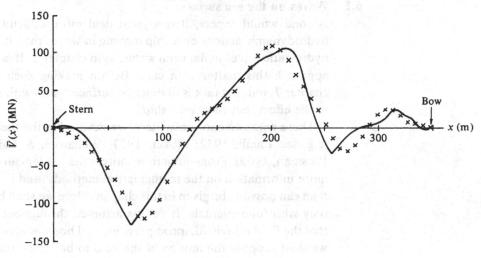

Fig. 5.11. The still water shearing force for the tanker in its loaded condition. The crosses represent approximations found by adding contributions from modes 2 to 7 inclusive.

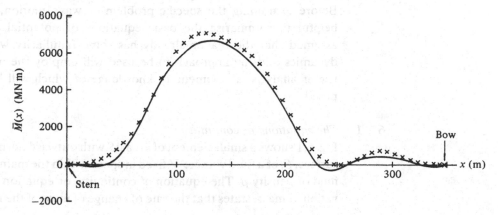

Fig. 5.12. The bending moment curve corresponding to that of shearing force shown in fig. 5.11.

6 Wave theory

You cannot eat breakfast all day,
  Nor is it the act of a sinner,
When breakfast is taken away,
  To turn your attention to dinner;
Trial by Jury

6.1 Waves on the sea surface

As one would expect, it is a great deal more difficult to estimate hydrodynamic actions on a ship moving in waves than it is to find the hydrostatic forces in flat calm water, as in chapter 5. It is necessary to approach this matter with care. Before making such estimates in chapter 7, our first task is to describe surface waves without reference to the effect they have on a ship.

There exists an enormous literature devoted to the theory of waves (e.g. see Lamb, 1932; Stoker, 1957; Wehausen & Laitone, 1960; Kinsman, 1965). From numerous sources the reader can obtain much more information on the mathematical methods used in wave theory than can possibly be given in this chapter. Here we shall be concerned only with fundamentals. It will be assumed throughout this chapter that the fluid is inviscid, incompressible and homogeneous. Moreover we shall suppose the motion of the fluid to be irrotational; that is to say the particles do not rotate during their motion along streamlines (although they can move along closed paths).

6.2 Fundamentals of hydrodynamics

Before examining the specific problem of wave motion, it will be helpful to summarise the basic equations of potential flow. It is assumed that the reader already has some familiarity with hydrodynamics and the approach to be used will employ the methods of vector analysis, an elementary knowledge of which will be presupposed.

6.2.1 *The equations of continuity*

Fig. 6.1 shows a small element of area dS with outward normal $\mathbf{n}$ in the surface, S, of a fluid volume V fixed in space within the main volume of fluid of density ρ. The equation of continuity, or equation of conservation of mass, states that the rate of change of mass of the fluid inside

V results solely from the flow of fluid across the boundary S. That is to say the equation denies the possibility of spontaneous generation or destruction of fluid, as would occur if sources or sinks were present within V.

Let the velocity of flow through the surface S at any point be $\mathbf{q}$. In a small interval of time δt the total outward mass flow of fluid across the surface S is, by Gauss's theorem,

$$\delta t \int_S \rho \mathbf{q} \cdot \mathbf{n} \, dS = \delta t \int_V \text{div} \, (\rho \mathbf{q}) \, dV,$$

where dV is an elemental volume of V. The total mass of fluid within V at any instant is

$$M = \int_V \rho \, dV,$$

and the rate of change of mass is

$$\frac{dM}{dt} = \int_V \frac{\partial \rho}{\partial t} \, dV.$$

Thus in time δt, the increase of fluid mass in V is $\delta t \, dM/dt$ which equals the mass of the net inward flow, i.e.

$$\delta t \int_V \frac{\partial \rho}{\partial t} \, dV = -\delta t \int_V \text{div} \, (\rho \mathbf{q}) \, dV,$$

which must be true for any fixed volume V and so must be true for dV. Hence

$$\frac{\partial \rho}{\partial t} + \text{div} \, (\rho \mathbf{q}) = 0$$

at all points in the fluid. For a fluid of constant density then,

$$\text{div} \, \mathbf{q} = 0. \tag{6.1}$$

Fig. 6.1. A volume of fluid V, with surface S. The outward unit vector to some small element dS of the surface is $\mathbf{n}$.

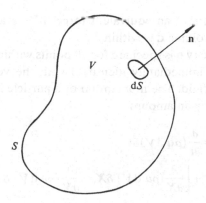

6.2.2 *Laplace's equation*

Rectangular coordinates $AXYZ$ are fixed so that the plane AXY lies in the horizontal, undisturbed free surface and AZ points vertically upwards. If coordinates (X, Y, Z) define the position of a fluid particle and the corresponding components of the velocity $\mathbf{q}$ are u, v, w, then equation (6.1) may be written

$$\frac{\partial u}{\partial X} + \frac{\partial v}{\partial Y} + \frac{\partial w}{\partial Z} = 0.$$

Moreover the necessary condition for the motion to be irrotational is

$$\text{curl } \mathbf{q} = \left(\frac{\partial w}{\partial Y} - \frac{\partial v}{\partial Z}, \quad \frac{\partial u}{\partial Z} - \frac{\partial w}{\partial X}, \quad \frac{\partial v}{\partial X} - \frac{\partial u}{\partial Y} \right) = \mathbf{0}. \tag{6.2}$$

There exists a velocity potential function $\Phi(X, Y, Z, t)$ which satisfies equation (6.2) when

$$u = -\frac{\partial \Phi}{\partial X}, \quad v = -\frac{\partial \Phi}{\partial Y}, \quad w = -\frac{\partial \Phi}{\partial Z}. \tag{6.3}$$

If these expressions are now substituted into equation (6.1) it is found that

$$\left(\frac{\partial^2}{\partial X^2} + \frac{\partial^2}{\partial Y^2} + \frac{\partial^2}{\partial Z^2} \right) \Phi \equiv \nabla^2 \Phi = 0. \tag{6.4}$$

This is the Laplace equation of hydrodynamics.

6.2.3 *Equation of motion*

If p is the pressure acting over the elemental surface dS shown in fig. 6.1, the force on this area is $-p\mathbf{n}\,dS$ and the total force on the surface S of V is

$$-\int_S p\mathbf{n}\,dS = -\int_V \text{grad } p\,dV.$$

This result may be deduced from Gauss's theorem. In addition, if the body forces acting throughout V are expressible in the form $\mathbf{F}$ per unit mass then the total resultant force acting on the body of fluid is

$$\int_V (\rho\mathbf{F} - \text{grad } p)\,dV.$$

This means that an equivalent force $(\rho\mathbf{F} - \text{grad } p)dV$ acts on every elemental volume dV within V.

The velocity $\mathbf{q}$ is specified for all points within V and for all instants and is not immediately identified with the velocity of a particular particle of fluid. The momentum of a particle is $\rho\mathbf{q}\,dV$ and in time δt it changes by an amount

$$\delta(\rho\mathbf{q}\,dV) = \frac{\partial}{\partial t}(\rho\mathbf{q}\,dV)\delta t$$

$$+ \left[\frac{\partial}{\partial X}(\rho u\,dV)\,\delta X, \quad \frac{\partial}{\partial Y}(\rho v\,dV)\,\delta Y, \quad \frac{\partial}{\partial Z}(\rho w\,dV)\,\delta Z \right],$$

where $\mathbf{q} = (u, v, w)$ and $(\delta X, \delta Y, \delta Z)$ is the movement of the particle in question. If we divide by δt and proceed to the limit as $\delta t \to 0$, identifying $\delta X/\delta t$ with u, $\delta Y/\delta t$ with v and $\delta Z/\delta t$ with w, we find that the total derivative is

$$\frac{D}{Dt}(\rho\mathbf{q}\,dV) = \left[\frac{\partial}{\partial t} + \mathbf{q}\cdot\text{grad}\right](\rho\mathbf{q}\,dV).$$

When Newton's laws of motion are now applied to the particle, it is found that

$$\left[\frac{\partial}{\partial t} + \mathbf{q}\cdot\text{grad}\right](\rho\mathbf{q}\,dV) = (\rho\mathbf{F} - \text{grad}\,p)\,dV.$$

It follows that

$$\frac{D\mathbf{q}}{Dt} = \frac{\partial\mathbf{q}}{\partial t} + (\mathbf{q}\cdot\text{grad})\,\mathbf{q} = \mathbf{F} - \frac{1}{\rho}\,\text{grad}\,p. \tag{6.5}$$

6.2.4 Bernoulli's equation

For an applied gravitational force, $\mathbf{F} (= 0, 0, -g)$, the equation of motion (6.5) can be written in the component form

$$\frac{\partial u}{\partial t} + \left(u\frac{\partial}{\partial X} + v\frac{\partial}{\partial Y} + w\frac{\partial}{\partial Z}\right)u = \frac{\partial u}{\partial t} + \tfrac{1}{2}\frac{\partial}{\partial X}(u^2 + v^2 + w^2) = -\frac{1}{\rho}\frac{\partial p}{\partial X},$$

$$\frac{\partial v}{\partial t} + \left(u\frac{\partial}{\partial X} + v\frac{\partial}{\partial Y} + w\frac{\partial}{\partial Z}\right)v = \frac{\partial v}{\partial t} + \tfrac{1}{2}\frac{\partial}{\partial Y}(u^2 + v^2 + w^2) = -\frac{1}{\rho}\frac{\partial p}{\partial Y},$$

$$\frac{\partial w}{\partial t} + \left(u\frac{\partial}{\partial X} + v\frac{\partial}{\partial Y} + w\frac{\partial}{\partial Z}\right)w = \frac{\partial w}{\partial t} + \tfrac{1}{2}\frac{\partial}{\partial Z}(u^2 + v^2 + w^2) = -\frac{1}{\rho}\frac{\partial p}{\partial Z} - g,$$

in which the results of equation (6.2) are used. Substitution of equation (6.3) into these equations and then integration gives

$$\frac{\partial\Phi}{\partial t} - \tfrac{1}{2}(u^2 + v^2 + w^2) - \frac{p}{\rho} - gZ = H \tag{6.6}$$

and this is a form of Bernoulli's equation. In general the quantity H is a function of time, but for steady, irrotational flow $\partial\Phi/\partial t = 0$ while H is independent of time.

For a fluid at rest under the action of gravity,

$$-\frac{p_h}{\rho} - gZ = H,$$

since the pressure p $(= p_h)$ is only hydrostatic, and it follows that H is a constant. If this result is substituted into equation (6.6), we find that

$$\frac{\Delta p}{\rho} = \frac{\partial\Phi}{\partial t} - \tfrac{1}{2}(u^2 + v^2 + w^2)$$

where Δp denotes departures from the hydrostatic value due to the fluid motion. Any time-dependent variation in the quantity H is included in the potential function.

6.3 Application to gravity waves
6.3.1 *Equation of the free surface*

Consider waves of amplitude $\zeta(X, Y, t)$ whose elevation is measured by reference to the plane AXY. The waves are travelling in the direction of the AX axis on the surface of water of depth d. The position of the rigid boundary is $Z = -d(X, Y)$, while at the free surface

$$Z = \zeta(X, Y, t).$$

Since a fluid particle located in the free surface remains in the surface throughout its motion† and the free surface moves with the fluid then, for any particle in the free surface,

$$\frac{DZ}{Dt} = \frac{D\zeta(X, Y, t)}{Dt}.$$

It follows that

$$w = \left(\frac{\partial}{\partial t} + \mathbf{q} \cdot \text{grad}\right) \zeta(X, Y, t),$$

from which we find that

$$\frac{\partial \zeta}{\partial t} + u\frac{\partial \zeta}{\partial X} + v\frac{\partial \zeta}{\partial Y} - w = 0 \tag{6.7}$$

on the free surface $Z = \zeta(X, Y, t)$.

6.3.2 *Boundary conditions*

For ocean waves, only rigid and free boundaries need be considered. At a *rigid* boundary there must be a zero fluid velocity component normal to the fixed surface. Hence on $Z = -d(X, Y)$,

$$w = -\frac{\partial \Phi}{\partial Z} = 0.$$

If there is any fluid velocity at such a boundary it must be tangential to the surface so that the boundary is always a streamline.

At the *free* boundary, we have the kinematic condition that a fluid particle at the free surface remains in the surface throughout its motion. Hence not only must equation (6.7) be satisfied on the surface $Z = \zeta(X, Y, t)$ but also the velocity potential, $\Phi(X, Y, Z, t)$ must satisfy the non-linear equation

$$\frac{\partial \Phi}{\partial t} - \tfrac{1}{2}(u^2 + v^2 + w^2) - \frac{p}{\rho} - g\zeta = 0$$

on $Z = \zeta(X, Y, t)$. In this last equation the arbitrary constant H of equation (6.6) is assumed to be accounted for in the quantity $\partial \Phi/\partial t$ and p denotes the change in pressure relative to the pressure at $Z = 0$. It

† For a proof of this result, see Lamb (1932).

will be seen that there exists an equilibrium condition represented by
$\zeta = 0 = u = v = w$.

6.4   Linearised theory of gravity waves
6.4.1   *Equations of wave motion*
The wave theory presented so far is non-linear and arrival at any
general solution is very difficult. By incorporating certain limitations
that are suggested by the physical nature of the problem, however,
simplifications can be made. A linearisation of the theory is permissible if it is assumed that the maximum wave amplitude is much smaller
than the wavelength. That is, the slope of the wave profile, $\partial\zeta/\partial X$, is
everywhere small. We shall confine our attention to the linearised
wave theory in which the squares and products of terms of all the
quantities and their differentials may be ignored.

Under these conditions the velocity potential Φ must satisfy the
Laplace equation

$$\nabla^2\Phi = 0.$$

The linearised conditions at the free surface are such that

$$\frac{\partial\Phi}{\partial t} - g\zeta = 0.$$

Thus, in equation (6.7),

$$\frac{\partial^2\Phi}{\partial t^2} + g\frac{\partial\Phi}{\partial Z} = 0 \quad \text{on} \quad Z = \zeta(X, Y, t) \doteq 0. \tag{6.8}$$

Finally

$$\frac{\partial\Phi}{\partial Z} = 0 \quad \text{on} \quad Z = -d(X, Y).$$

6.4.2   *Sinusoidal progressive waves*
The Laplace equation and rigid boundary condition at a constant
depth $Z = -d$ are satisfied by the solution

$$\Phi(X, Z, t) = A \cosh k(Z + d) \sin (kX - \omega t + \alpha),$$

whilst the free surface condition at $Z = 0$ in equation (6.8) gives the
'dispersion relationship'

$$\omega^2 = gk \tanh kd. \tag{6.9}$$

The corresponding wave elevation is

$$\zeta(X, t) = \frac{1}{g}\frac{\partial\Phi}{\partial t} = a \cos (kX - \omega t + \alpha), \tag{6.10}$$

where

$$a = -\frac{\omega A}{g} \cosh kd.$$

The wavenumber of this wave is k and it is related to the wavelength λ (see fig. 6.2) by the expression

$$k = \frac{2\pi}{\lambda}.$$

Similarly the frequency ω is related to the period T by

$$\omega = \frac{2\pi}{T}.$$

It is convenient to assume that the phase angle α is zero.

The velocity potential and stream functions may be written

$$\left.\begin{array}{l} \Phi(X, Z, t) = -\dfrac{ag}{\omega} \dfrac{\cosh\left[k(Z+d)\right]}{\cosh kd} \sin\left(kX - \omega t\right), \\[3mm] \Psi(X, Z, t) = -\dfrac{ag}{\omega} \dfrac{\sinh\left[k(Z+d)\right]}{\cosh kd} \cos\left(kX - \omega t\right), \end{array}\right\} \tag{6.11}$$

since

$$\frac{\partial \Phi}{\partial X} = \frac{\partial \Psi}{\partial Z}.$$

The factor $ag/\omega \cosh kd$ is sometimes replaced by $a\omega/k \sinh kd$ as is suggested by equation (6.9). Alternatively it is sometimes replaced by the factor $ac/\sinh kd$, where

$$c = \frac{\lambda}{T} = \frac{\omega}{k}$$

is the wave velocity.

6.4.3 Wave velocity

The wave velocity c is given by

$$c = \frac{\lambda}{T} = \frac{\omega}{k} = \left[\frac{g\lambda}{2\pi} \tanh\left(\frac{2\pi d}{\lambda}\right)\right]^{\frac{1}{2}}. \tag{6.12}$$

When $d/\lambda \to 0$ or $kd \to 0$, the wave velocity reduces to

$$c = (gd)^{\frac{1}{2}}. \tag{6.13}$$

Fig. 6.2. A sinusoidal wave which moves in the direction AX. Its amplitude is a and its wavelength is λ. The water depth is d.

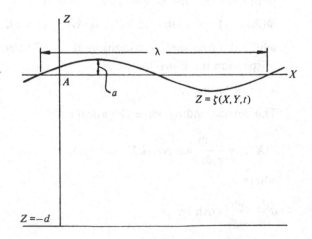

It thus becomes independent of the wavelength when the depth of water is small compared to the wavelength. This result applies to shallow water waves which are non-dispersive.

When $d/\lambda \to \infty$ or $kd \to \infty$, so that the depth of water is infinite,

$$c = \left(\frac{g\lambda}{2\pi}\right)^{\frac{1}{2}}$$

and so

$$\omega^2 = gk. \tag{6.14}$$

This 'dispersion equation' shows that waves on deep water are propagated with speeds which depend on their wavelength. The longer waves will be transmitted more rapidly from a storm area than short ones because they possess higher speeds, and at some distance away from the area only waves of similar characteristics are observed.

The theory of deep water waves can be assumed to apply when $\tanh kd \doteq 1$; thus a satisfactory approximation is obtained when $kd = 2.65$ so that $\tanh kd = 0.99$. Very little error arises when $d/\lambda > \frac{1}{2}$, but errors increase with shallower depths and the theory for shallow water applies when $d/\lambda < \frac{1}{8}$. In the intermediate range of values, that is to say when $\frac{1}{8} < d/\lambda < \frac{1}{2}$, equation (6.12) must be used. It has also to be remembered that the theory is valid only for a wave whose amplitude is much smaller than its wavelength; that is, the theory holds good provided some such criterion as $2a/\lambda < \frac{1}{20}$ is satisfied.

6.4.4 Fluid particle motions

When the velocity potential equation (6.10) is differentiated, it is found that the component velocities of the fluid particles are

$$\left. \begin{aligned} u &= -\frac{\partial \Phi}{\partial X} = \frac{akg}{\omega} \frac{\cosh\left[k(Z+d)\right]}{\cosh kd} \cos\left(kX - \omega t\right), \quad v = 0, \\ w &= -\frac{\partial \Phi}{\partial Z} = \frac{akg}{\omega} \frac{\sinh\left[k(Z+d)\right]}{\cosh kd} \sin\left(kX - \omega t\right). \end{aligned} \right\} \tag{6.15}$$

At the rigid boundary, $Z = -d$, the normal component of velocity, w, is everywhere zero. In shallow water, the horizontal component u has a non-zero amplitude and the fluid particle moves to and fro along a horizontal path. In deep water, the amplitude of the velocity component u decreases with increase of depth and a fluid particle on the bottom remains stationary.

Integration with respect to time of the component velocities of the fluid gives the displacements (X', Y', Z') of the fluid particles. That is

$$\left. \begin{aligned} X' &= -\frac{akg}{\omega^2} \frac{\cosh\left[k(Z+d)\right]}{\cosh kd} \sin\left(kX - \omega t\right), \quad Y' = 0, \\ Z' &= \frac{akg}{\omega^2} \frac{\sinh\left[k(Z+d)\right]}{\cosh kd} \cos\left(kX - \omega t\right), \end{aligned} \right\} \tag{6.16}$$

and these quantities are such that

$$\frac{X'^2}{\alpha^2}+\frac{Z'^2}{\beta^2}=1,$$

where

$$\alpha = \frac{akg}{\omega^2}\frac{\cosh\left[k(Z+d)\right]}{\cosh kd}, \qquad \beta = \frac{akg}{\omega^2}\frac{\sinh\left[k(Z+d)\right]}{\cosh kd}$$

and

$$\alpha^2 - \beta^2 = \frac{a^2}{\cosh^2 kd}\left(\frac{kg}{\omega^2}\right)^2 = \frac{a^2}{\sinh^2 kd}.$$

Thus the trajectory of the fluid particles is an ellipse with semi-major axis α and semi-minor axis β. In water of shallow and intermediate depth the orbital paths have the same distance between the foci, but the lengths of the axes decrease with depth. At $Z = -d$, $\beta = 0$ and the ellipses degenerate into straight lines, as illustrated in fig. 6.3(a).

In deep water $\alpha = a\,e^{kZ} = \beta$ and the trajectory of the fluid particles is a circle of radius $a\,e^{kZ}$. Thus for increasing depth the radius decreases so that at $Z = -d$ the particle is stationary as shown in fig. 6.3(b). By comparing equations (6.15) and (6.16) with (6.10) it will be seen that a surface particle moves in the direction of travel when it is at a crest and in the opposite direction when it is in a trough, as illustrated in fig. 6.3(c); moreover this is true of water having any depth, deep or shallow.

Fig. 6.3. The fluid particles move in elliptical paths during the propagation of sinusoidal waves. In shallow water, and water of intermediate depth, the paths are as sketched in (a). In deep water the ellipses become circles as in (b). The path of a particle in the surface of water of any depth is circular and the motion is as shown sketched in (c).

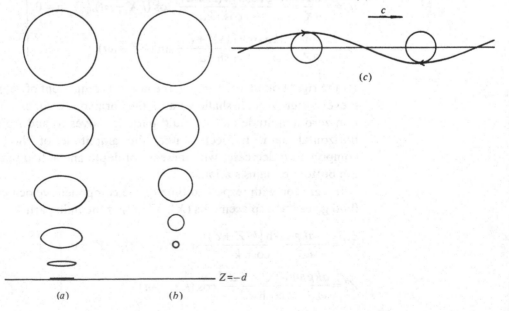

$Z = -d$

(a)   (b)

6.4.5 *Wave pressure*

As discussed in section 6.2.4 the term $-gZ$ represents hydrostatic pressure whereas $-\Delta p/\rho$ represents pressure changes due to other agencies. When sinusoidal waves travel on the surface of a fluid, pressure variation due solely to the wave disturbance is therefore given by Δp in the equation

$$\frac{\Delta p}{\rho} = \frac{\partial \Phi}{\partial t} - \tfrac{1}{2}(u^2 + v^2 + w^2).$$

In a linear analysis the kinetic energy term is neglected and from equation (6.11) we have

$$\Delta p = \rho \frac{\partial \Phi}{\partial t} = a\rho g \frac{\cosh [k(Z+d)]}{\cosh kd} \cos (kX - \omega t)$$

$$= \frac{\rho g \cosh [k(Z+d)]}{\cosh kd} \zeta(X, t).$$

This fluctuation is due to the passage of the wave.

In shallow water, the pressure variation is approximately

$$\Delta p = \rho g \zeta(X, t).$$

On the other hand in deep water,

$$\Delta p = \rho g\, e^{kZ} \zeta(X, t). \tag{6.17}$$

Thus, in this latter case, the amplitude of pressure fluctuations due to the passage of waves diminishes exponentially as depth increases. This is known as the 'Smith effect' (Smith, 1883). At $Z = -\lambda/2$, the pressure variation in deep water is a small fraction of that at $Z = 0$. Therefore the total pressure at $Z = -\lambda/2$ is mainly hydrostatic.

Fig. 6.4. A wave of amplitude a and unit width in water of depth d. A single wavelength λ is shown.

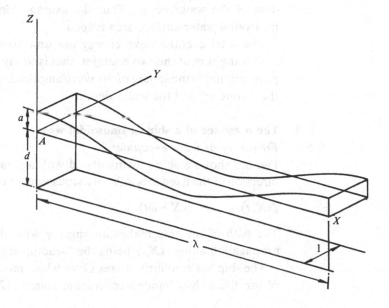

6.4.6   *Energy of a harmonic wave*

The potential energy of the harmonic wave

$$\zeta(X, t) = a\cos(kX - \omega t)$$

is the work needed to distort the calm water surface into this shape and it is independent of the motions of the fluid particles. Fig. 6.4 shows a sinusoidal wave of unit width bounded by two vertical planes at a distance of one wavelength apart. In a small interval dX, the weight of water displaced is $\rho g\zeta dX$ having a centre of mass at $\zeta/2$ above AXY. Thus the total work done in one wavelength is

$$\tfrac{1}{2}\rho g\int_0^\lambda \zeta^2\,dX = \tfrac{1}{2}\rho ga^2\int_0^\lambda \cos^2(kX - \omega t)\,dX = \tfrac{1}{4}\rho ga^2\lambda$$

and the average potential energy per unit horizontal water-surface area is $\tfrac{1}{4}\rho ga^2$.

The kinetic energy of the wave of unit width bounded by the two vertical planes is given approximately by

$$\tfrac{1}{2}\rho\int_0^\lambda\int_{-d}^0 (u^2 + v^2 + w^2)\,dX\,dZ.$$

Provided the elevation is small, the integration with respect to Z may be approximated to the range $(-d, 0)$. Substitution from equation (6.15) shows the kinetic energy to be

$$\tfrac{1}{2}\rho\frac{a^2k^2g^2}{\omega^2\cosh^2 kd}\int_0^\lambda\int_{-d}^0 \{\cosh^2[k(Z+d)]\cos^2(kX - \omega t)$$

$$+ \sinh^2[k(Z+d)]\sin^2(kX - \omega t)\}\,dX\,dZ$$

$$= \tfrac{1}{4}\rho\frac{a^2kg^2\lambda}{\omega^2}\tanh kd = \tfrac{1}{4}\rho ga^2\lambda$$

since, by equation (6.9), $\omega^2 = gk\tanh kd$. The expression is independent of the water depth. Thus the average kinetic energy per unit horizontal water-surface area is $\tfrac{1}{4}\rho ga^2$.

The total average wave energy per unit horizontal water-surface area is the sum of the two energies, that is to say $\tfrac{1}{2}\rho ga^2$. This energy is proportional to the square of the wave amplitude and is independent of the frequency and the water depth.

6.5   The presence of a ship in sinusoidal waves

6.5.1   *Frequency of wave encounter*

Fig. 6.5 shows a ship moving ahead with a steady velocity $\bar{U}$ in the direction of the fixed axis AX_0 in a sinusoidal sea having a wave profile

$$\zeta(X, t) = a\cos(kX - \omega t).$$

The path of the ship makes an angle χ with the direction of wave propagation along AX, χ being the 'heading angle'.

The ship has equilibrium axes $Oxyz$ which move along the axis AX_0 of the fixed right-handed coordinate frame $AX_0Y_0Z_0$ such that O

coincides with A at $t = 0$. This right-handed system of equilibrium axes has the undisturbed motion of the ship and the origin O occupies the position of trisection of the calm water surface, the stern and the plane of port and starboard symmetry; it would move along AX_0 if there were no parasitic motions. The frames $AX_0 Y_0 Z_0$ and $Oxyz$ always remain parallel to one another.

At time t, the coordinate X in $AXYZ$ may be transformed into coordinates in the $Oxyz$ system of axes. Thus by the transformation

$$X = X_0 \cos \chi + Y_0 \sin \chi = (\bar{U}t + x) \cos \chi + y \sin \chi$$

the wave elevation becomes

$$\zeta(x, y, t) = a \cos [kx \cos \chi + ky \sin \chi - (\omega - \bar{U}k \cos \chi)t]$$

$$= a \cos (kx \cos \chi + ky \sin \chi - \omega_e t) \tag{6.18}$$

in the equilibrium axes $Oxyz$, where the frequency of encounter is

$$\omega_e = \omega - \bar{U}k \cos \chi.$$

Substitution for k from equation (6.14) shows that

$$\omega_e = \omega - \frac{\bar{U}\omega^2}{g} \cos \chi \tag{6.19}$$

in deep water.

An alternative way of deriving this expression for frequency of encounter may be seen by reference to fig. 6.5. A deep water wave of velocity c travelling in the AX direction has a velocity $c/\cos \chi$ or $\omega/k \cos \chi$ in the direction AX_0. This is the speed with which a crest (say) moves in the direction of the ship's steady motion. The relative speed with which the wave overtakes the ship is therefore

$$\frac{\omega}{k \cos \chi} - \bar{U} = \frac{\omega - \bar{U}k \cos \chi}{k \cos \chi}.$$

Now the distance between successive crests in the direction AX_0 is $\lambda/\cos \chi$ so that the time taken for one complete wave to overtake the

Fig. 6.5. Fixed axes $AX_0 Y_0 Z_0$ with a ship, having equilibrium axes $Oxyz$, moving along AX_0. Sinusoidal waves are propagated in the direction AX of the frame $AXYZ$ so that the ship has a heading angle χ.

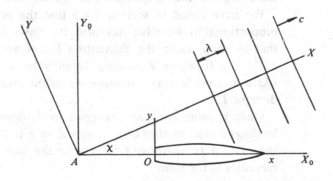

ship is given by this distance divided by the relative speed; that is,

$$T_e = \left| \frac{\lambda}{\cos \chi} \times \frac{k \cos \chi}{\omega - \bar{U}k \cos \chi} \right| = \frac{2\pi}{|\omega - \bar{U}k \cos \chi|} = \frac{2\pi}{|\omega_e|},$$

where, again,

$$\omega_e = \frac{2\pi}{T_e} = \omega - \bar{U}k \cos \chi.$$

The modulus signs are introduced to ensure that the time T_e is positive even though 'overtaking' does not occur. This result gives rise to three special cases (assuming that $\bar{U}$ is positive):

Value of χ	Value of $\bar{U}$	Nature of encounter	$\omega - \bar{U}k \cos \chi$	Overtaking
$\frac{\pi}{2} < \chi < \frac{3\pi}{2}$	any	Bow	+ ve	Head sea
$-\frac{\pi}{2} < \chi < \frac{\pi}{2}$	$\bar{U}k \cos \chi > \omega$	Bow	− ve	Waves overtaken by ship (head sea)
$-\frac{\pi}{2} < \chi < \frac{\pi}{2}$	$\bar{U}k \cos \chi < \omega$	Stern	+ ve	Waves overtaking ship (following sea)

6.5.2 The 'Smith correction'

Strictly speaking, procedure from wave elevation to wave force requires some form of integration of pressure fluctuations with immersed depth, the limits of integration being $0 > Z > -T$ (or $0 > z > -T$), where T is the local draught of the hull.

It is assumed that the presence of the ship does not interfere with the motions of the fluid particles so that the pressure distribution on the ship is the pressure of the undisturbed incident wave evaluated at the ship's hull. This is commonly known as the 'Froude–Krylov hypothesis'. The fluid loading applied to a ship may thus be found from a description of the wave disturbance as it is seen by an observer travelling with the equilibrium axes, i.e. by means of equation (6.18).

We have noted in section 6.4.5 that the pressure fluctuation is proportional to the wave elevation. It follows, from equation (6.17), that in deep water the fluctuating forces are all proportional to $e^{-k\bar{T}}\zeta(x, y, t)$, where $\bar{T}$ is some 'mean' value in the range $0 < \bar{T} < T$. (As a first crude approximation we might take $\bar{T}$ as half the local draught T.)

Once a value has been assigned to $\bar{T}$, dependence of the fluid loadings on the vertical coordinates Z or z is eliminated. This being the case, it is convenient to rewrite the expression for the wave elevation in the form

$$\zeta(x, y, t) = a \, e^{-k\bar{T}} e^{i(kx \cos \chi + ky \sin \chi - \omega_e t)} \tag{6.20}$$

and to assume that this variation at the surface prevails over the whole draught of the ship. When an average of this nature is used, the 'Smith correction' is said to be made.

A value may be found for $\bar{T}$ by the following rough and ready argument. Suppose that *body* axes $Oxyz$ are fixed to the ship with Oxz the plane of port and starboard symmetry and Oxy the mean water plane, as shown in fig. 6.6. The mean vertical force per unit length fore and aft due to pressure fluctuations Δp in a wave is approximately

$$F(x, y, t) = \int_{\text{keel}}^{\text{surface}} \Delta p(x, y, z, t) 2 \, dy.$$

If this expression is integrated by parts it gives

$$F(x, y, t) = [2y \, \Delta p]_{\text{keel}}^{\text{surface}} - \int_{\text{keel}}^{\text{surface}} 2y \, d(\Delta p).$$

Since $y = 0$ at the keel and $\Delta p = \rho g a \exp[i(kx \cos \chi + ky \sin \chi - \omega_e t)]$ $(= \xi$ say$)$ at the surface, by equation (6.17), the integrated term reduces to $B(x)\xi$ and

$$F(x, y, t) = B(x)\xi - \int_{-T}^{0} 2y \frac{d\Delta p(x, y, z, t)}{dz} \, dz$$

$$= B(x)\xi - 2k\xi \int_{-T}^{0} y \, e^{kz} \, dz.$$

The 'equivalent wave' of equation (6.20) would give a vertical force per unit length of $B(x)\xi \, e^{-k\bar{T}}$ so that

$$e^{-k\bar{T}} = 1 - \frac{2k}{B(x)} \int_{-T}^{0} y \, e^{kz} \, dz,$$

whence

$$\bar{T} = -\frac{1}{k} \ln \left[1 - \frac{2k}{B(x)} \int_{-T}^{0} y \, e^{kz} \, dz \right]. \tag{6.21}$$

6.5.3 Approximate symmetric and antisymmetric fluid actions

The wave elevation across a given section of hull is

$$\zeta(x, y, t) = a \, e^{-k\bar{T}} e^{i(kx \cos \chi + ky \sin \chi - \omega_e t)}.$$

Fig. 6.6. Axes $Oxyz$ fixed to a ship with Oxz coincident with the plane of port and starboard symmetry and Oxy in the mean water surface.

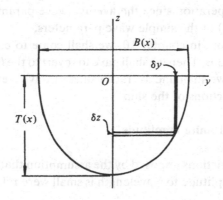

The rate of change of wave elevation is

$$\frac{D\zeta}{Dt} = \left(\frac{\partial}{\partial t} - \bar{U}\frac{\partial}{\partial x}\right)\zeta = -i\omega\zeta(x, y, t),$$

this being the vertical component of the velocity of the fluid particles as in equation (6.15) when the Smith correction is applied. Similarly,

$$\frac{D^2\zeta}{Dt^2} = -\omega^2\zeta(x, y, t).$$

The average wave elevation across a section of the ship is approximately

$$\bar{\zeta} = \frac{1}{B(x)}\int_{-0.5B}^{0.5B} \zeta(x, y, t)\,dy = a\alpha(x)\,e^{-k\bar{T}}\,e^{i(kx\cos\chi - \omega_e t)}, \tag{6.22}$$

where

$$\alpha(x) = \frac{\sin\left[0.5B(x)k\sin\chi\right]}{0.5B(x)k\sin\chi}. \tag{6.23}$$

The parameter $\alpha(x)$ accounts for the variation of fluid motions across the beam of the ship and also provides a means of allowing for the influence of short waves. The average rates of change of wave elevation may be determined in a similar manner and are given by

$$\frac{D\bar{\zeta}}{Dt} = -i\omega\bar{\zeta}, \qquad \frac{D^2\bar{\zeta}}{Dt^2} = -\omega^2\bar{\zeta}. \tag{6.24}$$

These expressions may reasonably be assumed to give those characteristics of the wave elevation which determine symmetric responses of the hull to the sinusoidal sea illustrated in fig. 6.5.

An alternative, and even simpler, hypothesis is that symmetric response is governed by the wave elevation characteristics along the centre line, $y = 0$, of the ship. In other words, we merely consider the ship as moving in a system of waves

$$\zeta(x, t) = a\,e^{-k\bar{T}}\,e^{i(kx\cos\chi - \omega_e t)}. \tag{6.25}$$

In this chapter and the following one, this simplified standpoint is adopted, while in the calculations of chapter 8 the average wave elevation parameters are used. The change to $\bar{\zeta}$ and its derivatives is a simple operation since the average wave parameters are only a multiple $\alpha(x)$ of the simple wave parameters.

Later on, in chapter 10, we shall come to consider antisymmetric fluid actions. Then we shall have to revert to the former approach. That is to say we shall find average values of wave elevation and wave slope across sections of the ship.

6.6   Waves of finite amplitude
6.6.1   *Stokes waves*
The restrictions imposed by the assumption that the ratio of maximum wave amplitude to wavelength is small were relaxed by Stokes (1880)

in his studies of waves of small but finite amplitudes. He retained the assumption that the fluid motion is irrotational, unlike Gerstner (1809) whose theory of trochoidal waves admits finite amplitudes. Stokes suggested the use of expansions about the mean water level when the water depth is not small compared with a typical wavelength. For example, the velocity potential function may be expressed in the form

$$\Phi = \varepsilon \Phi_1 + \varepsilon^2 \Phi_2 + \varepsilon^3 \Phi_3 + \ldots ,$$

where $\Phi_1, \Phi_2, \Phi_3, \ldots$ are suitable 'component' potentials and the arbitrary parameter ε is of the order of magnitude of the wave slope, which is small. The success of this approach depends upon the fact that the maximum value of the ratio of the non-linear terms to the linear ones in the wave equation is of order ε in deep water.

The solution of the non-linear wave equations is obtained by successive approximations, though a general convergence criterion has not been fully established. Nor is the technique satisfactory when the wave slope approaches the order of unity, as in the case where waves are about to become unstable and break. Various solutions up to the third order are given by Wiegel (1964). To this order the profiles of the Stokes wave and the trochoidal wave are identical.

The symmetry of crest and trough about the mean water level which is observed in the simple sinusoidal, or Airy, wave theory breaks down with waves of finite amplitude. The crests are now steeper than the troughs and are much higher above the calm water surface level than the troughs are below it. The difference is roughly as sketched in fig. 6.7.

Skjelbreia & Hendrickson (1961) have shown that the profile of a fifth-order Stokes wave is of the form

$$\zeta(X, t) = a \cos \theta + a^2 k (B_{22} + a^2 k^2 B_{24}) \cos 2\theta$$
$$+ a^3 k^2 (B_{33} + a^2 k^2 B_{35}) \cos 3\theta$$
$$+ a^4 k^3 B_{44} \cos 4\theta + a^5 k^4 B_{55} \cos 5\theta.$$

Fig. 6.7. A rough comparison of wave forms. When the amplitude of a sinusoidal wave ceases to be adequately 'small', the wave has a form more like that predicted by the Stokes theory.

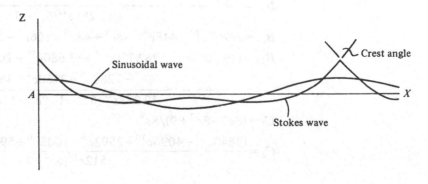

The associated velocity potential function is

$$\Phi(X, t) = \bar{c}[a(A_{11} + a^2 k^2 A_{13} + a^4 k^4 A_{15}) \cosh kd \sin \theta$$
$$+ a^2 k (A_{22} + a^2 k^2 A_{24}) \cosh 2kd \sin 2\theta$$
$$+ a^3 k^2 (A_{33} + a^2 k^2 A_{35}) \cosh 3kd \sin 3\theta$$
$$+ a^4 k^3 A_{44} \cosh 4kd \sin 4\theta$$
$$+ a^5 k^4 A_{55} \cosh 5kd \sin 5\theta].$$

In this expression

$$\theta = kX - \omega t,$$

and the wave velocity is

$$\bar{c} = \left[\frac{g}{k} (1 + a^2 k^2 C_1 + a^4 k^4 C_2) \tanh kd \right]^{\frac{1}{2}}$$

The coefficients are defined as follows, with

$$s = \sinh kd, \qquad c = \cosh kd:$$

$$A_{11} = 1/s,$$

$$A_{13} = -c^2(5c^2 + 1)/8s^5,$$

$$A_{15} = -(1184c^{10} - 1440c^8 - 1992c^6 + 2641c^4 - 249c^2 + 18)/1536s^{11},$$

$$A_{22} = 3/8s^4,$$

$$A_{24} = (192c^8 - 424c^6 - 312c^4 + 480c^2 - 17)/768s^{10},$$

$$A_{33} = (13 - 4c^2)/64s^7,$$

$$A_{35} = (512c^{12} + 4224c^{10} - 6800c^8 - 12\,808c^6 + 16\,704c^4$$
$$- 3154c^2 + 107)/4096s^{13}(6c^2 - 1),$$

$$A_{44} = (80c^6 - 816c^4 + 1338c^2 - 197)/1536s^{10}(6c^2 - 1),$$

$$A_{55} = \frac{-(2880c^{10} - 72\,480c^8 + 324\,000c^6 - 432\,000c^4 + 163\,470c^2 - 16\,245)}{61\,440s^{11}(6c^2 - 1)(8c^4 - 11c^2 + 3)},$$

$$B_{22} = c(2c^2 + 1)/4s^3,$$

$$B_{24} = c(272c^8 - 504c^6 - 192c^4 + 322c^2 + 21)/384s^9,$$

$$B_{33} = 3(8c^6 + 1)/64s^6,$$

$$B_{35} = \frac{(88\,128c^{14} - 208\,224c^{12} + 70\,848c^{10} + 54\,000c^8 - 21\,816c^6 + 6264c^4 - 54c^2 - 81)}{12\,288s^{12}(6c^2 - 1)},$$

$$B_{44} = c(768c^{10} - 448c^8 - 48c^6 + 48c^4 + 106c^2 - 21)/384s^9(6c^2 - 1),$$

$$B_{55} = \frac{(192\,000c^{16} - 262\,720c^{14} + 83\,680c^{12} + 20\,160c^{10} - 7280c^8 + 7160c^6 - 1800c^4 - 1050c^2 + 225)}{12\,288s^{10}(6c^2 - 1)(8c^4 - 11c^2 + 3)},$$

$$C_1 = (8c^4 - 8c^2 + 9)/8s^4,$$

$$C_2 = \frac{(3840c^{12} - 4096c^{10} + 2592c^8 - 1008c^6 + 5944c^4 - 1830c^2 + 147)}{512s^{10}(6c^2 - 1)}.$$

For deep water in which $kd \to \infty$, the surface profile of the fifth-order Stokes wave reduces to

$$\zeta(x, t) = a\cos\theta + a^2k(\tfrac{1}{2} + \tfrac{17}{24}a^2k^2)\cos 2\theta + a^3k^2(\tfrac{3}{8} + \tfrac{153}{128}a^2k^2)\cos 3\theta$$

$$+ \frac{a^4k^3}{3}\cos 4\theta + \tfrac{125}{384}a^5k^4\cos 5\theta.$$

De (1955) extended Stokes's theory (for deep water waves) to the fifth order and showed that even this fifth-order analysis should not be used when $d/\lambda < \tfrac{1}{8}$. Furthermore, the larger the values of $2a/\lambda$, the greater is the value of d/λ at which Stokes's theory becomes unreliable.

Expressions can be obtained for the stream function and for the components of particle velocity and acceleration in deep and shallow water. The fifth-order wave theory predicts (a) higher values of the fluid particle velocity components, as illustrated in fig. 6.8 (which is adapted from the work of Skjelbreia & Hendrickson, 1961), and (b) that the acceleration components do not differ much from those predicted by the simple theory of sinusoidal waves. For these two reasons a wave theory of higher order should be used in calculations of wave force where drag forces (proportional to the square of the velocity) are important, as with the forces on an offshore structure or a pile. When inertia (i.e. acceleration) forces predominate, however, as in ships, a first-order (simple sinusoidal wave) theory is probably sufficiently accurate for most purposes.

6.6.2 Wave breaking

Stokes made the assumption that, if wave breaking is to occur, the fluid particle speed at the very crest must be at least equal to the wave

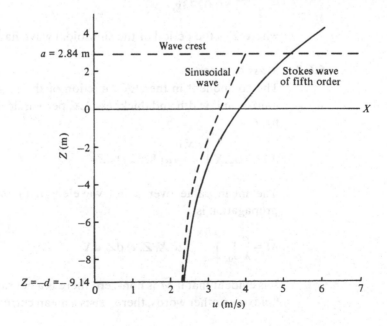

Fig. 6.8. Variations of the horizontal component u of particle velocity in a section through the crest of a sinusoidal wave and a Stokes fifth-order wave, for a wave height $h = 5.68$ m and depth $d = 9.14$ m.

propagation speed so that $u/c \geqslant 1$. For waves of small amplitude in deep water, equation (6.15) gives a maximum particle speed of $u = a\omega = akc$. But from the basic assumption of small amplitude it follows that $u/c = ak \ll 1$. Therefore breaking is never predicted by the theory of waves having small amplitude and the possibility arises only in the theory of finite-amplitude waves. Stokes showed that in deep water the maximum total crest angle (made by assuming that the water surface on either side of the peak can be approximated by two straight lines as sketched in fig. 6.7) as the wave begins to break is 120°, though this may be an over-estimate for real waves.

Michell (1893) found that the maximum possible steepness for a Stokes wave is

$$\left(\frac{2a}{\lambda}\right)_{max} = 0.142 \simeq \tfrac{1}{7}.$$

Should this limit be exceeded, the speed of the fluid particles at the crest would exceed the wave speed and so the wave would break. The maximum wave propagation speed c_{max} of a Stokes wave to third order is given by

$$c_{max} \simeq 1.1c,$$

where $c = g/\omega$ is the propagation speed of the sinusoidal wave in deep water. Thus, the maximum correction to be applied to the wave propagation speed to allow for wave steepness is about 10% of the speed for small amplitude waves. An alternative form of the Michell breaking condition is

$$\frac{(2a)_{max}}{T^2} \simeq 0.0273g,$$

where T is the period of the sinusoidal wave having small amplitude.

6.6.3   *Mass transport*

The component in the AX direction of the momentum of a slice of fluid of unit width and thickness ΔX, perpendicular to the axis AX in fig. 6.4 is

$$\Delta M = \rho \Delta X \int_{-d}^{\zeta(X, t)} u(X, Z, t)\, dZ.$$

The mean value over a full wavelength in the direction of wave propagation is

$$\bar{M} = \frac{\rho}{\lambda} \int_0^\lambda \int_{-d}^{\zeta(X, t)} u(X, Z, t)\, dZ\, dX.$$

It is evident that if $\bar{M}$ is non-zero there is a net rate of mass transfer of fluid or, in other words, there exists a mean current or drift in the fluid.

Substitution from equations (6.15) and (6.10) gives

$$\bar{M} = \frac{\rho}{\lambda} \int_0^{\lambda} \int_{-d}^{\zeta(X,\,t)} \frac{akg \cosh[k(Z+d)]}{\omega} \, \frac{}{\cosh kd} \cos(kX - \omega t) \, dZ \, dX$$

$$= \frac{\rho}{\lambda} \int_0^{\lambda} \frac{g}{\omega} \frac{\sinh\{k[\zeta(X,t)+d]\}}{\cosh kd} \zeta(X,t) \, dX.$$

Expansion of the sinh term as a power series in $\zeta(X, t)$ gives

$$\bar{M} = \frac{\rho g}{\lambda \omega} \tanh kd \int_0^{\lambda} \left[\left(1 + \frac{k^2\zeta^2}{2!} + \dots \right) \zeta \right.$$

$$\left. + k\zeta^2 \left(1 + \frac{k^2\zeta^2}{3!} + \dots \right) \coth kd \right] dX.$$

The only non-zero contribution to the mean rate of fluid momentum in the direction of wave propagation arises from the second term in the square bracket. (The first term, being an odd sinusoidal function, on integration over a wavelength and period is zero.) After integration, and the use of equation (6.9) the mean momentum per unit area in the direction of wave propagation is found to be

$$\bar{M} = \tfrac{1}{2}\rho a^2 \omega \coth kd + O(a^4).$$

An alternative more general derivation of this result may be obtained using vector methods. The mean momentum per unit width of section is

$$\bar{\mathbf{M}} = \rho \overline{\int_{-d}^{\zeta(X,\,t)} \mathbf{q} \, dZ} = -\rho \overline{\int_{-d}^{\zeta(X,\,t)} \operatorname{grad} \Phi(X, Y, Z, t) \, dZ},$$

where the overbar denotes the mean value and the results of equation (6.3) have been used so that we assume the motion to be irrotational. Now by Leibnitz's rule for differentiation of integral signs, it follows that

$$\operatorname{grad} \int_{-d}^{\zeta(X,\,t)} \Phi(X, Y, Z, t) \, dZ$$

$$= \Phi(X, Y, \zeta, t) \operatorname{grad} \zeta + \int_{-d}^{\zeta(X,\,t)} \operatorname{grad} \Phi(X, Y, Z, t) \, dZ,$$

in which the two terms on the right-hand side arise because the wave elevation is a function of the horizontal coordinate. The term on the left is the gradient of an oscillatory function whose mean value is zero. Thus we find that

$$\bar{\mathbf{M}} = \overline{\rho \Phi(X, Y, \zeta(X, t), t) \operatorname{grad} \zeta} \simeq \overline{\rho \Phi(X, Y, 0, t) \operatorname{grad} \zeta},$$

which is a second-order term being produced by the products of terms. Substitution into this result from equations (6.10) and (6.11) gives the mean momentum per unit area in the direction of wave propagation to order a^2 to be

$$\bar{M} = \tfrac{1}{2}\rho a^2 \omega \coth kd.$$

(Note that the mean momentum per unit area in each of the directions AY and AZ is zero.)

Since $\bar{M}$ can be interpreted as the net mass flux per unit width of the wave there must exist a forward translation or drift of the fluid particles in the direction of the travelling wave. That is to say there is a net wave current or 'mass transport velocity' which is of order a^2. In these circumstances the particle motion cannot be along closed paths as shown in fig 6.3. However, in the linear small-amplitude theory, the mass transport velocity is of the second order and therefore negligibly small. It follows that in the linear theory of sinusoidal waves it may be ignored, and we may regard the fluid particles as travelling in closed orbits.

Finite-amplitude irrotational waves, on the other hand, must be accompanied by a mass transport of water in the direction of the wave motion. There exists a mass transport velocity u_c and the particles do not travel around closed paths.

Longuet-Higgins (1953) has derived a general theory of mass transport taking account of viscosity. The resultant mass transport velocity can differ markedly from that predicted by Stokes on the assumption of a perfect, non-viscous fluid in irrotational flow and it leads to results that agree well with observation. Longuet-Higgins expresses the mass transport velocity as

$$u_c = \frac{a^2 \omega k \cosh\left[2k(Z+d)\right]}{2 \sinh^2 kd} + C,$$

where C is an arbitrary constant. This expression can be shown to be equivalent to the one initially derived by Stokes.

If it is assumed that there is a zero total mass flow across any plane between the bottom and mean water surface, it is found that

$$C = -\frac{a^2 \omega}{2d} \coth kd.$$

This assumption is appropriate in wave experiments undertaken in a tank and, for such a case, Longuet-Higgins derives a suitable expression for the mass transport velocity. In a viscous fluid of finite depth d, it is

$$u_c = \frac{a^2 \omega k}{4 \sinh^2 kd}\left[3 + 2(\mu+1)\cosh 2kd + kd(3\mu^2 + 4\mu + 1)\sinh 2kd\right.$$

$$\left. + 3(\mu^2 - 1)\left(\frac{\sinh 2kd}{2kd} + \frac{3}{2}\right)\right],$$

where $\mu = Z/d$ and the distribution is illustrated in fig. 6.9, for $a^2 \omega k = 1$.

It will be seen from fig 6.9 that there exists a circulatory flow in the tank, this being superimposed on the particle motion. This is a conclusion that should be borne in mind when tests are made in waves

in a towing tank. Such consideration may also apply when the influence
is examined of large waves (such as the so-called '100 year wave') on
models of marine or offshore structures. Direct scaling of the experimental results of such tests to full size may lead to erroneous
conclusions since the mass transport velocity may not be so significant
in the open sea.

In deep water, where $kd \to \infty$, the mass transport velocity simplifies
to

$$u_c = a^2 \omega k\, e^{2kZ}.$$

This quantity may be large at the surface $Z = 0$, but it decreases rapidly
with increase of depth. Thus wind-induced currents which run with the
waves are only expected to be significant in the sea surface.

Fig. 6.9. The distribution of mass transport velocity u_c of a viscous fluid
when small, though finite, waves travel along the surface of a closed tank.
The assumption is made that there is no net mass transport across any
section perpendicular to the direction of wave propagation. (Longuet-
Higgins, 1953.)

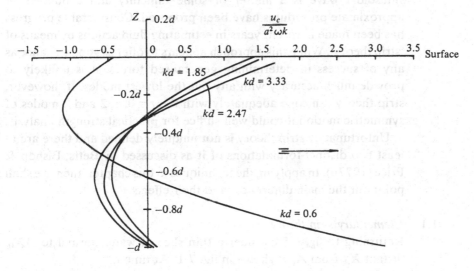

7 Symmetric generalised fluid forces

> It is purely a matter of skill,
> Which all may attain if they will:
> But every Jack,
> He must study the knack
> If he wants to be sure of his Jill!
> *The Yeomen of the Guard*

7.1 Strip theory

The estimation of hydrodynamic loading applied to the hull by a sinusoidal wave is a matter of some difficulty and a number of approximate procedures have been proposed. Considerable progress has been made in recent years in estimating fluid actions by means of 'strip theory'. While this approach appears to offer as good a chance as any of success in determining generalised forces, it is unlikely to provide much accuracy with any but the lowest modes. If, however, strip theory can cope adequately with the $r = 0, 1, 2$ and 3 modes of symmetric motion it could well suffice for practical strength analysis.

Unfortunately strip theory is not uniquely defined and there are at least two distinct formulations of it as discussed by Betts, Bishop & Price (1977b). In applying the techniques in this chapter, then, we shall point out the main differences and their effects.

7.1.1 *Elementary strip theory*

Returning to fig. 6.5 consider a thin slice of water normal to AX_0, distant X_A from A, as shown in fig. 7.1. At time t,

$$X_A = \bar{U}t + x,$$

Fig. 7.1. At some instant t, a slice of the hull lies in the plane of a slice of water. The water slice is normal to AX_0, distant X_A from the origin A of the fixed axes and x from the origin O of the equilibrium axes. Strip theory predicts the force exerted by the fluid slice on its instantaneously coincident hull slice.

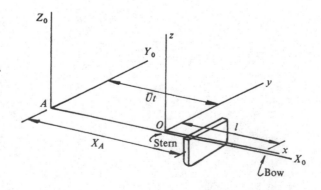

O having coincided with A at the instant $t = 0$, where

$$\frac{dx}{dt} = -\bar{U}.$$

Strip theory seeks to predict the force applied by the strip of fluid (fig. 7.1) to the hull.

The relative displacement of the ship and water surface is

$$\bar{z}(x, t) = w(x, t) - \zeta(x, t), \tag{7.1}$$

where $w(x, t)$ is the upward displacement of the section of the hull coincident with the strip and $\zeta(x, t)$ is the local surface elevation. The quantity $\bar{z}(x, t)$ is thus a measure of hull emergence at the water strip. The upward force per unit length exerted by the fluid on the hull, $F(x, t)$, is dependent upon $\bar{z}(x, t)$ and its total derivatives with respect to time. According to Gerritsma & Beukelman (1964) and to Lewis (1967),

$$F(x, t) = -\left\{ \frac{D}{Dt}\left[m(x) \frac{D\bar{z}(x, t)}{Dt} \right] + N(x) \frac{D\bar{z}(x, t)}{Dt} + \rho g B(x) \bar{z}(x, t) \right\}. \tag{7.2}$$

In this expression, $m(x)$ is the local 'added mass' per unit length, $N(x)$ is the local 'fluid damping coefficient' and the operator D/Dt is the total derivative with respect to time; that is

$$\frac{D}{Dt} = \frac{\partial}{\partial t} + \frac{\partial}{\partial x}\frac{dx}{dt} = \frac{\partial}{\partial t} - \bar{U}\frac{\partial}{\partial x}.$$

The origins of this expression for $F(x, t)$ are semi-empirical even though it has been employed with considerable success in practice. Notice that, as it stands, the expression is such that *an* added mass per unit length $m(x)$ and *a* fluid damping coefficient $N(x)$ are referred to, each associated with the appropriate ship section and no more.

It may be shown that the fluid force $F(x, t)$ also depends on frequency of the relative displacement. The dependence is not an easy one to express and, indeed, unless $\bar{z}(x, t)$ varies sinusoidally it is necessary to identify its sinusoidal components through the Fourier transform and to assemble the contributions that the various components make to $F(x, t)$. It is common to ignore this frequency-dependence, however, and to employ added mass and damping functions at their face value, i.e. as functions $m(x)$ and $N(x)$ of x only. In the interests of brevity we shall ignore frequency dependence and merely write $m(x)$ and $N(x)$ in the development of the theory. Later, when we come to calculations for ships we shall reinstate the frequency-dependence.

The total derivatives of $\bar{z}(x, t)$ are

$$\frac{D\bar{z}(x, t)}{Dt} = \left(\frac{\partial}{\partial t} - \bar{U}\frac{\partial}{\partial t} \right)\bar{z}(x, t)$$

and

$$\frac{D^2 \bar{z}(x, t)}{Dt^2} = \left(\frac{\partial}{\partial t} - \bar{U}\frac{\partial}{\partial x}\right)^2 \bar{z}(x, t).$$

It follows that

$$F(x, t) = -m(x)\frac{D^2 \bar{z}(x, t)}{Dt^2} - \left[N(x) - \bar{U}\frac{dm(x)}{dx}\right]\frac{D\bar{z}}{Dt} - \rho g B(x)\bar{z}(x, t).$$

(7.3)

The relative displacement $\bar{z}(x, t)$ can now be eliminated by introducing the quantity $[w(x, t) - \zeta(x, t)]$. In this way it is found that

$$F(x, t) = -H(x, t) + Z(x, t),$$ (7.4)

where

$$H(x, t) = m(x)\frac{D^2 w(x, t)}{Dt^2} + \left[N(x) - \bar{U}\frac{dm(x)}{dx}\right]\frac{Dw(x, t)}{Dt}$$

$$+ \rho g B(x)w(x, t),$$ (7.5)

$$Z(x, t) = m(x)\frac{D^2 \zeta(x, t)}{Dt^2} + \left[N(x) - \bar{U}\frac{dm(x)}{dx}\right]\frac{D\zeta(x, t)}{Dt}$$

$$+ \rho g B(x)\zeta(x, t).$$ (7.6)

The force $H(x, t)$ per unit length depends only on the motion of the ship and it would be applied to the hull by flat calm water. The expression for the force $Z(x, t)$ per unit length is exactly analogous to that for $H(x, t)$ and so $Z(x, t)$ would be applied to the ship if it ploughed straight through a sinusoidal 'hydrostatic' sea without responding to the excitation. It is to allow for the inevitable difference between real waves and 'hydrostatic' ones that we make use of the Smith correction referred to in section 6.5.2; indeed when we substitute for $\zeta(x, t)$ using equation (6.24) we find that

$$Z(x, t) = \{-m(x)\omega^2 - i\omega[N(x) - \bar{U}m'(x)] + \rho g B(x)\}\zeta(x, t).$$ (7.7)

Here and henceforth we shall signify differentiation with respect to x by means of a prime.

7.1.2   Alternative form of strip theory

As we have mentioned, strip theory is not a single, generally accepted formulation of hydrodynamic action, and alternative forms have been proposed. In alternative approaches (e.g. see Salvesen, Tuck & Faltinsen, 1970; Vugts, 1971) the hydrodynamic actions are initially found from potential flow theory in which the fluid is assumed irrotational, incompressible and inviscid. The strip theory approximation is devised as a local two-dimensional approximation in the last stages of this general analysis. (This is in marked contrast to the previous analysis in which the 'strip' concept is adopted from the outset.) It has been shown by Flokstra (1974) that the fluid action per unit length can be expressed in the form

$$F(x, t) = -\left(\frac{D}{Dt}\left\{\left[m(x) + \frac{i}{\omega_e}N(x)\right]\frac{D\bar{z}(x, t)}{Dt}\right\} + \rho g B(x)\bar{z}(x, t)\right). \quad (7.8)$$

Here, ω_e is the frequency of encounter, and the relative displacement is

$$\bar{z}(x, t) = w(x, t) - \zeta(x, t)$$

$$= e^{-i\omega_e t}\sum_{r=0}^{\infty} p_r w_r(x) - a\, e^{-k\bar{T}}e^{i(kx\cos\chi - \omega_e t)}, \quad (7.9)$$

$p_0, p_1, p_2, \ldots$ being a sequence of complex constants.

On comparing this expression for $F(x, t)$ with that of equation (7.2), it is seen that the term involving the fluid damping function $N(x)$ is different. The difference apparently originates in the interpretation of the 'momentum of the fluid motion' in the underlying hydrodynamic theories. In this theory the effect of the outgoing waves has been included in the expression for the momentum whereas in the theory quoted earlier it is not. For zero forward velocity, there is no difference between the two expressions for $F(x, t)$.

With this formulation we find that

$$H(x, t) = \left[m(x) + \frac{iN(x)}{\omega_e}\right]\frac{D^2 w(x, t)}{Dt^2}$$

$$- \bar{U}\left[m'(x) + \frac{iN'(x)}{\omega_e}\right]\frac{Dw(x, t)}{Dt} + \rho g B(x)w(x, t), \quad (7.10)$$

$$Z(x, t) = \left[m(x) + \frac{iN(x)}{\omega_e}\right]\frac{D^2\zeta(x, t)}{Dt^2}$$

$$- \bar{U}\left[m'(x) + \frac{iN'(x)}{\omega_e}\right]\frac{D\zeta(x, t)}{Dt} + \rho g B(x)\zeta(x, t), \quad (7.11)$$

and, with waves as specified in chapter 6, the latter result reduces to

$$Z(x, t) = \left\{-\omega^2\left[m(x) + \frac{\bar{U}N'(x)}{\omega\omega_e}\right] - i\omega\left[\frac{\omega N(x)}{\omega_e} - \bar{U}m'(x)\right]\right.$$

$$\left. + \rho g B(x)\right\}\zeta(x, t). \quad (7.12)$$

7.1.3 *The added mass coefficient*

Both of the formulations of strip theory that we have introduced refer to the distribution of added mass $m(x)$, and to the fluid damping coefficient $N(x)$. These quantities have therefore to be determined for a given hull. (Notice that there is no suggestion that mass is really 'added' in the sense that an identifiable body of water is 'entrained' by the hull; in referring to 'added mass' we are merely abiding by common usage. The 'virtual mass' is the combination of the actual mass and the added mass.)

Ideally the hydrodynamic properties of the ship should be determined by means of a three-dimensional analysis. Strip theory simplifies the theory by making it two dimensional. That is to say, any disturbances in the fluid caused by the motions of the ship section concerned only propagate in directions perpendicular to the axis Ox in the plane of symmetry, disturbances parallel to the longitudinal axis being ignored.

The added mass may be deduced for spheres, ellipsoids and similar bodies (such as cylinders, etc.) whose cross-sectional shapes can be specified. They are assumed to be moving in an infinite, incompressible, irrotational and inviscid fluid of density ρ and the appropriate velocity potential function is constructed. But determination of the necessary velocity potential function is not easy for prismatic bodies with arbitrary ship-shaped sections. For them, more refined analytic methods have to be employed – a matter that is discussed by de Jong (1973).

Consider a totally immersed circular cylinder of beam (diameter) B and infinite length moving in a direction perpendicular to its longitudinal axis with a velocity U, so that its kinetic energy per unit length is $mU^2/2$, m being the mass per unit length. Because of the cylinder's motion, particles of fluid are forced to move out of its way to the rear, thereby acquiring kinetic energy. The appropriate two-dimensional velocity potential function of the circular cylinder is

$$\Phi = \frac{UB^2 x}{4(x^2 + z^2)} = \frac{UB^2 \cos \theta}{4r},$$

where $x = r \cos \theta$ and the components of the fluid velocity are

$$u = -\frac{\partial \Phi}{\partial x} = \frac{UB^2 \cos 2\theta}{4r^2}, \qquad v = 0, \qquad w = -\frac{\partial \Phi}{\partial z} = \frac{UB^2 \sin 2\theta}{4r^2}.$$

An element of fluid has kinetic energy

$$\tfrac{1}{2}\rho(u^2 + v^2 + w^2)\, dx\, dy\, dz$$

and the total kinetic energy per unit length of the fluid is

$$\tfrac{1}{2}\rho \iint (u^2 + w^2)\, dx\, dz,$$

(assuming that $v = 0$ everywhere so that the flow is two dimensional). It follows that the kinetic energy of fluid per unit length of cylinder at any instant is

$$\tfrac{1}{2}\rho \int_{\frac{1}{2}B}^{\infty} \int_0^{2\pi} \frac{U^2 B^4}{16 r^3}\, dr\, d\theta = \rho\pi \frac{B^2 U^2}{8} = \tfrac{1}{2}\bar{m}U^2,$$

where $\bar{m}$ is the mass of fluid displaced by unit length of cylinder. The total kinetic energy of the fluid and cylinder is $(m + \bar{m})U^2/2$ per unit axial length. If an external force F per unit length acts on the cylinder in the direction of motion then

$$FU = \frac{d}{dt}\left[(m + \bar{m})\frac{U^2}{2}\right]$$

and so

$$F = (m + \bar{m})\dot{U},$$

assuming there is no change in potential energy. The quantity $(m + \bar{m})$ is the virtual mass and $\bar{m}$ is the added mass, both reckoned per unit length.

For ship-shaped sections, the added mass is quoted in terms of the added mass of a cylinder having 'comparable' size. That is to say, a constant C is defined such that

$$C = \frac{\text{added mass per unit length of ship-shaped section}}{\text{added mass per unit length of 'comparable' infinite cylinder}}.$$

For a ship executing vertical motion in the free surface, the non-dimensional added mass coefficient is

$$C \equiv C_V = \frac{m_V}{\rho\pi B^2/8}.$$

The apparent halving of the added mass per unit length of the infinite 'comparable cylinder' is due to the fact that the cylinder is now half immersed at a free surface and no longer totally immersed. For horizontal motions the non-dimensional added mass coefficient is

$$C \equiv C_H = \frac{m_H}{\rho\pi T^2/2}.$$

Here, B is the beam and T is the draught of the ship section.

Since the added moment of inertia per unit length of a circular cylinder rotating in an inviscid fluid is nil, the added moment of inertia $\bar{I}$ per unit length of a hull is compared arbitrarily with the quantity $\rho\pi T^4$. Thus

$$C \equiv C_T = \frac{\bar{I}}{\rho\pi T^4}.$$

Tables 7.1 and 7.2 illustrate results for a number of two-dimensional and three-dimensional geometric shapes. These can all be determined analytically. They are to be assumed fully submerged except where it is stated otherwise. The data contained in Tables 7.1 and 7.2 are adapted from results quoted by Saunders (1957).

Although the derivation of the added mass of ship-shaped sections is nowadays becoming a routine matter in naval hydrodynamics, it is, in general, based on the pioneering work of F. M. Lewis (1929). He developed a method whereby, in the absence of a free surface, the added mass of an infinitely long ship-shaped section oscillating in a fluid of infinite depth could be obtained from that of a semi-circle of unit radius by means of the conformal transformation

$$K = Y + iZ = a_0(\eta + \eta^{-1}a_1 + \eta^{-3}a_3).$$

In this expression $\eta = e^{i\beta}$ describes the semi-circle which is mapped into the ship section described by the coordinates (Y, Z) in the

Table 7.1. *The added masses of some two-dimensional bodies*

Form of two-dimensional body	Description of body	Added mass	Added moment of inertia
	Rod of circular cross-section	$m_V = \dfrac{\rho \pi B^2}{4}$	$\bar{I} = 0$
	Rod of elliptic cross-section	$m_H = \dfrac{\rho \pi}{4}(T^2 \cos^2 \alpha + B^2 \sin^2 \alpha)$	$\bar{I} = \dfrac{\rho \pi (B^2 - T^2)^2}{32}$
	Long flat plate	$m_V = \dfrac{\rho \pi B^2}{4}$	$\bar{I} = \dfrac{\rho \pi B^4}{128}$
	Rod of rectangular cross-section	$m_V = \dfrac{k_1 \rho \pi B^2}{4}$	$\bar{I} = \dfrac{k_2 \rho \pi B^4}{16}$

B/T	0.1	0.2	0.5	1	2	5	10
k_1	2.23	1.98	1.70	1.51	1.36	1.21	1.14
k_2	1470	94	2.4	0.234	0.15	0.15	0.147

Form	Description	Added mass	Added moment of inertia
	Rod of square section with fins on the corners	$m_V = \dfrac{k_3 \rho \pi B^2}{4}$	$\bar{I} = \dfrac{k_4 \rho \pi B^4}{16}$

$2d/B$	0.05	0.1	0.25
k_3	1.61	1.72	2.19
k_4	0.31	0.40	0.69

Form	Description	Added mass	Added moment of inertia
	Floating rectangular box	$m_V = 0.19 \rho \pi B^2$ $m_H = 0.0625 \rho \pi B^2$	$\bar{I} = 0.007 \rho \pi B^4$
	Floating rectangular box in water of limited depth	$m_V = \dfrac{k_5 \rho \pi B^2}{4}$	

$2e/B$	∞	2.6	1.8	1.5	0.5	0.25
k_5	0.75	0.83	0.89	1.0	1.35	2.0

Table 7.1. (*Continued*)

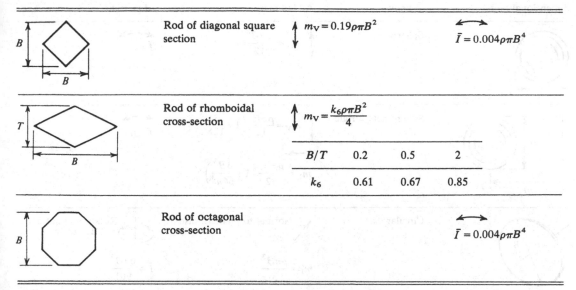

Rod of diagonal square section	$m_V = 0.19\rho\pi B^2$	$\bar{I} = 0.004\rho\pi B^4$

Rod of rhomboidal cross-section	$m_V = \dfrac{k_6\rho\pi B^2}{4}$			

B/T	0.2	0.5	2
k_6	0.61	0.67	0.85

Rod of octagonal cross-section	$\bar{I} = 0.004\rho\pi B^4$

Note: m_V, m_H are added mass for vertical and horizontal motion respectively.

K-plane and a_0 is a scale factor having the dimension of length. The coefficients a_1 and a_3 are constants which must be determined for a given section. Initially the method was developed for use in vibration calculations and, strictly speaking, the results apply only to symmetric ship sections having vertical tangents at the waterline and a horizontal tangent at the line of symmetry, which oscillate with infinite frequency.

After substitution for η in the transformation, the real and imaginary terms are found to be given by

$$Y = a_0[(1 + a_1)\cos\beta + a_3\cos 3\beta],$$

$$Z = a_0[(1 - a_1)\sin\beta - a_3\sin 3\beta].$$

It follows that $\partial Y/\partial\beta = 0$ at $\beta = 0$, so the ship section has vertical sides at the water line, while $\partial Z/\partial\beta = 0$ at $\beta = \pi/2$, so the ship section has a horizontal tangent at the line of symmetry. Now the boundary conditions, $\beta = 0$ when $Y = B/2$ and $\beta = \pi/2$ when $Z = T$, show that

$$\frac{B}{2} = a_0(1 + a_1 + a_3), \qquad T = a_0(1 - a_1 + a_3).$$

Table 7.2. *The added masses of some three-dimensional bodies*

Form of three-dimensional body	Description of body	Added mass	Added moment of inertia

Sphere

$$m_V = \frac{\rho \pi B^3}{12}$$

$$\bar{I} = 0$$

Sphere near a rigid wall

$$m_V = \frac{\rho \pi B^3}{12}\left(1 + \frac{3B^3}{128h^3}\right)$$

$$\bar{I} = 0$$

$$m_H = \frac{\rho \pi B^3}{12}\left(1 + \frac{3B^3}{64h^3}\right)$$

Circular disc

Normal to plane of disc

$$m_L = \frac{\rho \pi B^3}{3}$$

Rotation about a diameter

$$\bar{I} = \frac{\rho \pi B^5}{901}$$

Prolate ellipsoid

$$m_V = \frac{k_1 \rho \pi B T^2}{6}$$

$$m_L = \frac{k_2 \rho \pi B T^2}{6}$$

$$\bar{I}_L = \frac{k_3 \rho \pi B T^2}{90}(B^2 + T^2)$$

B/T	1.0	1.5	2.0	2.51	2.99	3.99	4.99	6.01	6.97	8.01	9.02	9.97	∞
k_1	0.5	0.621	0.702	0.763	0.803	0.860	0.895	0.918	0.933	0.945	0.954	0.960	1.0
k_2	0.5	0.305	0.209	0.156	0.122	0.082	0.059	0.045	0.036	0.029	0.024	0.021	0
k_3	0	0.094	0.240	0.367	0.465	0.608	0.701	0.764	0.805	0.840	0.865	0.883	1.0

Oblate spheroid

$$m_L = \frac{\rho \pi T^3}{6}\left[\frac{e - (\sin^{-1} e)(1 - e^2)^{1/2}}{\sin^{-1} e - e(1 - e^2)^{1/2}}\right]$$

$$\text{where} \quad e = \left(1 - \frac{B^2}{T^2}\right)^{1/2}$$

Elliptical disc

Normal to plane of disc

$$m_L = \frac{k_4 \rho \pi B^2 T}{6}$$

B/T	1	2	5	10
k_4	0.637	0.41	0.19	0.098

Note: m_V, m_H, m_L are added mass for vertical, horizontal and longitudinal motion respectively.

The coefficients a_1 and a_3 are related to the sectional area coefficient σ_s and beam–draught ratio Λ of a ship section through the equations

$$\sigma_s = \frac{S}{BT} = \frac{\pi(1-a_1^2-3a_3^2)}{4[(1+a_3)^2-a_1^2]}, \qquad \Lambda = \frac{B}{2T} = \frac{1+a_1+a_3}{1-a_1+a_3},$$

or

$$a_0 = \frac{B}{2(1+a_1+a_3)},$$

$$a_1 = \frac{(1+a_3)(\Lambda-2)}{(\Lambda+2)},$$

$$a_3 = \frac{-C_0+3+(9-2C_0)^{\frac{1}{2}}}{C_0},$$

where

$$C_0 = 3 + \frac{4\sigma_s}{\pi} + \left(\frac{\Lambda-2}{\Lambda+2}\right)^2\left(1-\frac{4\sigma_s}{\pi}\right).$$

Here, the cross-sectional area of the ship section is

$$S = 2\int_0^{\pi/2} Y\,dZ = a_0^2\frac{\pi}{2}(1-a_1^2-3a_3^2).$$

The family of ship sectional shapes generated by such transformations is commonly referred to as that of 'Lewis forms'.

Landweber & Macagno (1957) have shown that for vertical oscillations,

$$m_V = \frac{a_0^2\pi\rho}{2}[(1+a_1)^2+3a_3^2] \quad\text{and}\quad C_V = \frac{(1+a_1)^2+3a_3^2}{(1+a_1+a_3)^2},$$

and for horizontal motions

$$m_H = \frac{a_0^2\pi\rho}{2}[(1-a_1)^2+3a_3^2] \quad\text{with}\quad C_H = \frac{(1-a_1)^2+3a_3^2}{(1-a_1+a_3)^2}.$$

Fig. 7.2 illustrates the form of the added mass coefficients for vertical and horizontal motions, neglecting free surface effects.

It has been shown by von Kerczek & Tuck (1969) that Lewis forms of symmetric ship-shaped sections are only valid provided that the sectional area coefficient lies in a range defined by a minimum

$$\min(\sigma_s)\begin{cases} = \dfrac{3\pi}{32}\left(2-\dfrac{B}{2T}\right) & \text{(for } B/2T \leqslant 1\text{),} \\[2ex] = \dfrac{3\pi}{32}\left(2-\dfrac{2T}{B}\right) & \text{(for } B/2T \geqslant 1\text{),} \end{cases}$$

and an overall maximum

$$\max(\sigma_s) = \frac{\pi}{2}\left(\frac{B}{2T}+\frac{2T}{B}+10\right).$$

For conventional ship-shaped Lewis form sections (in contrast to bulbous ship-shaped sections) the upper limit on the sectional area coefficient is given by Landweber & Macagno (1957) as

$$\max(\sigma_s)_{con} \begin{cases} = \dfrac{3\pi}{128}\left(12+\dfrac{B}{2T}\right) & \text{(for } B/2T \leqslant 1), \\[2ex] = \dfrac{3\pi}{128}\left(12+\dfrac{2T}{B}\right) & \text{(for } B/2T \geqslant 1). \end{cases}$$

It is not recommended, however, that Lewis forms be used for sections with area coefficients close to the maximum value; Lewis forms close to the upper limit differ considerably from common bulb shapes. Fig. 7.3 illustrates the range of $B/2T$ and σ_s for different Lewis forms. Other two-parameter mappings have been suggested by Prohaska (1947) which give better approximations to some ship sections. Landweber & Macagno (1959) proposed a three-parameter mapping in which the third geometrical quantity is taken as the second moment of area of the cross-section.

By a suitable choice of N in the more general conformal transformation

$$K = a_0\left[\eta + \sum_{n=1}^{N} a_{2n-1}\eta^{2n-1}\right],$$

greater accuracy and definition may be obtained for a ship section than is possible with the simple Lewis transformation in which $N = 2$. The

Fig. 7.2. Non-dimensional added masses, neglecting surface effects for (a) vertical motion, (b) horizontal motion. (Note that $\Lambda = B/2T$ and $\sigma_s = S/BT$.)

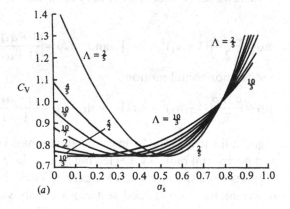

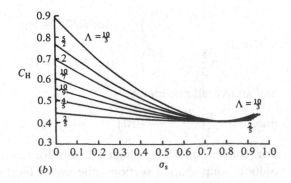

constants a_{2n-1} are the transformation coefficients. The vertical and horizontal added masses are now

$$m_V = \frac{a_0^2 \pi \rho}{2} \left[1 + 2a_1 + \sum_{n=1}^{N} (2n-1)a_{2n-1}^2 \right],$$

$$m_H = \frac{a_0^2 \pi \rho}{2} \left[1 - 2a_1 + \sum_{n=1}^{N} (2n-1)a_{2n-1}^2 \right].$$

The theory discussed so far is an 'ideal' one in the sense that the effects of the presence of the free surface are ignored and the frequency of oscillation is assumed to be infinite. Now the frequency range of ship motions is such that these two factors generally reduce the 'ideal' added mass significantly. Vertical motion of the ship produces a standing wave system which, being in phase, does not involve dissipation of energy but only modifies the inertial forces.

The nature of this effect can perhaps best be seen by referring to the work of Ursell (1949a,b). He studied, mathematically, the motion of the fluid when an infinitely long circular cylinder (beam ≪ length) oscillates harmonically in the vertical direction in the free surface of an infinite body of fluid. In the analysis he assumed that

(a) the amplitude of the oscillations is small compared with the diameter of the cylinder and length of waves generated,
(b) the mean position of the horizontal axis coincides with the mean surface position of the fluid,
(c) the fluid is inviscid, incompressible and the motion of the fluid particles is irrotational so that the appropriate velocity potential satisfies Laplace's equation (equation 6.4),
(d) the oscillation is such that a linearised analysis is valid.

Under these assumptions, the motions may be found as the solution of a linear velocity potential boundary value problem. The solution is obtained by superimposing suitably chosen functions in such a way that

Fig. 7.3. Restrictions have to be placed on the use of Lewis forms and the suggested ranges of their applicability are as shown.

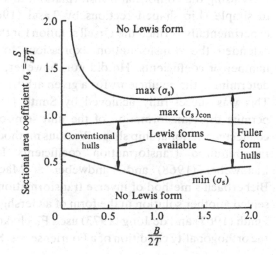

each separate function satisfies the Laplace equation and the linearised free surface boundary condition, while a linear combination of these functions satisfies the remaining boundary conditions on the cylinder. The appropriate velocity potential function, Φ_V, is composed of a source potential and a sum of multipole potentials both placed in the origin giving

$$\Phi_V = C \operatorname{Re} \left\{ \left[\Phi_{\text{source}} + \sum_{m=1}^{\infty} (p_{2m} + iq_{2m}) \Phi_{2m} \right] e^{-i\omega t} \right\},$$

where C is a constant, ω is the frequency of vertical oscillation and the coefficients p_{2m}, q_{2m} are chosen in such a way that the linearised boundary condition on the surface of the cylinder is satisfied. The source potential, Φ_{source}, is complex but the multipole potential Φ_{2m} is real. In real form, the velocity potential of vertical motion may be expressed as

$$\Phi_V = C \left[\left(\Phi_c + \sum_{m=1}^{\infty} p_{2m} \Phi_{2m} \right) \cos \omega t + \left(\Phi_s + \sum_{m=1}^{\infty} q_{2m} \Phi_{2m} \right) \sin \omega t \right],$$

where $\Phi_{\text{source}}(Y, Z) = \Phi_c + i\Phi_s = \Phi_{\text{source}}(-Y, Z)$ and

$$\Phi_c = \pi e^{+kZ} \cos kY,$$

$$\Phi_s = \pi e^{+kZ} \sin kY - \int_0^{\infty} \frac{e^{-\nu Y}}{k^2 + \nu^2} (\nu \cos \nu Z + k \sin \nu Z) \, d\nu,$$

$$\Phi_{2m} = \cos 2m\beta + k \left\{ \frac{\cos [(2m-1)\beta]}{(2m-1)} \right.$$
$$\left. + \sum_{n=1}^{N} \frac{(-1)^{n+1}(2n-1)a_{2n-1} \cos [(2m+2n-1)\beta]}{(2m+2n-1)} \right\},$$

$$k = \frac{\omega^2}{g}.$$

A full derivation and description of the terms involved in this expression is given by de Jong (1973).

By using the conformal transformation, this analysis was extended to simple ship-shaped sections by Tasai (1960a,b). Porter (1960) experimentally verified the Ursell solution for the circular cylinder and extended the transformation expressions to include an arbitrary number of coefficients. He did not, however, provide a method for determining the coefficients for a given arbitrary ship-shaped section. This was successfully achieved by Smith (1967) and the method permits the transformation of the unit semi-circle into any simply-connected sectional ship shape. Various methods are used to calculate the required transformation coefficients for a given section. Landweber (1968) and Landweber & Macagno (1967) applied Bieberbach's method of inverse transformation and later (1975) presented another solution in the form of a Gershgorin integral equation. Smith (1967) and de Jong (1973) used Fil'chakova's method based on the orthogonality condition of a Fourier series. Solution by an iterative procedure similar to the Newton–Raphson method was employed by Smith (1967) and von Kerczek & Tuck (1969).

The results of Ursell's analysis showed that the free-surface effects can be taken into account by modifying the C_V coefficient to $k_4 C_V$. He computed values of this factor k_4 for a circular cylinder of beam B in terms of the frequency parameter $\omega^2 B/2g$. Porter (1960) extended this calculation to Lewis form shapes and a selection of these findings is shown in fig. 7.4. The curves vary significantly for small values of $\omega^2 B/2g$ but tend to unity for increasing values of that parameter. This implies that the added mass of the ship-shaped sections are frequency-dependent – strongly so for motions whose frequency approximates to dominant wave frequencies. But the dependence decreases with increasing frequency and is far less pronounced at frequencies associated with hull vibrations due to mechanical excitation.

Newman (1971, 1977) discusses the form of the coefficients whose variation is shown in fig. 7.4 by reference to the high- and low-frequency limits of the free surface condition given in equation (6.8). For progressive sinusoidal waves,

$$-\omega^2 \Phi + g \frac{\partial \Phi}{\partial Z} = 0 \qquad \text{(on } Z = 0\text{)}.$$

Fig. 7.4. The non-dimensional added mass C_V associated with a Lewis form can be modified to account for free-surface effects by replacing it with $k_4 C_V$. This curve shows how the multiplier varies with the dimensionless driving frequency $\omega^2 B/2g$.

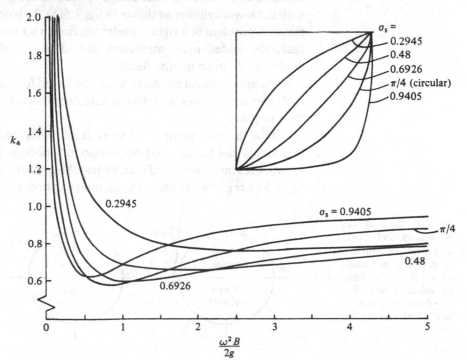

In the limit as $\omega \to 0$, this reduces to

$$\frac{\partial \Phi}{\partial Z} = 0 \qquad \text{(on } Z = 0),$$

which corresponds to the conditions at a rigid boundary with zero normal velocity. Thus the motion of the heaving cylinder can be thought of as if a second cylinder moves in fluid occupying the upper half space, as in fig. 7.5(a); the image cylinder above the plane $Z = 0$ is 180° out of phase with the lower cylinder so that when the lower cylinder is moving down, the upper cylinder is moving up, and vice versa. The motion is therefore symmetrical about the plane $Z = 0$ and the boundary condition of a 'rigid' interface is satisfied. This is no longer a free-surface problem, but corresponds instead to the pulsation of a dilating body whose volume fluctuates. Now, for a two-dimensional body the added mass coefficient must be infinite because continuity requires that fluid must oscillate back and forth all the way out to infinity, the compressibility being nil. In three dimensions, however, the fluid can distribute itself spatially in any direction and the magnitude of the added mass is merely finite, though still large.

In the limit as $\omega \to \infty$, the free surface condition reduces to

$$\Phi = 0 \qquad \text{(on } Z = 0).$$

To satisfy such a condition, the normal velocities on the upper and lower surfaces must be of opposite sign so that the potential is an odd function of Z. Therefore the image cylinder must now move in phase with the lower cylinder as shown in fig. 7.5(b). This corresponds to the classical problem of a rigid cylinder moving in an infinite fluid. In this case, the added mass coefficient should be equal to that of the double-body in an infinite fluid.

No simple physical explanation can be found for the character of the hydrodynamic forces at intermediate frequencies. Calculations are then required.

Although only vertical motions have been considered, similar arguments can be adduced for horizontal motions (sway, roll). The symmetries involved would then be just the opposite of those shown in fig. 7.5 as regards the phase of the image cylinders.

Fig. 7.5. Illustration of Newman's (1971, 1977) argument concerning the form of the coefficients shown in fig. 7.4. Diagram (a) refers to the limit $\omega \to 0$ and (b) refers to the limit $\omega \to \infty$.

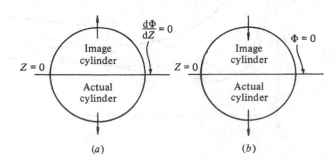

The coefficients for two-dimensional ship-shaped (rather than semi-circular) sections are widely used for the computation of ship motions in waves. It is assumed that the fluid motion at each transverse section can be analysed separately and is identical to the motion which would result from the vertical oscillations of the same section in a purely two-dimensional flow. This is the basis of 'strip theory'.

Fig. 7.6 shows curves of $m(x, \omega)$ for the destroyer hull, each relating to a particular value of frequency of encounter. Likewise the curves of figs. 7.7 and 7.8 relate to the tanker in the loaded and ballast conditions respectively. It is of interest to compare these curves with those of the actual mass distributions of the hulls (see figs. 4.2 and 4.12, though the latter shows weight).

By way of an alternative to Ursell's analytical approach, a direct numerical evaluation of the velocity potential may be made, as discussed by Frank (1967). The section contour is approximated by a series of straight line elements, while the velocity potential is represented by a distribution of wave sources over the contour. The strength of the sources is constant along each segment and is determined from a set of integral equations derived from the boundary condition on the surface of the cylinder. A minor difficulty occurs at certain frequencies when the integral equation method fails to give a solution, as discussed by Faltinsen (1969). This defect may be overcome by applying a smoothing technique to the hydrodynamic coefficients as functions of frequency so as to remove such singularities from the results. A numerical method of this sort may be used for

Fig. 7.6. The distribution of added mass $m(x, \omega)$ along the hull of the destroyer, for some values of frequency if the ship were stationary or encounter frequency if the ship were moving. The hull sections are assumed to be Lewis forms.

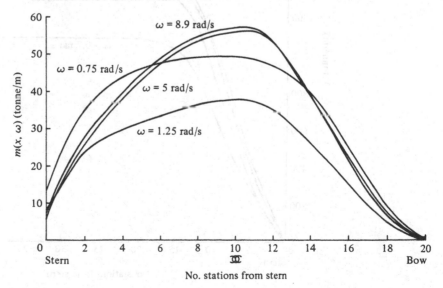

sections of greater complexity than those envisaged in the conformal transformation method. But this is only at the expense of a much greater investment of effort.

Another purely numerical method using a finite element approach has been proposed by Bai & Yeung (1974). This has the advantage that the method of solution does not depend on any particular choice of body geometry but can, in principle, be applied to two- or three-dimensional bodies of any shape in any water depth. The disadvantage is the amount of computer effort involved. It is interesting to note that the added mass of a two-dimensional circular cylinder in heave at low frequency in deep water tends to infinity whereas this method suggests that it approaches a finite value in shallow water as $\omega \to 0$.

7.1.4   *Three-dimensional reduction factor*
Three-dimensional calculations of the added mass and damping coefficients have been made for submerged and floating ellipsoids and some idealised forms as shown in fig. 7.9. The data contained in this figure are adapted from results quoted by Newman (1971). In practice, however, to account for the three-dimensional nature of flow around a ship's hull, the distributed sectional added masses along the ship's

Fig. 7.7. The distribution of added mass $m(x, \omega)$ along the hull of the loaded tanker, for some values of frequency if the ship were stationary or encounter frequency if the ship were moving. The hull sections are assumed to be Lewis forms.

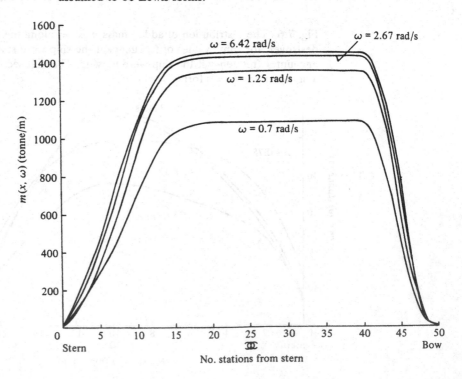

length are artificially reduced by a constant fraction. A three-dimensional reduction factor, J, used for ships is derived from calculations for ellipsoids having the same length-to-breadth ratio and overall dimensions as the hull. This factor is defined as the ratio between the kinetic energy of the exact three-dimensional flow and the kinetic energy of the approximate two-dimensional flow, both taken for the whole ship. Approximate values of the three-dimensional added mass of the whole ship are obtained by multiplying the added mass of the two-dimensional flow by the corresponding reduction factor. (For example, in some ship studies, Lamb's accession correction factors are used.)

Alternatively such factors may be based on the empirical interpretation of experimental ship model tests. A factor that is sometimes used is

$$J_n = 1.02 - 3\left(1.2 - \frac{1}{n}\right)\frac{B}{l},$$

where n is the number of nodes and B is the waterplane breadth amidships. Such a formulation was proposed by Townsin (1969), though the assumptions upon which it is based remain a source of speculation.

Fig. 7.8. The distribution of added mass $m(x, \omega)$ along the hull of the tanker in ballast, for some values of frequency if the ship were stationary or encounter frequency if the ship were moving. The hull sections are assumed to be Lewis forms.

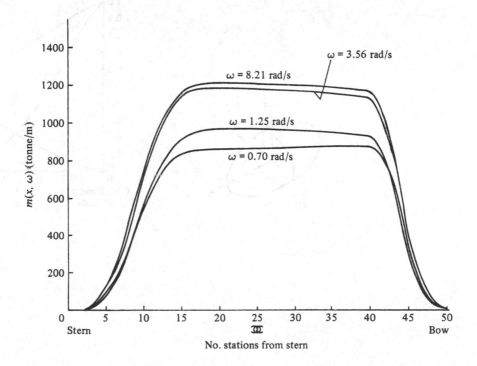

Fig. 7.9. Added masses and inertias of a prolate spheroid and ellipsoid. The major semi-axis Ox is of length a and the minor semi-axes Oy and Oz are of lengths b and c respectively, as shown in sketch (a).

For the prolate spheroid, $b = c$ and

$$\bar{m}_x = \frac{\text{added mass for surge acceleration along } Ox\text{-axis}}{\frac{4}{3}\pi\rho ab^2},$$

$$\bar{m}_y = \frac{\text{added mass for sway acceleration along } Oy\text{-axis}}{\frac{4}{3}\pi\rho ab^2},$$

$$\bar{I}_y = \frac{\text{added pitch moment of inertia about } Oy\text{-axis}}{\frac{4}{15}\pi\rho ab^2(a^2+b^2)}.$$

The results are presented in (b).

Curves (c)–(h) refer to the ellipsoid, for which

$$\bar{m}_x = \frac{\text{added mass for surge acceleration along } Ox\text{-axis}}{\frac{4}{3}\pi\rho abc},$$

$$\bar{I}_{xx} = \frac{\text{added roll moment of inertia about } Ox\text{-axis}}{\frac{4}{15}\pi\rho abc(b^2+c^2)},$$

$$\bar{m}_y = \frac{\text{added mass for sway acceleration along } Oy\text{-axis}}{\frac{4}{3}\pi\rho abc},$$

$$\bar{I}_{yy} = \frac{\text{added pitch moment of inertia about } Oy\text{-axis}}{\frac{4}{15}\pi\rho abc(a^2+c^2)},$$

$$\bar{m}_z = \frac{\text{added mass for heave acceleration along } Oz\text{-axis}}{\frac{4}{3}\pi\rho abc},$$

$$\bar{I}_{zz} = \frac{\text{added yaw moment of inertia about } Oz\text{-axis}}{\frac{4}{15}\pi\rho abc(a^2+b^2)}.$$

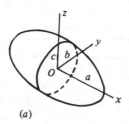

(a)

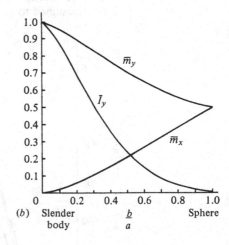

(b) Slender body $\dfrac{b}{a}$ Sphere

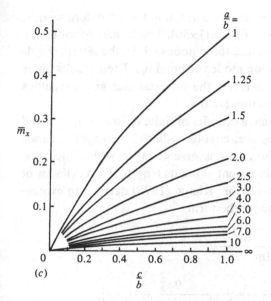

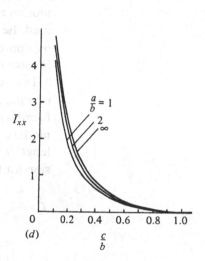

(c)

(d)

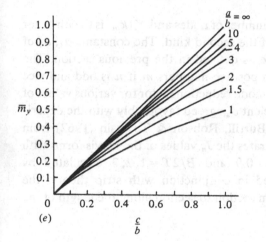

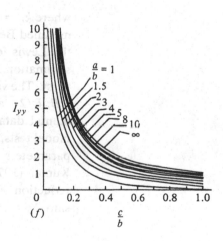

(e)

(f)

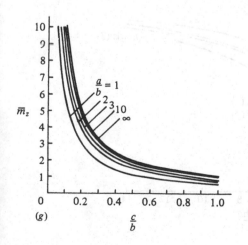

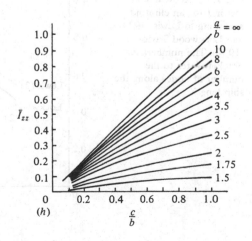

(g)

(h)

Fig. 7.10 illustrates the calculated correction factors determined by Lewis (1929) and Lockwood Taylor (1930). The former considered a motion of pure shear of two and three nodes whilst the latter considered the two- and three-node modes of bending. Their results differ by approximately 10% illustrating the uncertainties and difficulties involved in using such reduction factors.

In an alternative approach, a cylinder of finite or infinite length and of various sectional shapes (i.e. circular, elliptic, rectangular, Lewis form) is considered to perform a transverse vibration with a sinusoidal modal distribution of displacement along its length. For a cylinder of length l with a Lewis form section, Kumai (1975) derived an expression for the reduction factor in the form

$$J_n = \frac{16}{\pi^2[(1+a_1)^2 + 3a_3^2]} \sum_m \frac{m^2}{(m^2-n^2)^2}$$

$$\times \left[\frac{(1+a_1)^2}{1+k_m Y_0(k_m)/Y_1(k_m)} + \frac{9a_3^2}{3+k_m Y_2(k_m)/Y_3(k_m)} \right],$$

where $k_m = mB/l$, n is the number of nodes and $Y_i(k_m)$ is the ith order modified Bessel function of the second kind. The constants a_1, a_3 of the Lewis form section are as defined in the previous section. The summation is over all even positive numbers m if n is odd and vice versa. The value of the two-node reduction factor for various values of l/B, $B/2T$ and area coefficient σ_s agreed favourably with the experimental data obtained by Burrill, Robson & Townsin (1962) from model tests. Fig. 7.11 illustrates the J_n values of the Lewis forms with parameters $l/B = 7.0$, $\sigma_s = 0.9$ and $B/2T = 1, 2, 3$ calculated by Kumai (1977). When used in conjunction with strip theory, the reduction factor varies from section to section along the length of the ship.

Fig. 7.10. Reduction factor J for an ellipsoid according to Lewis (1929) and Lockwood Taylor (1930). The numbers in brackets refer to the number of nodes along the ship length.

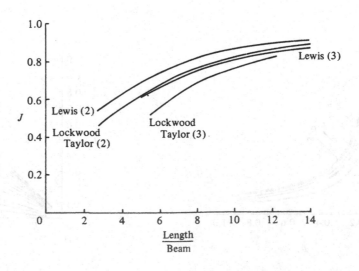

Other methods have been proposed for evaluating the added mass for three-dimensional flow. These are based on variational techniques employing a Lagrangean formulation for the coupling of the energies in both the elastic body and surrounding fluid. It is always an inherent assumption, however, in all the methods employed to calculate the reduction factor, that cross-sections do not deform. This is only tenable with the lower modes of vibration and becomes untenable for the higher mode numbers. It also appears that the factor is only applicable to the diagonal terms of the added mass matrix and not to the off-diagonal terms.

7.1.5 *The fluid damping coefficient*
The fluid damping of vertical motions is associated with the generation of dissipative waves. When a forced harmonic oscillation is executed by the two-dimensional cylinder (or hull), a progressive wave is set up on the surface which travels to infinity on both sides of the cylinder, in a direction perpendicular to the axis. These latter waves carry energy away from the oscillating hull, causing the fluid to have a damping influence on the motion of the cylinder. It can be shown by an extension of Ursell's theory to cylinders with arbitrary cross-section

Fig. 7.11. Three-dimensional reduction factors for a three-dimensional cylindrical body having a sectional shape of Lewis form.

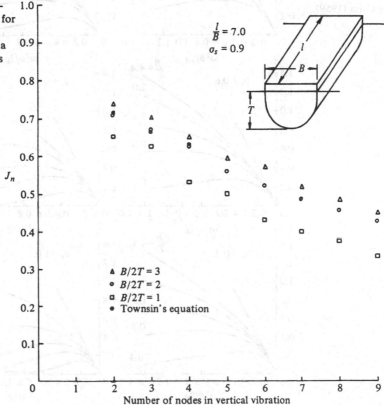

$\frac{l}{B} = 7.0$

$\sigma_s = 0.9$

▲ $B/2T = 3$
∘ $B/2T = 2$
□ $B/2T = 1$
• Townsin's equation

J_n

Number of nodes in vertical vibration

(e.g. see Porter, 1960; de Jong, 1973) that the damping coefficient of a ship-shaped section is given by

$$N(x) = \frac{\rho g^2 \bar{A}^2}{\omega_e^3},$$

where ω_e is the frequency of the radiated wave (which is equal to the frequency of wave encounter) and the non-dimensional wave amplitude ratio is

$$\bar{A} = \frac{\text{Amplitude of radiated or progressive wave at infinity}}{\text{Amplitude of section's forced vertical oscillation}}$$

The quantity $N(x)$, too, varies with frequency. The nature of the dependence is illustrated in fig. 7.12, which is based on the original computations of Grim (1959).

Fig. 7.12. The variation of dimensionless wave-amplitude ratio $\bar{A}$ with dimensionless driving frequency $\omega^2 B/2g$ for a range of sectional area coefficients σ_s and beam-draught ratios. The curves refer to Lewis forms and were computed by Grim (1959).

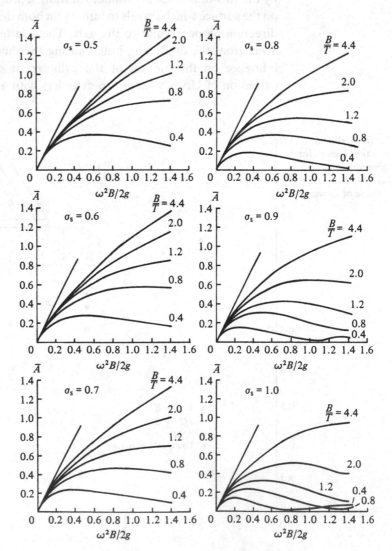

Newman (1971, 1977) also discussed the limiting values of fluid damping as $\omega \to 0$ and $\omega \to \infty$. He concluded that when the driving frequency is very small (fig. 7.5(a)) the damping coefficient and the wave amplitude ratio both tend to zero because no waves are generated. But when the semi-circular cylinder moves at the interface with infinite frequency the conditions are those of fig. 7.5(b) which would prevail if the cylinder moved in an infinite fluid; consequently no waves are set up and so the damping and wave amplitude ratio must vanish for high frequencies.

Figs. 7.13, 7.14 and 7.15 show distributions of the damping coefficient $N(x, \omega)$ along the ships' hulls which correspond to those of $m(x, \omega)$ shown in figs. 7.6, 7.7 and 7.8 respectively.

7.1.6   *Comparison of Lewis, conformal and Frank's methods*
Three general approaches have been mentioned by means of which added masses and damping coefficients may be estimated. First, the Lewis method employs a conformal transformation with two disposable coefficients. Secondly, a more comprehensive use may be made of conformal mapping. Finally, the Frank method seeks to model the flow directly.

To compare these methods of calculation of the added mass and damping coefficients for ship-shaped sections, Bishop, Price & Tam (1978a,b) obtained results over a wide range of non-dimensional frequency $\omega^2 B/2g$ for a number of sections. Where conformal transformations of higher order than those of Lewis's method were used,

Fig. 7.13. The distribution of damping coefficient $N(x, \omega)$ along the hull of the destroyer for some values of frequency (stationary ship) or encounter frequency (if ship under way). The hull sections are assumed to be Lewis forms.

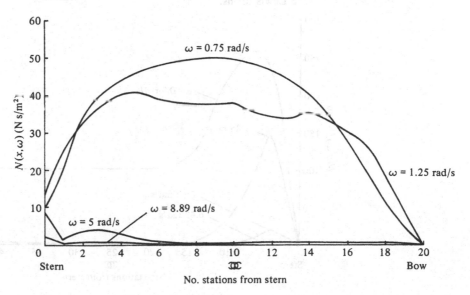

the transformation coefficients were determined by a quick computation method devised by Peckham (1970), which is essentially a variant of the Gauss–Newton method.

(*i*) *Chine section.* Fig. 7.16(*a*) shows the section under investigation; it has a pronounced knuckle. The ten-parameter fit reproduces the sectional shape with reasonable accuracy whereas the Lewis form and four-parameter fit give poor representations. The added mass calculated using the ten-parameter mapping method is greater than that found with the Lewis fit, as shown in fig. 7.16(*b*). The reverse is true for the damping coefficient, as illustrated in fig. 7.16(*c*). These findings are in agreement with results for a family of chine sections obtained by Maeda (1975) who used a close-fit method making use of stream functions.

(*ii*) *Rectangular section.* The eight-parameter conformal transformation generates a contour that is in close agreement with the rectangular ship section shown in fig. 7.17(*a*). The appropriate Lewis form is very much inferior. When compared with the Lewis form solution, the eight-parameter close-fit method again suggests a higher added mass and lower damping, as shown in figs. 7.17(*b*) and (*c*) respectively.

(*iii*) *Triangular section.* Close agreement is also found between the triangular ship section and the generated contour when an eight-parameter transformation is employed, as shown in fig. 7.18(*a*). The Lewis form is, once more, a cruder approximation to the given shape.

Fig. 7.14. The distribution of damping coefficient $N(x, \omega)$ along the hull of the loaded tanker for some values of frequency (stationary ship) or encounter frequency (if ship under way). The hull sections are assumed to be Lewis forms.

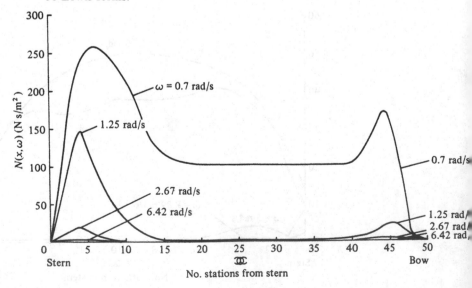

In contrast to the rectangular section, the Lewis form solution produces larger added mass and considerably lower damping values (see in figs. 7.18(b) and (c) respectively) when compared with the close-fit conformal solution.

(iv) *Fine section.* The fine section shown by the full line in fig. 7.19(a) is a reproduction of 'section 7' used by Faltinsen (1969) in an investigation of two-dimensional added mass and damping coefficients by Frank's close-fit method. Fig. 7.19(a) shows that the conformal method produces a far better generated contour than the Lewis fit which even fails to distinguish the basic characteristics of the original section. The conformal mapping fails to fit the original shape in the region $Y = 7.4$ m, $Z = 0.0$ m, but this is to be expected from the theory.

For $\omega^2 B/2g < 1$, the values of non-dimensional added mass calculated by the Frank and the multi-parameter conformal methods are in close agreement and are higher than those of the Lewis form fit, as shown in fig. 7.19(b). In the higher frequency range, too, the Lewis form fit produces lower values than does the conformal fit, but it will be seen that the results derived by Frank's method are then plagued with

Fig. 7.15. The distribution of damping coefficient $N(x, \omega)$ along the hull of the tanker in ballast for some values of frequency (stationary ship) or encounter frequency (if ship under way). The hull sections are assumed to be Lewis forms.

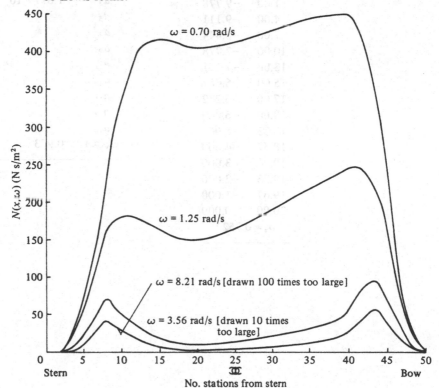

Fig. 7.16(a). A chine section determined by the offsets quoted and represented by the marks •. The full-line curve and chained curve are obtained with a ten-parameter and a four-parameter conformal mapping respectively, while the broken curve is the appropriate Lewis form.

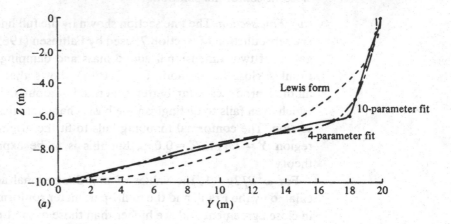

(a)

Chine section

Offsets	
Y (m)	Z (m)
0.00	−10.000
1.00	−9.778
4.00	−9.111
7.00	−8.444
10.00	−7.778
13.00	−7.111
15.00	−6.667
17.00	−6.222
18.00	−6.000
18.33	−5.000
18.67	−4.000
19.00	−3.000
19.33	−2.000
19.67	−1.000
20.00	0.000
$\sigma_s = 0.7522$	

N	Lewis	Conformal
a_0	14.724	14.963
a_1	3.396×10^{-1}	3.646×10^{-1}
a_3	1.877×10^{-1}	-2.895×10^{-2}
a_5		-3.647×10^{-2}
a_7		1.530×10^{-2}
a_9		9.381×10^{-3}
a_{11}		1.137×10^{-2}
a_{13}		-9.688×10^{-4}
a_{15}		-6.401×10^{-4}
a_{17}		-1.816×10^{-3}
a_{19}		3.194×10^{-3}
$C_V(\infty)$	0.973	1.052

Fig. 7.16(*b*). The variation of dimensionless added mass with the quantity $\omega^2 B/2g$. The full-line curve corresponds to the ten-parameter fit and the broken curve to the Lewis form.

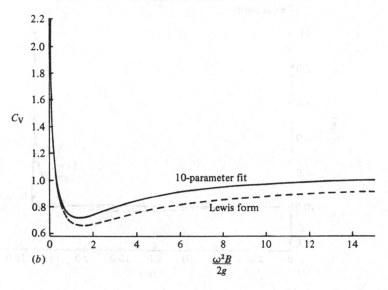

(b)

Fig. 7.16(*c*). The variation of $\bar{A}^2$ with $\omega^2 B/2g$.

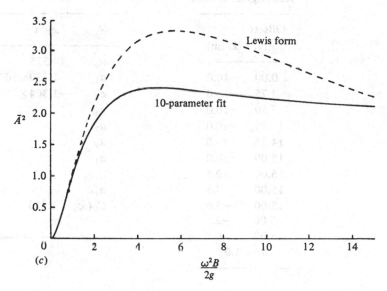

(c)

Fig. 7.17(a). A rectangular section determined by the offsets quoted and represented by the marks •. The full-line curve is obtained with an eight-parameter conformal mapping, while the broken curve is the appropriate Lewis form.

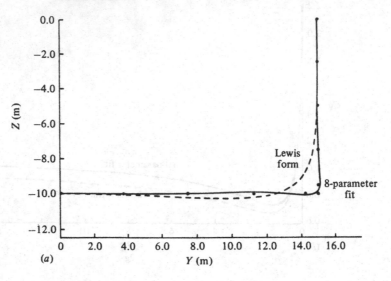

(a)

Rectangular section

Offsets		N	Lewis	Conformal
Y (m)	Z (m)			
0.00	−10.0	a_0	14.375	14.700
3.75	−10.0	a_1	1.739×10^{-1}	1.835×10^{-1}
7.50	−10.0	a_3	-1.304×10^{-1}	-1.620×10^{-1}
11.25	−10.0	a_5		-1.786×10^{-2}
14.25	−10.0	a_7		1.469×10^{-2}
15.00	−10.0	a_9		8.010×10^{-3}
15.00	−9.5	a_{11}		-3.788×10^{-3}
15.00	−7.5	a_{13}		-4.667×10^{-3}
15.00	−5.0	a_{15}		1.880×10^{-3}
15.00	−2.5	$C_V(\infty)$	1.313	1.427
15.00	0.0			
$\sigma_s = 1.0$				

Fig. 7.17(b). The variation of dimensionless added mass with the quantity $\omega^2 B/2g$. The full-line curve corresponds to the eight-parameter fit and the broken curve to the Lewis form.

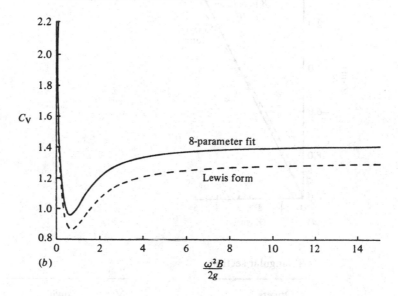

(b)

Fig. 7.17(c). The variation of $\bar{A}^2$ with $\omega^2 B/2g$.

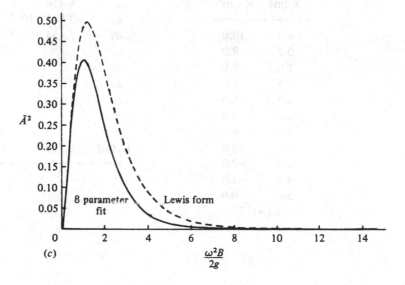

(c)

Fig. 7.18(a). A triangular section determined by the offsets quoted and represented by the marks •. The full-line curve is obtained with an eight-parameter fit and the broken curve is the appropriate Lewis form.

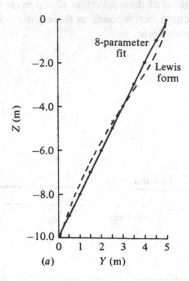

(a)

Triangular section

Offsets ($Y = 0.5Z + 5$)		N	Lewis	Conformal
Y (m)	Z (m)			
0.0	−10.0	a_0	6.436	6.422
0.5	−9.0	a_1	-3.885×10^{-1}	-4.064×10^{-1}
1.0	−8.0	a_3	1.654×10^{-1}	1.401×10^{-1}
1.5	−7.0	a_5		1.256×10^{-2}
2.0	−6.0	a_7		1.734×10^{-2}
2.5	−5.0	a_9		3.512×10^{-3}
3.0	−4.0	a_{11}		6.898×10^{-3}
3.5	−3.0	a_{13}		9.150×10^{-4}
4.0	−2.0	a_{15}		3.424×10^{-3}
4.5	−1.0	$C_V(\infty)$	0.756	0.685
5.0	0.0			
$\sigma_s = 0.5$				

Fig. 7.18(b). The variation of dimensionless added mass with the quantity $\omega^2 B/2g$. The full-line curve corresponds to the eight-parameter fit and the broken curve to the Lewis form.

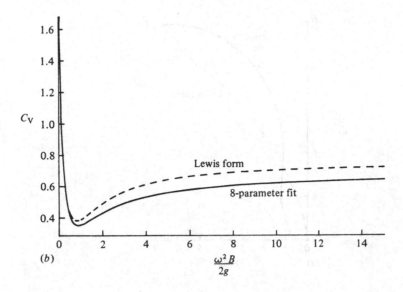

(b)

Fig. 7.18(c). The variation of $\bar{A}^2$ with $\omega^2 B/2g$.

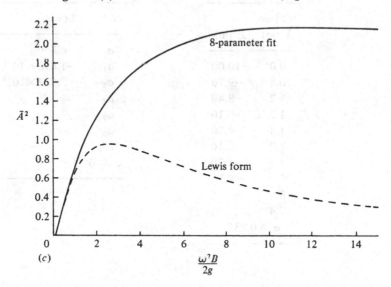

(c)

Fig. 7.19(a). A fine section determined by the offsets quoted and
represented by the marks •. The full-line curve is obtained with a six-
parameter fit and the broken curve is the appropriate Lewis form.

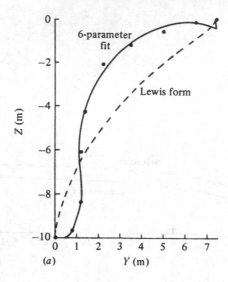

(a)

Fine section

Offsets	
Y (m)	Z (m)
0.0	−10.00
0.8	−9.70
1.2	−8.40
1.2	−6.10
1.4	−4.30
2.2	−2.10
3.5	−1.20
5.0	−0.55
6.5	−0.15
7.4	0.00
$\sigma_s = 0.235$	

N	Lewis	Conformal
a_0	6.850	6.957
a_1	-1.898×10^{-1}	-2.881×10^{-1}
a_3	2.701×10^{-1}	2.945×10^{-1}
a_5		1.531×10^{-1}
a_7		-1.228×10^{-2}
a_9		-5.553×10^{-2}
a_{11}		-3.357×10^{-2}
$C_V(\infty)$	0.750	0.729

Fig. 7.19(b). The variation of dimensionless added mass with the quantity $\omega^2 B/2g$. The full-line curve corresponds to a six-parameter fit and the broken curve to the Lewis form. The chained curve is found using Frank's method.

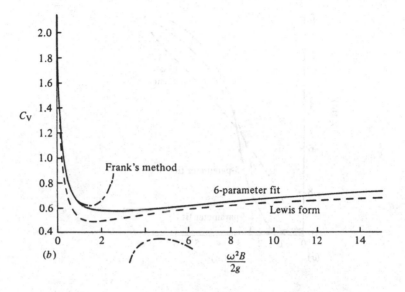

Fig. 7.19(c). The variation of $\bar{A}^2$ with $\omega^2 B/2g$. The chained curve is found using Frank's method.

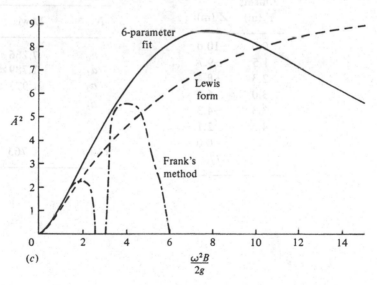

Fig. 7.20(a). A bulbous section determined by the offsets quoted and represented by the marks •. The full-line curve is obtained with a five-parameter fit and the chained curve with a three-parameter fit. The broken curve is the appropriate Lewis form.

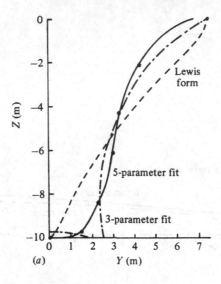

(a)

Bulbous section

Offsets	
Y (m)	Z (m)
0.0	−10.0
1.5	−9.7
2.3	−8.4
3.0	−6.1
3.3	−4.3
4.3	−2.1
7.4	0.0
$\sigma_s = 0.477$	

N	Lewis	Conformal
a_0	7.266	7.445
a_1	-1.789×10^{-1}	-3.081×10^{-1}
a_3	1.973×10^{-1}	1.200×10^{-1}
a_5		1.093×10^{-1}
a_7		4.863×10^{-2}
a_9		2.407×10^{-2}
$C_V(\infty)$	0.763	0.611

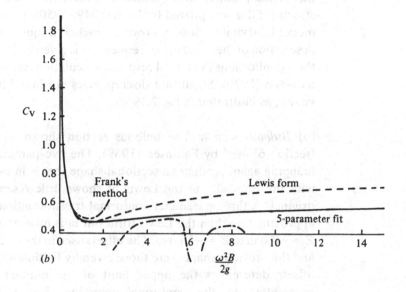

Fig. 7.20(b). The variation of dimensionless added mass with the quantity $\omega^2 B/2g$. The full line corresponds to the five-parameter fit and the broken curve to the Lewis form, while the chained curve is found using Frank's method.

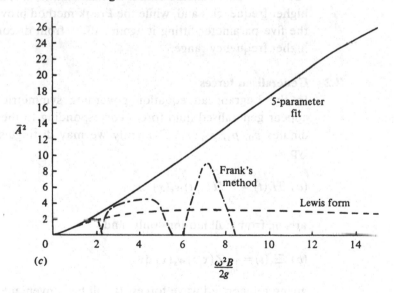

Fig. 7.20(c). The variation of $\bar{A}^2$ with $\omega^2 B/2g$. The chained curve is found using Frank's method.

discontinuities at discrete irregular frequencies. Faltinsen indicated that this latter defect may be overcome by interpolation over the irregularities. Unfortunately these frequencies are an inherent basic theoretical problem which cannot be overcome by using a finer mesh structure; this was proved by John (1949, 1950). Eventually Frank's method inevitably breaks down at higher frequencies. A detailed description of these irregular frequencies is given by Frank (1967) and their implications in vertical response calculations is given by Frank & Salvesen (1970). Significant discrepancies are found in the damping values, as illustrated in fig. 7.19(c).

(v) *Bulbous section.* The bulbous section shown in fig. 7.20(a) is 'section 6' used by Faltinsen (1969). The five-parameter conformal mapping again produces a sectional shape that is in close agreement with the original, but the Lewis fit shows little resemblance to the original. A three-parameter conformal transformation gives a better approximation than the Lewis form but fails miserably in the region $Z = -8.0$ m to $Z = -10.0$ m. The offsets used in the calculation for this and the previous shape were those given by Faltinsen. The number of offsets determines the upper limit of the number of disposable parameters in the conformal mapping. For $\omega^2 B/2g < 1$, the differences in the added mass and damping values shown in fig. 7.20(b) and (c) respectively are small. But substantial disparities appear at higher frequencies and, while the Frank method provides support for the five-parameter fitting it again suffers from discontinuities in the higher frequency range.

7.2 Generalised forces

In the Lagrangean equation governing symmetric motions there appear generalised fluid forces corresponding to the principal coordinates $p_0, p_1, p_2, \ldots$. Evidently we may distinguish between two types:

$$(a) \quad H_s(t) = \int_0^l H(x, t) w_s(x) \, dx, \tag{7.13}$$

arising from hull motions only, and

$$(b) \quad \Xi_s(t) = \int_0^l Z(x, t) w_s(x) \, dx, \tag{7.14}$$

giving the applied wave forces. It will be convenient to discuss these two sorts of generalised fluid force separately.

7.2.1 *Generalised forces due to hull motion*

If the vertical deflection of the hull is

$$w(x, t) = e^{-i\omega_e t} \sum_{r=0}^{\infty} p_r w_r(x),$$

we find that

$$H_s(t) = \sum_{r=0}^{\infty} [A_{rs}\ddot{p}_r(t) + B_{rs}\dot{p}_r(t) + C_{rs}p_r(t)] \qquad (s = 0, 1, 2, 3, \dots),$$

(7.15)

where the values of A_{rs}, B_{rs} and C_{rs} depend upon which of the two forms of strip theory is used. If the theory described in section 7.1.2 is employed, it is found that

$$A_{rs} = \int_0^l \left\{ m(x)w_r(x)w_s(x) + \frac{\bar{U}^2}{\omega_e^2} m(x)w_r'(x)w_s'(x) \right.$$

$$\left. + \frac{\bar{U}}{\omega_e^2} N(x)[w_r'(x)w_s(x) - \underline{w_s'(x)w_r(x)}] \right\} dx$$

$$- \frac{\bar{U}^2}{\omega_e^2} [m(x)w_s(x)w_r'(x)]_0^l + \frac{\bar{U}}{\omega_e^2} [N(x)w_s(x)w_r(x)]_0^l,$$

(7.16)

$$B_{rs} = \int_0^l \left\{ N(x)w_r(x)w_s(x) + \frac{\bar{U}^2}{\omega_e^2} N(x)w_r'(x)w_s'(x) \right.$$

$$\left. + \bar{U}m(x)[w_r(x)w_s'(x) - w_s(x)w_r'(x)] \right\} dx$$

$$- \frac{\bar{U}^2}{\omega_e^2} [N(x)w_s(x)w_r'(x)]_0^l - \bar{U}[m(x)w_s(x)w_r(x)]_0^l,$$

(7.17)

$$C_{rs} = \rho g \int_0^l B(x)w_s(x)w_r(x) \, dx,$$

(7.18)

for $s = 0, 1, 2, \dots$. In this grouping the term C_{rs} contains only 'hydrostatic' effects and all dynamic contributions are grouped in either A_{rs} or B_{rs} by noting that $\ddot{p}_r(t)$ is real and equal to $-\omega_e^2 p_r(t)$ while $\dot{p}_r(t)$ is imaginary and equal to $-i\omega_e p_r(t)$.

If the simpler form of strip theory (i.e. that of section 7.1.1) is used, it is found that equation (7.5) leads to the same results except that the underlined terms are absent.

The nature of the coefficients A_{rs}, B_{rs}, C_{rs} is of the utmost importance. They appear in the equations of motion along with the coefficients a_{rs}, b_{rs}, c_{rs} of the dry hull but they are nothing like as straightforward. Unlike the inertia and stiffness coefficients, the A_{rs} and C_{rs} do not form a symmetric array, let alone a diagonal one. Further, the B_{rs} do not form a symmetric or positive definite array and – most important – they do not even remotely produce a diagonal one.

The values of the elements A_{rs}, B_{rs} and C_{rs} depend on a number of factors. They are partially determined by the added mass $m(x)$ and damping factors $N(x)$ and so by

(a) the shape and size of the immersed part of the hull, and
(b) the encounter frequency;

they depend upon the operating conditions, as may be seen from equations (7.16) and (7.17), and so upon

(c) the operating speed, $\bar{U}$;

they also depend on the characteristic functions of the hull, i.e. upon

(d) the principal modes, $w_s(x)$.

In the language of theoretical dynamics, a ship is a 'non-conservative linear system having frequency-dependent coefficients'. The various relationships are not simple and rather little can be said about the values that can be expected in practice, at least until experience with them has accumulated.

7.2.2 Generalised wave force

The generalised force at the sth principal coordinate of the dry hull is given by equation (7.14), where the applied wave force per unit length of hull, $Z(x, t)$, is given by equation (7.7) or by equation (7.12). If we use the latter form, corresponding to the 'alternative theory' of section 7.1.2 we find that

$$\Xi_s(t) = \int_0^l a\, e^{-k\bar{T}} e^{i(kx\cos\chi - \omega_e t)}\left\{-\left[m(x) + \frac{iN(x)}{\omega_e}\right]\omega^2\right.$$

$$\left. + i\bar{U}\left[m'(x) + \frac{iN'(x)}{\omega_e}\right]\omega + \rho g B(x)\right\}w_s(x)\,dx.$$

It follows from equation (6.6) that

$$\Xi_s(t) = \int_0^l a\, e^{-k\bar{T}} e^{i(kx\cos\chi - \omega_e t)}\left\{-\omega^2\left[m(x) + \frac{\underline{\bar{U}N'(x)}}{\omega\omega_e}\right]\right.$$

$$\left. - i\omega\left[N(x) - \bar{U}m'(x) + \frac{\underline{\bar{U}N(x)k}}{\omega_e}\cos\chi\right] + \rho g B(x)\right\}w_s(x)\,dx$$

$$= \Xi_s\, e^{-i\omega_e t} \quad (s = 0, 1, 2, 3, \dots). \tag{7.19}$$

In this form, the expression for the amplitude Ξ_s is readily distinguished from that found from the simpler strip theory of section 7.1.1; the underlined terms are absent in the simpler case.

The amplitude of the excitation of the sth principal coordinate, Ξ_s, evidently depends on a number of parameters. It will be seen to be affected by all the quantities which determine the generalised forces due to hull motion, plus those parameters which determine the waves:

the wave amplitude a,
the wave frequency ω,
the heading angle χ and
the wave number k.

7.3   Discussion of the generalised fluid forces

It is now possible to write down the equation governing motion at p_s, the sth generalised coordinate. We have found that

$$a_{ss}[\ddot{p}_s(t) + \omega_s^2 p_s(t)] + \sum_{r=0}^{\infty} [A_{rs}\ddot{p}_r(t) + (\alpha_{rs} + \beta_{rs} + B_{rs})\dot{p}_r(t) + C_{rs}p_r(t)]$$

$$= \Xi_s\, e^{-i\omega_e t} \qquad (s = 0, 1, 2, 3, \ldots) \tag{7.20}$$

and it has been shown that the form of the coefficients A_{rs}, B_{rs}, C_{rs} and of Ξ_s depends on the form of strip theory employed.

For a ship encountering a given sinusoidal wave, the steady state response is such that

$$p_r(t) = p_r\, e^{-i\omega_e t}$$

or, in matrix form when $n + 1$ generalised coordinates are thought to be sufficient so that $\mathbf{p} = \{p_0, p_1, p_2, \ldots, p_n\}$,

$$\mathbf{p}(t) = \mathbf{p}\, e^{-i\omega_e t}. \tag{7.21}$$

The assumption of this type of solution has the practical effect of reducing the generalised equations of motion to a matrix form:

$$[(\mathbf{c} + \mathbf{C}) - \omega_e^2(\mathbf{a} + \mathbf{A}) - i\omega_e(\mathbf{b} + \mathbf{B})]\mathbf{p} = \Xi. \tag{7.22}$$

Here

$\mathbf{a}$ is the diagonal inertia matrix of the dry hull;

$\mathbf{A}$ is the matrix of elements of the type (7.16);

$\mathbf{b}$ is the symmetric structural damping matrix given by

$\mathbf{b} = \boldsymbol{\alpha} + \boldsymbol{\beta}$,

where $\boldsymbol{\alpha}$ is the symmetric shear damping matrix, and $\boldsymbol{\beta}$ is the symmetric bending damping matrix;

$\mathbf{B}$ is the matrix of elements of the type (7.17);

$\mathbf{c}$ is the diagonal stiffness matrix of the dry hull with elements $c_{ss} = \omega_s^2 a_{ss}$;

$\mathbf{C}$ is the symmetric matrix of elements of the type (7.18)

Ξ is the wave excitation matrix, equal to $\{\Xi_0, \Xi_1, \Xi_2, \ldots, \Xi_n\}$.

Equation (7.22) is readily written in an apparently familiar form by defining three square matrices

$\mathscr{A} = \mathbf{a} + \mathbf{A}$,

$\mathscr{B} = \mathbf{b} + \mathbf{B}$,

$\mathscr{C} = \mathbf{c} + \mathbf{C}$;

i.e.

$$(\mathscr{C} - \omega_e^2 \mathscr{A} - i\omega_e \mathscr{B})\mathbf{p} = \Xi,$$

and it is now a very short step to an affirmation that $\mathscr{A}$ is an 'inertia matrix' comprising a structural component and an 'added fluid inertia' component. Likewise (one might say that) $\mathscr{B}$ is a damping matrix consisting of a structural portion and a portion of hydrodynamic origin and $\mathscr{C}$ is a combined structural and fluid stiffness matrix. Such an approach is entirely in line with common usage and we shall adopt it since it is very convenient. But it may be misleading.

The equation governing forced sinusoidal vibration of a simple oscillator is familiar. It is

$$a\ddot{x} + b\dot{x} + cx = F\,e^{i\omega t},$$

where a is the mass, b the damping coefficient, c the stiffness, x the displacement, F the amplitude of excitation and ω the driving frequency. If we seek a solution of the form $x = X\,e^{i\omega t}$, we find that

$$(c - \omega^2 a + i\omega b)X = F.$$

Now it is well known that a passive system with several degrees of freedom modifies the equation by requiring the use of mass, damping and stiffness *matrices* **a**, **b** and **c**, X and F being replaced by column vectors **X** and **F**. The theory is given, for example, in Bishop, Gladwell & Michaelson (1965). Had we taken the excitation and response proportional to $e^{-i\omega t}$, rather than $e^{i\omega t}$, the analogy with the present theory would appear to be obvious. The trouble is that **a**, **b** and **c** are all symmetric and positive definite (or positive semi-definite) for a passive system, whereas our $\mathscr{A}$, $\mathscr{B}$ and $\mathscr{C}$ do not possess this restricted form. We have no real assurance, for instance, that $\mathscr{B}$ does indeed account for the 'damping' of the response.

Although we shall continue to refer to **A**, **B** and **C** as the 'added inertia', 'fluid damping' and 'fluid stiffness' matrices respectively, we shall do so with the understanding that this is a rough and ready usage. The ship proceeding through waves constitutes a 'non-conservative system' and a general theory of the dynamics of such systems exists only for the simpler case in which $\mathscr{A}$, $\mathscr{B}$ and $\mathscr{C}$ are frequency-*in*dependent.[†] (In the terms of naval hydrodynamics this means that only quasi-steady flow is catered for.) This is not an insuperable obstacle to progress in the present context however since, practically, one is committed to the use of a computer; but it does mean that the extraction of *general* conclusions from numerical results is something that must be approached with great care.

7.3.1 *Hydrodynamic damping*

According to the simpler form of strip theory, the damping coefficient derived from equation (7.17) is

$$B_{rs} = \int_0^l \{N(x)w_r(x)w_s(x) + \bar{U}m(x)[w_r(x)w_s'(x) - w_r'(x)w_s(x)]\}\,\mathrm{d}x$$
$$- \bar{U}[m(x)w_r(x)w_s(x)]_0^l.$$

The damping at p_s due to motion at p_r is $B_{rs}\dot{p}_r$. Similarly, the damping at p_r due to motion at p_s is $B_{sr}\dot{p}_s$, where B_{sr} is obtained by reversing the suffices in the above expression for B_{rs}.

[†] See Wahed & Bishop (1976) and Fawzy & Bishop (1976).

For a slender ship with pointed ends, the integrated term vanishes since $m(0) = 0 = m(l)$. We then have

$$B_{rs} = \int_0^l N(x) w_r(x) w_s(x) \, \mathrm{d}x$$
$$+ \bar{U} \int_0^l m(x) [w_r(x) w_s'(x) - w_r'(x) w_s(x)] \, \mathrm{d}x,$$

$$B_{sr} = \int_0^l N(x) w_r(x) w_s(x) \, \mathrm{d}x$$
$$- \bar{U} \int_0^l m(x) [w_r(x) w_s'(x) - w_r'(x) w_s(x)] \, \mathrm{d}x.$$

These two damping terms both depend on speed $\bar{U}$ in the same way, but the dependence is of opposite sign. It follows that $(B_{rs} + B_{sr})$ is independent of speed. This feature was first established by Timman & Newman (1962) for rigid body modes $(r, s) = (0, 1)$ and we now see that it is of more general validity.

It will be seen that if $r = s$, and the hull lines are still such that $m(0) = 0 = m(l)$, the damping terms become independent of ship speed. That is

$$B_{ss} = \int_0^l N(x) [w_s(x)]^2 \, \mathrm{d}x \qquad (s = 0, 1, 2, \ldots).$$

Thus while the off-diagonal terms $(r \neq s)$ of the hydrodynamic damping matrix remain speed-dependent, obeying the Timman–Newman relations, the leading diagonal terms are speed-independent.[†] Now it may be cogently argued that, in general, it is only these latter terms which represent dissipative damping, the off-diagonal terms serving only to couple the modes (Betts, 1975). Thus dissipative hydrodynamic damping remains speed-independent in the sense that $\bar{U}$ does not appear explicitly, though it must be remembered that the damping coefficient $N(x)$ does depend on frequency of encounter so that there is an implicit speed-dependence. This fact was pointed out à propos the rigid-body modes by Korvin-Kroukovsky (1961), but its continuing validity for all modes has not been generally recognised.

The implications of this are considerable, if only because some writers have attributed very substantial wave damping to speed effects (e.g. see Goodman, 1971; Aertssen, 1971; Kaplan & Sargent, 1972; Hoffman & van Hooff, 1973).

It is interesting to speculate on the transom-sterned ship, for which presumably $m(0) \neq 0$. In this case a mathematical speed-dependence remains, even in the leading diagonal terms. Yet can this in reality lead to such different damping behaviour in a ship with a transom as opposed to a canoe – or rounded – stern? From a physical standpoint

[†] In this we find that the strip theory approach is consistent with a direct three-dimensional approach to the generalised fluid forces developed by Eatock Taylor (1972).

one suspects not, but the mathematical implications are certainly intriguing. At present one can only hazard that the speed-dependent terms arising are unlikely to be of much significance in practice, for distortion modes.

Most evidence suggests that all hydrodynamic dissipative damping is negligible (compared to internal sources of hull damping) in distortion modes of high order at zero speed; see, for example, the theories of Sezawa & Watanabe (1936), Kumai (1958), Borg (1960), and the model experiments of Lockwood Taylor (1930), Ochi (1961) and Kuo (1963). Such hydrodynamic damping could surely not, therefore, increase sufficiently to attain levels comparable to that of structural damping at the relatively low speeds of displacement ships. Furthermore, this same argument would imply that any transfer of energy between (distortion) modes, via the coupling afforded by the off-diagonal speed-dependent terms, would also be very small, whatever the form of stern.†

If the alternative form of strip theory is used, there are extra contributions to the coefficients B_{rs} and B_{sr}. These are the ones corresponding to the underlined terms of equation (7.17). For the rigid body modes $r = 0, 1$ and $s = 0, 1$ the Timman–Newman relationship remains valid provided that we assume $m(0) = m(l) = 0 = N(0) = N(l)$. For higher modes, however, the Timman–Newman relationships do not apply and the sum of the cross-coupling coefficients is not independent of speed. Further, for $r = s$ the damping terms are now dependent on ship speed, i.e.

$$B_{ss} = \int_0^l N(x)\left\{[w_s(x)]^2 + \frac{\bar{U}^2}{\omega_e^2}[w_s'(x)]^2\right\} dx \qquad (s = 0, 1, 2, \ldots).$$

We see, then, that the predictions of the two forms of strip theory as regards generalised hydrodynamic damping actions differ markedly. The position may be summed up as in Table 7.3. This alternative analysis does give a speed-dependent term, but it does so only because we have used a different form of strip theory. Perhaps the speed dependence does indeed exist – though not for reasons that have been adduced in the literature.

To give some idea of orders of magnitude of the generalised fluid damping it is of interest to refer to the 250 000 DWT tanker of length 350 m in the ballast condition operating with a speed of 6 m/s in head seas. The values shown in Table 7.4 were found taking the first four

---

† Note that despite the interest shown over the years in the possibility of a speed-dependence of the dissipative damping of ship hulls, no clear evidence for it has ever been published as far as we are aware. A review of available trials data by Betts (1975) found no correlation of damping of distortion modes with ship speed in those trials (admittedly few) covering a reasonable range of ship speed.

modes (the principal coordinates being scaled as usual to 1 m deflection at the stern):

$r = 0$, heave,

$r = 1$, pitch,

$r = 2$, 2-node flexure dry,

$r = 3$, 3-node flexure dry.

It will be seen that the damping actions couple the motions in the principal modes; moreover the coupling is by no means negligible, mainly because of the effects of a large contribution to the overall fluid damping by the influence of added mass. On the other hand, the magnitudes of the non-dimensional diagonal elements are little affected by the different influence of forward speed in the two theories.

Table 7.5 illustrates the relative non-dimensional magnitudes of the structural damping and the diagonal hydrodynamic fluid damping terms at various non-dimensional frequencies $\omega_e (l/g)^{\frac{1}{2}}$. Over the frequency range considered it is seen that the influences of hydrodynamic damping and structural damping are reversed with increasing mode number. For example, in mode 2 the structural damping is of the same order of magnitude as the hydrodynamic damping at $\omega_e(l/g)^{\frac{1}{2}} = 17$, whereas in mode 7 structural damping is dominant even for $\omega_e(l/g)^{\frac{1}{2}} = 4$. It does appear from this limited evidence that, for certain frequencies and mode numbers the magnitudes of structural and hydrodynamic damping are comparable.

Table 7.3. *Comparison of hydrodynamic coefficients predicted by the two forms of strip theory for a ship with a pointed stern* $(m(0) = 0 = m(l) = N(0) = N(l))$

Coefficient	Theory due to Gerritsma & Beukelman (1964)	Theory due to Salvesen *et al.* (1970) and Vugts (1971)
Added mass	$A_{rs}(+\bar{U}) \neq A_{sr}(-\bar{U})$	$A_{rs}(+\bar{U}) = A_{sr}(-\bar{U})$ for all modes r, s
Damping	$B_{rs}(+\bar{U}) = B_{sr}(-\bar{U})$ for all modes r, s	$B_{rs}(+\bar{U}) = B_{sr}(-\bar{U})$ for all modes r, s
	$B_{rs}(+\bar{U}) + B_{sr}(+\bar{U})$ is independent of $\bar{U}$ for all modes r, s	$B_{rs}(+\bar{U}) + B_{sr}(+\bar{U})$ is independent of $\bar{U}$ for modes 0, 1 only
	B_{ss} is independent of $\bar{U}$ as suggested by three-dimensional theory of Eatock Taylor (1972)	B_{ss} is speed-dependent

7.3.2 Fluid stiffness

When a response $p_s(t)$ is assumed of the form $p_s \exp(-i\omega_e t)$ and generalised forces due to hull motion are then sought, these forces may be in phase with $p_s(t)$ or in quadrature with it. This is because we are only concerned with sinusoidally varying parameters. The generalised fluid forces that are in phase with $p_s(t)$ account for 'stiffness' and 'inertia' effects and the division between them is to some extent arbitrary. The quantity

$$A_{rs}\ddot{p}_r(t) + C_{rs}p_r(t) = (C_{rs} - \omega_e^2 A_{rs})p_r\, e^{-i\omega_e t}$$

is determined by whichever form of strip theory is used, but the theory is not concerned with the division between C_{rs} and $-\omega_e^2 A_{rs}$.

As quoted in equation (7.18), the stiffness terms correspond to purely hydrostatic effects. If the hull suffers a distortion $p_r(t)w_r(x)$ in the rth mode, the increment of buoyancy force applied to a slice of thickness Δx is $-\rho g B(x)p_r(t)w_r(x)\,\Delta x$. This represents a contribution $-\rho g B(x)p_r(t)w_r(x)w_s(x)\,\Delta x$ to the generalised force at p_s and so

$$C_{rs}p_r(t) = p_r(t)\rho g \int_0^l B(x)w_r(x)w_s(x)\,\mathrm{d}x.$$

This does not imply that the conditions of hydrostatic loading prevail, for fluid accelerates in the vicinity of the hull. It simply means that we choose to regard departures from hydrostatic loading as being associated with 'added inertia'. This option is available to us with both forms of the strip theory.

Table 7.4. *Generalised mass and hydrodynamic damping matrices of tanker in ballast*

$$\text{Matrix } [a_{ss}] = \begin{bmatrix} 7.8 \times 10^4 & 0 & 0 & 0 \\ 0 & 2.4 \times 10^4 & 0 & 0 \\ 0 & 0 & 2.2 \times 10^4 & 0 \\ 0 & 0 & 0 & 7.9 \times 10^3 \end{bmatrix} (\text{tonne m}^2)$$

$$\text{Matrix } \left[\frac{B_{rs}(l/g)^{\frac{1}{2}}}{a_{ss}}\right] = \begin{bmatrix} 8.98 & -0.09 & -1.97 & -1.49 \\ -4.12 & 6.21 & -0.88 & -1.52 \\ -4.33 & -5.42 & 9.64 & 3.90 \\ -15.70 & -6.10 & 0.92 & 10.63 \end{bmatrix} \begin{array}{l} \text{at } \omega_e(l/g)^{\frac{1}{2}} = 4.0 \\ \text{(including under-} \\ \text{lined terms} \\ \text{in definition of } B_{rs}) \end{array}$$

$$\text{Matrix } \left[\frac{B_{rs}(l/g)^{\frac{1}{2}}}{a_{ss}}\right] = \begin{bmatrix} 8.98 & -0.09 & -1.97 & -1.49 \\ -4.12 & 6.13 & -0.83 & -1.54 \\ -4.33 & -5.36 & 9.24 & 3.92 \\ -15.70 & -6.12 & 0.97 & 10.07 \end{bmatrix} \begin{array}{l} \text{at } \omega_e(l/g)^{\frac{1}{2}} = 4.0 \\ \text{(excluding under-} \\ \text{lined terms} \\ \text{in definition of } B_{rs}) \end{array}$$

7.3.3 Added inertia

It can legitimately be argued that, by grouping all dynamic effects in the coefficients A_{rs} and B_{rs} we make things a little more complicated. For it will be seen from equation (7.16) that, since we only make an arbitrary distinction between C_{rs} and $-\omega_e^2 A_{rs}$, all the terms in A_{rs} other than

$$\int_0^l m(x) w_r(x) w_s(x) \, \mathrm{d}x$$

can be simplified slightly if multiplied by $-\omega_e^2$ and transferred to C_{rs}. This is quite true, but in adopting our convention we are following a custom that is quite well established and which does have the advantage of representing all static effects in the stiffness terms $C_{rs} p_r$.

Even if the lines of the hull are such that $m(0) = 0 = m(l) = N(0) = N(l)$, so that there is some simplification of the elements, the added inertia matrix is non-symmetric, dependent on speed and on the distribution of added mass and fluid damping.

7.3.4 Wave excitation

The column vector of wave excitation Ξ consists of two parts. Equation (7.19) shows that the sth element has a 'hydrostatic' portion

$$\int_0^l a\rho g \, \mathrm{e}^{-k\bar{T}} \mathrm{e}^{\mathrm{i}(kx\cos\chi - \omega_e t)} B(x) w_s(x) \, \mathrm{d}x.$$

Table 7.5. *Comparison of non-dimensional diagonal damping coefficients of tanker in ballast (column under — includes underlined terms in theory)*

Mode number m	Structural damping[a]	Non-dimensional encounter frequency = 4 Hydrodynamic damping		Non-dimensional encounter frequency = 9.6 Hydrodynamic damping		Non-dimensional encounter frequency = 17 Hydrodynamic damping	
0	–	8.98	8.98	1.85	1.85	0.21	0.21
1	–	6.21	6.13	1.70	1.69	0.23	0.23
2	0.63	9.64	9.24	2.13	2.11	0.25	0.25
3	15.83	10.63	10.07	1.79	1.76	0.16	0.16
4	29.90	10.57	9.69	1.79	1.74	0.17	0.16
5	44.40	13.44	11.90	2.38	2.32	0.25	0.25
6	57:80	13.10	11.36	2.44	2.38	0.27	0.27
7	69.50	9.13	7.27	1.72	1.65	0.21	0.21

[a] $\pi^{-1}\omega_s \delta_s (l/g)^{\frac{1}{2}}$, where ω_s is the sth natural frequency of the dry hull and δ_s is the logarithmic decrement of the sth mode. The method of rendering the damping dimensionless is as follows:

for fluid damping the quantity $\dfrac{B_{ss}(l/g)^{\frac{1}{2}}}{a_{ss}}$ is used;

for structural damping, therefore, the quantity $\dfrac{b_{ss}(l/g)^{\frac{1}{2}}}{a_{ss}} = \dfrac{(a_{ss}\omega_s \delta_s/\pi)(l/g)^{\frac{1}{2}}}{a_{ss}} = \pi^{-1}\omega_s \delta_s (l/g)^{\frac{1}{2}}.$

To this must be added a correction whose form depends on which form of strip theory is employed. This latter correction depends on:

operating speed, $\bar{U}$;
distribution of added mass, $m(x)$;
distribution of damping coefficient, $N(x)$;
wave frequency, ω.

Further, if the alternative form of strip theory is used it will depend on

wave number, k;
heading angle, χ.

7.3.5 *Matrices for generalised fluid forces*

It is tempting to quote the values of specific matrices, giving the numerical values of A_{rs}, B_{rs} and C_{rs}, in order to try to show the order of the changes brought about by alterations of operating conditions and wave parameters. This can of course readily be done by extracting the required data in the programs employed. It will be appreciated, however, that so great is the number of variables involved, it turns out to be virtually impossible to give a comprehensive picture of the ways in which the matrices vary for different ships in different operating conditions and seaways.

No doubt some impression will be made in the fullness of time on the problem of imparting some sort of 'picture'. But to attempt to resolve the matter at the present juncture is simply too great an undertaking. In any event, this is likely to remain a complicated problem for the out-and-out specialist.

8 Symmetric response

Oh, sweet surprise – oh, dear delight,
To find it undisputed quite,
All musty, fusty rules despite,
That Art is wrong and Nature right!
Utopia Limited

8.1 Receptances and response functions

Consider the output response, $q(t)$, of a linear system having a single degree of freedom. This output depends on an input excitation, $Q(t)$, being governed by the differential equation

$$\ddot{q}(t) + 2\nu\omega_1\dot{q}(t) + \omega_1^2 q(t) = Q(t), \tag{8.1}$$

where ν is the damping factor and ω_1 is the (one and only) natural frequency. For a sinusoidal input

$$Q(t) = Q_0 e^{i\omega t},$$

the steady state output is

$$q(t) = q_0 e^{i\omega t} = H(\omega) Q_0 e^{i\omega t} = H(\omega) Q(t), \tag{8.2}$$

where the complex receptance is given by

$$H(\omega) = \frac{1}{\omega_1^2 - \omega^2 + 2i\nu\omega_1} = \frac{\exp\left[-i \tan^{-1}\left(\dfrac{2\nu\omega_1}{\omega_1^2 - \omega^2}\right)\right]}{[(\omega_1^2 - \omega^2)^2 + 4\nu^2\omega_1^2]^{\frac{1}{2}}}. \tag{8.3}$$

It is convenient to identify a 'response amplitude operator' (RAO)

$$|H(\omega)| = \frac{1}{[(\omega_1^2 - \omega^2)^2 + 4\nu^2\omega_1^2]^{\frac{1}{2}}} = \left|\frac{\text{amplitude of output}}{\text{amplitude of input}}\right|. \tag{8.4}$$

The system being linear, a sinusoidal input produces a steady sinusoidal output of the same frequency. A generalisation of this approach for systems with multiple degrees of freedom is commonplace (e.g. see Bishop & Johnson, 1960 or Bishop, Gladwell & Michaelson, 1965). The theory has also been adopted for the solution of the matrix response equations derived in chapter 7. This present chapter is largely devoted to receptances like $H(\omega)$ but in chapter 9 we shall examine a second type of receptance and relate it to the present sort.

We shall show in chapter 11 that the output mean square spectral density of a linear time-invariant system, $\Phi_{oo}(\omega)$, is related to that of input $\Phi_{ii}(\omega)$ by the equation

$$\Phi_{oo}(\omega) = |H_{oi}(\omega)|^2 \Phi_{ii}(\omega),$$

where $H_{oi}(\omega)$ is the appropriate receptance function. In the present context, it is convenient to use encounter frequency, ω_e, as the independent variable and to identify the 'input' as wave elevation $\zeta(x, t)$. That is to say

$$\Phi_{oo}(\omega_e) = |H_{o\zeta}(\omega_e)|^2 \Phi_{\zeta\zeta}(\omega_e). \tag{8.5}$$

Outputs 'o' of interest are the displacement, bending moment, shearing force or any other appropriate response at any section of the hull. In this chapter we shall refer only to symmetric responses.

The quantity $\Phi_{\zeta\zeta}(\omega_e)$ is derived from a wave spectrum $\Phi_{\zeta\zeta}(\omega)$ using standard methods that are discussed in chapter 11. It represents a part of the 'given' operating condition of a ship. By contrast, $\Phi_{oo}(\omega_e)$ is what has to be found in order that the statistics of the output random process of interest can be assessed. Such matters as proneness to deck wetting, the likelihood of large vertical accelerations, liability to frequent slamming or the possible occurrence of excessive stresses can be examined once $\Phi_{oo}(\omega_e)$ has been determined. Evidently, then, it is necessary to find $|H_{o\zeta}(\omega_e)|$, whatever 'o' may be. That will be our main object in this chapter.

By definition, $H_{o\zeta}(\omega_e)$ is the (complex) response to waves of unit amplitude. In what follows, therefore, we shall take a, the wave amplitude, to equal unity. Two other parameters relating to the sinusoidal waves remain open, however, though in deep water they are related; they are the wavelength λ and wave frequency ω. The operating conditions are determined by the ship speed $\bar{U}$ and heading χ, both of which are assumed to be constant and prescribed. Now

$$\omega_e = \omega - \frac{\bar{U}\omega^2}{g} \cos \chi \tag{8.6}$$

in deep water and so we can calculate the encounter frequency for given operating conditions in a given sinusoidal sea. Knowing ω_e we can calculate $H_{o\zeta}(\omega_e)$ for a given ship.

As previously discussed in section 7.3, the principal coordinates p_r ($r = 0, 1, 2, \ldots, n$) of the ship are governed by the matrix equation

$$(\mathscr{C} - \omega_e^2 \mathscr{A} - i\omega_e \mathscr{B})\mathbf{p} = \mathbf{D}\mathbf{p} = \Xi \, e^{-i\omega_e t}. \tag{8.7}$$

It is therefore possible to evaluate $\mathbf{p}$ since

$$\mathbf{Ip} = \frac{\text{adj } \mathbf{D}}{\det \mathbf{D}} \, \Xi, \tag{8.8}$$

where $\mathbf{I}$ is a unit matrix. In this way we find the p_r and, hence, the responses

$$
\left.\begin{array}{l}
w(x, t) = e^{-i\omega_e t} \sum_{r=0}^{n} p_r w_r(x), \\[2mm]
M(x, t) = e^{-i\omega_e t} \sum_{r=2}^{n} p_r M_r(x), \\[2mm]
V(x, t) = e^{-i\omega_e t} \sum_{r=2}^{n} p_r V_r(x),
\end{array}\right\} \tag{8.9}
$$

since $M_0(x) = 0 = M_1(x) = V_0(x) = V_1(x)$. The quantities p_r are in general complex while the $w_r(x)$, $M_r(x)$, $V_r(x)$ are not. Provided the wave amplitude $a = 1$, the sums are the appropriate complex response functions $H_{o\zeta}(\omega_e)$.

It will be noted that $H_{o\zeta}(\omega_e)$ depends on $\mathbf{D}$ and $\boldsymbol{\Xi}$ for a given ship, as well as upon ω_e. But the matrices $\mathbf{D}$ and $\boldsymbol{\Xi}$ themselves depend, in general, on sea state, on the operating conditions and on ω_e. In short, the response function $H_{o\zeta}(\omega_e)$ of a given ship is not likely to behave in a particularly simple way as ω_e is changed. The quantity ω_e cannot be regarded as a simple independent variable as it commonly can elsewhere in dynamics.

Despite this difficulty it is possible to deduce certain general features of the response functions from the equations of motion.

(a) It is clear that responses at the coordinates are coupled. Generally speaking for a given value of ω_e, all of them vary sinusoidally with this frequency.

(b) If, as ω_e is varied, the quantity $|\det \mathbf{D}|$ goes through a minimum, all the coordinates take maximum ('resonant encounter') values together at the appropriate 'resonant encounter frequencies'.

(c) The values of the generalised fluid forces at the generalised coordinates p_s are found by forming terms

$$
\Xi_s = \int_0^l Z(x) w_s(x) \, dx,
$$

where the function $Z(x)$ is dependent on the hydrodynamic theory employed. Now the value of this integral will depend on the extent to which $Z(x)$ and the characteristic modal function $w_s(x)$ reinforce or cancel each other along the ship's length. In other words Ξ_s depends upon the ratio

$$
\frac{l}{\lambda} = \frac{\text{ship length}}{\text{wavelength}},
$$

for the heading concerned.

To sum up then, the coupled responses at the generalised coordinates $p_0, p_1, p_2, \ldots$ will be large if 'resonant encounter' or if 'ship-wave matching' occurs or if both occur together.

So low are the levels of damping involved, ship responses are usually magnified much more by resonant encounter than they are by ship-wave matching. The exceptions to this general rule are the 0, 1 modes which suffer heavy hydrodynamic damping. Were it not for the effects of fluid actions, these two modes could not be brought to resonance anyway since they are associated with zero natural frequencies.

8.1.1 Dependence of response functions on encounter frequency

Suppose that the operating conditions, $\bar{U}$ and χ are held constant and that ω_e is 'large' over 'most' of the ship's length. Figs. 7.6, 7.14 for the destroyer and 7.7, 7.8, 7.15, 7.16 for the tanker show that in this event $\mathbf{D}$ will not depend sensitively on ω_e for a given sea state. Under these restrictions equation (8.7) becomes, in effect, an equation with constant coefficients in which ω_e is a simple independent variable.

Intuition suggests that, under these conditions, each coordinate p_r $(r = 2, 3, 4, \dots)$ will produce a curve like that sketched in fig. 8.1. Moreover one would expect that each coordinate will be identified with a distinct dominant peak. If this is so, equations (8.9) show that the values of $|H_{o\zeta}(\omega_e)|$ will similarly have peaks at the various resonance values of ω_e as indicated in fig. 8.2. Notice that a peak in $|H_{o\zeta}(\omega_e)|$ will be small even when the corresponding dominant $|p_r|$ peak is large if the output is reckoned at some section x which is so chosen as to make $w_r(x)$, $M_r(x)$ or $V_r(x)$ – whichever is relevant – very small.

Maxima in $|p_0|$ and $|p_1|$ are likely to occur at very low encounter frequencies since both are associated with zero natural frequencies of the dry hull. But they are likely to fall in the range in which fluid damping is large, as suggested in figs. 7.14, 7.15, 7.16. Resonant encounter is likely to be less important than it is with higher encounter frequencies.

It is not easy to see how the hypothetical curve in fig. 8.2 is likely to vary if ω, $\bar{U}$ and χ are allowed to change. A different wave frequency ω will produce a different value of ω_e for the same operating conditions and we might surmise that the curve would not change greatly and that

Fig. 8.1. The likely dependence of amplitude of a coordinate on encounter frequency. Peaks correspond to minima of $|\det \mathbf{D}|$ so that all p_r have the same 'resonant' encounter frequencies. Each p_r is likely to have a distinct dominant peak. Very low values of ω_e (to the left of the broken line) are excluded by the assumptions made in the text.

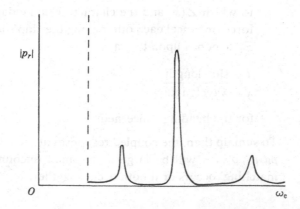

only the position of the representative point on it would alter. But if the operating conditions, $\bar{U}$ and/or χ, are altered significantly the curve is likely to be changed perceptibly – though it seems probable on physical grounds that even then the positions of the peaks along the ω_e axis are unlikely to change much. Although this is little more than conjecture, we shall find that it is borne out by calculated results.

8.1.2   Dependence of response functions on wavelength

It may seem at first sight that the dynamicist's traditional approach of regarding the encounter frequency ω_e as an independent variable has much to commend it. This is indeed the case but we shall nevertheless now examine an alternative. Consider the dimensionless ratio of ship length l to wavelength λ. By definition

$$\lambda = \frac{2\pi}{k},$$

where k is the wave number. But in deep water

$$k = \frac{\omega^2}{g}$$

and so

$$\lambda = \frac{2\pi g}{\omega^2}. \tag{8.10}$$

If ω is now eliminated between equations (8.6) and (8.10) it is found that

$$\omega_e = \left[\left(\frac{2\pi g}{l}\right)\frac{l}{\lambda}\right]^{\frac{1}{2}} - \left(\frac{2\pi \bar{U}}{l}\right)\frac{l}{\lambda}\cos\chi. \tag{8.11}$$

Fig. 8.2. The likely form taken by the modulus of a response function for fixed operating conditions. The curve has a sequence of peaks at the resonance values of encounter frequency. Each peak is labelled with the likely dominant resonant coordinate. To the left of the broken line (which would lie at about 1.3 rad/s for a very large ship to about 2.5 rad/s for a fairly small one) there will be maxima in which p_0 and p_1 dominate, but they will not conform to a simple pattern.

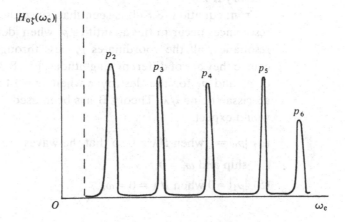

That is ω_e is related to l/λ for a given ship in a way that is determined by the operating conditions. Notice that only the positive square root is admissible in equation (8.11) since ω must be greater than zero in equation (8.6). There is thus a one-to-one correspondence between ω_e and l/λ for a given ship with given operating conditions.

At certain values of ω_e conditions of resonance exist. It follows that there are values of l/λ at which this is the case and so results like those sketched in figs. 8.1 and 8.2 are to be expected in which the abscissae are values of l/λ. But the positions of the maxima of $|H_{o\zeta}(\omega_e)|$ will then depend heavily on the operating conditions $\bar{U}$ and χ, as equation (8.11) shows. This means that, whereas one might on physical grounds expect a ship to possess inherent resonant encounter frequencies, it will certainly not possess inherent values of l/λ at which resonance occurs.

It is really a matter of taste which type of specification is used for a response function, i.e. whether ω_e or l/λ is used. It is certainly arguable that ω_e is preferable in the study of resonant encounter whereas l/λ is more useful in considerations of ship-wave matching. Perhaps ω_e will appeal more to the dynamicist and l/λ will seem the more useful to the naval architect.

8.1.3 *The location of peaks in response curves*

While it is intuitively to be expected on physical grounds that the $r = 2, 3, \ldots,$ resonance values of ω_e are almost independent of the operating conditions (i.e. on $\bar{U}$ and χ save insofar as they determine ω_e) this is not assured by the theory. On the other hand we should expect resonance values of l/λ to be heavily dependent on $\bar{U}$ and χ. It is worth while therefore to illustrate these points numerically. Here, and henceforth in this chapter, we shall refer to the ships whose properties are examined in chapter 4 (i.e. to the destroyer and to the 250 000 DWT tanker in ballast and fully loaded). In doing so we shall employ the two versions of strip theory that were introduced in chapter 7, referring to that of Gerritsma & Beukelman (1964) as 'Theory A' and that of Salvesen, Tuck & Faltinsen (1970) and Vugts (1971) as 'Theory B'.

From equation (8.8) it is seen that, since the matrix $\mathbf{D}$ is complex, resonances occur in the quantities p_r when $|\det \mathbf{D}|$ is a minimum. At resonance, all the coordinates p_r pass through maxima, though of course they are of different magnitudes. Fig. 8.3 shows the amplitudes of p_0 and p_1 for the destroyer when $\bar{U} = 14$ m/s and $\chi = 135°$ the abscissae being l/λ. Theory B has been used. We notice that, as one would expect,

(a) $|p_0| = 1$ when $l/\lambda = 0$ so that the waves are much longer than the ship and $\omega_e = 0$;

(b) $|p_1| = 0$ when $l/\lambda = 0 = \omega_e$.

But while $|p_0|$ and $|p_1|$ both have maxima at the same value l/λ – approximately 1.2 – it will be seen that there is no significant peak in which $|p_0|$ predominates and $|p_1|$ has a smaller hump by reason of coupling.

Turning next to the distortion modes, fig. 8.4 shows $|p_r|$ plotted against l/λ with $r = 2, 3, 4, 5$, for the same ship in the same operating conditions. Now we see that

(a) All the $|p_r|$ have maxima in the region of the maximum in $|p_0|$ and $|p_1|$ that we noted previously, i.e. around $l/\lambda = 1.2$, though there is some spreading along the l/λ axis;

(b) All the $|p_r|$ have maxima at approximately the same values of l/λ;

(c) Each coordinate provides one dominant peak and the arrangement is such that p_2 gives the first, p_3 the second and so on. Notice particularly that, due to coupling, p_2 produces a significant (though not dominant) peak at $l/\lambda = 1.2$, where $|p_0| \doteqdot 1.12$; $|p_1| \doteqdot 2.12$; $|p_2| \doteqdot 0.04$. The higher coordinates $p_3, p_4, \ldots$, too, produce peaks at about $l/\lambda = 1.2$.

(d) While we could detect evidence of ship-wave matching where the 'p_1 peak' was concerned, there is little or none for the $p_2, p_3, \ldots$ peaks since l/λ is generally far too large. This turns out to be generally true and so we conclude that peaks in the distortion responses are due to resonant encounter.

Table 8.1 shows values of l/λ and ω_e at which resonant encounter takes place. From it we see that, as we surmised, there appear to be characteristic values of encounter frequency which depend very little

Fig. 8.3. The amplitudes of p_0 and p_1 for small values of l/λ for the destroyer with $\bar{U} = 14$ m/s and $\chi = 135°$.

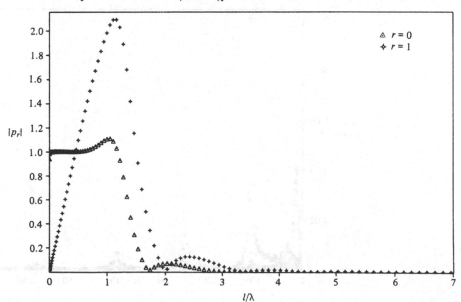

on operating conditions. As already explained, however, this is a feature that is unlikely to be shared by the heave and pitch modes, 0 and 1.

It is of interest to compare resonant encounter frequencies with natural frequencies of the dry hull. Thus, to take the '$|p_3|$-dominated peak' of the destroyer, the resonant encounter frequency is 18.3 rad/s (or 2.9 Hz), more or less independent of the operating conditions. Were the free–free hull of the destroyer *in vacuo* it would be found to have a natural frequency of 26.8 rad/s (or 4.3 Hz). The presence of the water has decreased what would be the resonant frequency of the dry free–free hull by about a third.

Turning to the l/λ values at resonance in Table 8.1, we may take it as a working hypothesis that the higher the value of l/λ the less important it is. Remembering that the heading angle χ is such that $\chi = 0$ represents a following sea and $\chi = 180°$ a head sea, it will be seen that for any given heading, resonances occur at decreasing values of l/λ as $\bar{U}$ increases. At a forward speed of 6 m/s in a head sea the 'two-node wet mode' (i.e. the resonant mode at which $|p_2|$ predominates) is excited in the destroyer, the tanker in ballast and the fully laden tanker by waves of lengths 6.6 m, 21 m and 30 m respectively (though at least the first of these figures raises questions as to the validity both of strip theory and of the structural representation). In practice, significant fluid loading seems more likely from waves of length 30 m than from waves having either of the other two lengths. It can therefore be

Fig. 8.4. The amplitudes of p_2, p_3, p_4 and p_5 for the destroyer with $\bar{U} = 14$ m/s and $\chi = 135°$; Theory B.

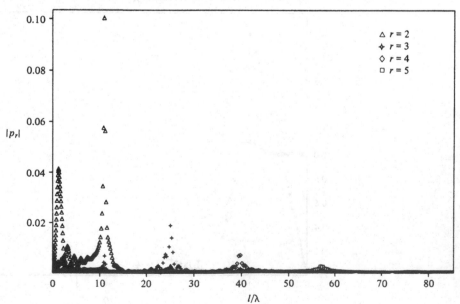

Table 8.1. Values of l/λ and frequencies of encounter at which resonance occurs

Destroyer — Ū = 6 m/s, Ū = 10 m/s, Ū = 14 m/s

Index of dominant mode	Dry hull ω_r (rad/s)	Dry hull equiv l/λ	Wet hull ω_e (rad/s)	Ū = 6 m/s χ=45°	χ=90°	χ=135°	χ=180°	Ū = 10 m/s χ=45°	χ=90°	χ=135°	χ=180°	Ū = 14 m/s χ=45°	χ=90°	χ=135°	χ=180°
2	13.3	320	8.79	60.8	139.1	22.0	16.9	32.6	139.1	14.8	11.1	21.9	139.1	11.2	8.4
3	26.8	1295	18.3	108.5	603.1	53.5	40.0	60.2	603.1	34.8	24.9	41.2	603.1	25.9	19.0
4	39.4	2793	27.8	154.3	1392.0	86.8	64.3	86.8	1392.0	55.6	40.7	59.9	1392.0	41.1	29.9
5	57.1	5867	39.0	207.0	2739.0	127.4	93.6	117.6	2739.0	80.7	58.8	81.6	2739.0	59.4	43.0

Tanker loaded — Ū = 3 m/s, Ū = 6 m/s, Ū = 9 m/s

Index of dominant mode	Dry hull ω_r (rad/s)	Dry hull equiv l/λ	Wet hull ω_e (rad/s)	Ū = 3 m/s χ=45°	χ=90°	χ=135°	χ=180°	Ū = 6 m/s χ=45°	χ=90°	χ=135°	χ=180°	Ū = 9 m/s χ=45°	χ=90°	χ=135°	χ=180°
2	4.0	90	2.67	239.9	40.3	20.3	17.1	85.7	40.3	14.2	11.5	48.8	40.3	11.1	8.8
3	10.0	565	6.42	382.5	232.7	73.5	58.9	151.4	232.7	46.4	36.0	90.8	232.7	34.4	26.2
4	16.6	1562	11.0	543.3	683.2	151.8	118.5	226.2	683.2	91.2	69.2	138.9	683.2	66.0	49.5
5	24.0	3248	16.1	714.0	1463.5	247.5	190.1	306.3	1463.5	144.2	108.2	190.7	1463.5	103.0	76.4

Tanker in ballast — Ū = 3 m/s, Ū = 6 m/s, Ū = 9 m/s

Index of dominant mode	Dry hull ω_r (rad/s)	Dry hull equiv l/λ	Wet hull ω_e (rad/s)	Ū = 3 m/s χ=45°	χ=90°	χ=135°	χ=180°	Ū = 6 m/s χ=45°	χ=90°	χ=135°	χ=180°	Ū = 9 m/s χ=45°	χ=90°	χ=135°	χ=180°
2	6.7	252	3.56	275.3	71.6	31.4	26.1	102.0	71.6	21.2	16.9	59.2	71.6	16.2	12.7
3	15.7	1390	3.21	446.6	380.6	102.9	81.5	181.1	380.6	59.3	48.7	109.8	380.6	46.5	35.2
4	29.6	4932	14.9	674.4	1253.4	224.4	172.9	287.7	1253.4	131.5	98.9	178.6	1253.4	94.1	70.0
5	43.8	10815	22.7	927.3	2934.9	378.8	287.4	407.2	2934.9	215.7	160.6	256.2	2934.9	152.6	112.2

expected that excitation of this mode is a far more serious matter with the loaded tanker than it is with the other two vessels.

Just as it is possible to compare resonant encounter frequencies with natural frequencies of the dry hull, so it is possible to compare l/λ values at resonance with an 'equivalent l/λ' determined by the dry hull. Since

$$\frac{l}{\lambda} = \frac{lk}{2\pi} = \frac{l\omega^2}{2\pi g}$$

we can define an 'equivalent l/λ' for the rth dry mode by using the quantity $l\omega_r^2/2\pi g$. The appropriate values are quoted in Table 8.1. The interpretation to be placed on this quantity is that it is the l/λ ratio for waves whose frequency ω is that of the rth dry mode.

8.1.4   *Resonance of a ship at rest*

We have seen that the values of resonant encounter frequency for a ship appear to be almost independent of operating conditions. It is natural to enquire whether or not they can be measured as resonant shaking frequency, ω_{res}, when the ship is stationary. Now it happens that the destroyer which we have referred to underwent a resonance test when it was stationary in calm water (Hicks, 1972); the results are those given and compared with the calculated resonant encounter frequencies in Table 8.2. The results given in brackets relate to a sister ship in a similar displacement condition and are thought by Hicks to reflect inaccuracy of the ship data – data which, perforce, we have used in this book.

It will be seen from Table 8.2 that the measured calm water resonance frequencies of the destroyer are at least comparable with the resonant encounter frequencies of the sinusoidal sea. Indeed the higher the modal order, the better is the agreement. While admittedly the evidence is very limited it does seem possible that we have here a straightforward method of estimating resonant encounter frequencies; either the calm water resonance frequencies can be measured or they may be estimated using fairly rudimentary rules which ignore the parameters $\bar{U}, \chi$ and ω altogether (e.g. see Hicks, 1972). This is

Table 8.2. *Resonance frequencies of the destroyer in calm water*

Index of dominant dry mode	Calm water resonance frequency	ω_{res} (rad/s)	Resonant encounter frequency (Table 8.1) ω_e (rad/s)
2	9.99	(9.42)	8.79
3	20.11	(19.42)	18.3
4	29.97		27.8
5	39.52		39.0

unquestionably an important matter and it should be investigated for a variety of ships.

8.1.5 Seakeeping

In effect, seakeeping theory is concerned with the responses at p_0 and p_1, the behaviour of the remaining coordinates being ignored. On the basis of curves such as those given in fig. 8.3 it is possible to speculate on the nature of seakeeping. In this section, then, we shall discuss the destroyer's performance in somewhat conjectural terms since it appears that the study of seakeeping still requires considerable development in terms of physical reasoning.

The fact that $|p_0| = 1$ when $l/\lambda = 0 = \omega_e$ is an outcome of ship-wave matching. There is no dynamic effect and so resonant encounter does not occur.

Consider now the question of whether or not resonant encounter takes place so as to produce a $|p_0|$-dominated peak. At the very low values of l/λ, the ship's added mass is very large and fluid damping will also be large. Resonant encounter, if it occurs, can be expected at a very low value of ω_e and, therefore, at a low value of l/λ. Moreover the resonance curve will not have a pronounced hump (if it has one at all); if the heave motion were subject to approximately critical damping the response curve would merely maintain its unity value up to the vicinity of the resonant value of ω_e, and hence over a range of l/λ. There is a suggestion in fig. 8.3 that something like this may indeed be the case. The $|p_0|$ curve remains almost flat up to $l/\lambda \doteqdot 0.7$ and since $\chi = 135°$ it follows that, at that value,

$$\frac{\text{ship length}}{\text{wavelength}} \doteqdot \frac{l \cos 45°}{l/0.7} \doteqdot 0.5,$$

where the wavelength is reckoned along the ship's axis. This ratio appears to be too large to be accounted for entirely in terms of ship-wave matching. With such a resonant response at p_0 it is not to be expected that a corresponding peak will be visible in the $|p_1|$ curve.

The heaving frequency of the stationary destroyer in calm water has been estimated by Hicks (1972) who found it to be 0.21 Hz or 1.32 rad/s. If this is assumed to be equal to, and independent of, the resonant encounter frequency for heave – and for near critical damping this estimate can be expected to be too high – equation (8.11) suggests that the relevant value of l/λ would be given by

$$1.32 = \left[\frac{2\pi \times 9.81}{111} \left(\frac{l}{\lambda} \right) \right]^{\frac{1}{2}} + \frac{2\pi \times 14}{111} \left(\frac{l}{\lambda} \right) \cos 45°,$$

whence $l/\lambda \doteqdot 1.02$ which is the only permissible solution. Although it is perhaps slender, this is one more piece of evidence which suggests that, where heave is concerned, ship-wave matching gradually gives way to resonant encounter as l/λ is increased from zero in fig. 8.3.

The $|p_1|$ curve has a maximum value of about 2.1 at $l/\lambda \doteqdot 1.2$. If this peak were due solely to ship-wave matching it could be expected to occur where the ratio

$$\frac{\text{ship length}}{\text{wavelength along ship's axis}}$$

is about 0.5 or (since the vessel has a pointed bow) a little larger – 0.6 say. Further, if, in the complete absence of dynamic effects, the vessel were to conform to the maximum slope of the sinusoidal wave of unit amplitude, the displacement at its stern would be of the order of $\pi/2$ or 1.6 approximately. That is to say, ship-wave matching alone would place a peak of magnitude 1.6 (at most) at the abscissa $l/\lambda = 0.6/\cos 45° = 0.85$ roughly.

Three features of the $|p_1|$ curve in fig. 8.3 suggest that resonant encounter now plays a significant part in determining the pitching response. First of all there is the perceptible peak at $l/\lambda \doteqdot 1.1$ in the $|p_0|$ curve due to coupling. Secondly, the peak in $|p_1|$ occurs at too high a value of l/λ (1.2 instead of 0.85). Thirdly, the peak is too high (2.1 instead of, at most, 1.6).

If we make the assumption that resonant encounter only accounts for the peak in $|p_1|$ we conclude from equation (8.7) that the appropriate frequency is

$$\omega_e = \left(\frac{2\pi \times 9.81 \times 1.2}{111}\right)^{\frac{1}{2}} + \frac{2\pi \times 14 \times 1.2 \cos 45°}{111} \doteqdot 1.5 \text{ rad/s}.$$

It is of interest to compare this result with the value of the pitching frequency of the stationary ship in calm water. This was calculated by Hicks (1972) and found to be 0.23 Hz or 1.45 rad/s.

In conclusion, we repeat that this is a subject which appears to merit more study. At best the conclusions we have drawn for one particular ship under one set of operating conditions can only be regarded as tentative. And there is an obvious need to examine results for other ships and other operating conditions.

8.2 Computations
8.2.1 *The ships*
Before examining response functions in detail it will be helpful to recall details of the ships. The relevant modal functions are summarised in figs. 8.5 and 8.6 while the mechanical properties are those given in Tables 4.9 and 4.11. The lengths and weights of the hulls are given in Table 8.3.

It will be remembered that, in computing the modal properties of the dry hulls, the destroyer was assumed to be divided into 20 slices and the tanker into 50. These divisions were also used in the computation of the fluid actions. For each section the added mass $m(x, \omega_e)$ and fluid damping $N(x, \omega_e)$ were calculated for a range of frequency using a Lewis-form fit at every section.†

The calculations referred to in this book have been made by means of a suite of programs. This suite is a comprehensive one and that portion of it relating to the dry hull was used to obtain the results given in chapter 4. As already explained, two alternative hydrodynamic formulations – Theory A or Theory B – are available.

8.2.2 Number of modes used

It is of interest to investigate the sensitivity of practical results on the value used for $n = 2, 3, 4, \ldots$ in the modal summation. Thus a preliminary investigation was made for both ships into the variation of the displacement, bending moment and shearing force amplitudes amidships using values $n = 2, 3, 4$ and 7. Little difference is made to the responses at low values of l/λ no matter which of these values of n was used.

† As suggested by Hoffman & van Hooff (1976), it would be more accurate to employ an alternative multi-coefficient 'close-fit' technique, particularly in regions near the bow and stern. Although provision is made for doing this in the programs employed in these studies, a Lewis-form fit was used at all sections for ease, speed and economy of calculation. The improvements that can be expected from more accurate representations are discussed in section 8.4.5.

Fig. 8.5. The first five principal modes of free symmetric vibration (a) for a destroyer and (b) for a tanker. In (b) the dotted curves refer to the ballast condition and the full-line curves to the fully loaded condition.

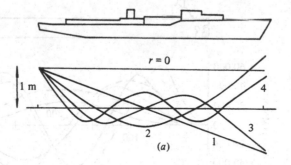

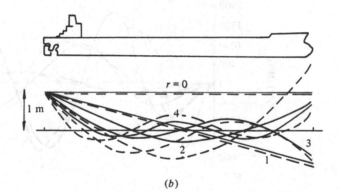

Fig. 8.6. (a) Bending moment and shearing force curves corresponding to the modal deflections of fig. 8.5(a). (b) Similar curves for the tanker, corresponding to fig. 8.5(b). Notice that the modal indices $r = 0, 1$ do not have associated $M_r(x)$ or $V_r(x)$ curves.

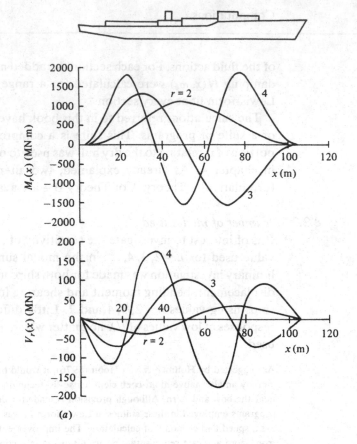

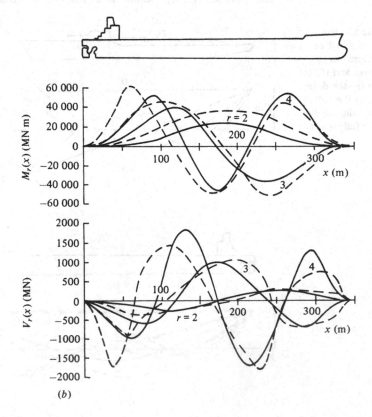

With increasing l/λ, differences can be expected to increase. This is readily shown to be the case, with the greatest discrepancies occurring in the heights of the peaks at resonance. Indeed for any given value of n, resonances will cease to be found if the values of l/λ examined are high enough. (The same can be said if ω_e, rather than l/λ, is taken as the independent variable.)

In this chapter we shall present results for the ships we have referred to. For the results to be quoted, the value $n = 4$ was used in most of the modal summations. Generally speaking, however, it is quite immaterial if $n = 7$ is used, say. It is not suggested, of course, that the principal mode $w_7(x)$ is accurately known – plainly it is *not* for the destroyer – but rather that there is safety in using the value $n = 7$. All significant modal distortion will be catered for and any errors in the highest three or four modes employed are neither here nor there. For practical purposes, however, $n = 4$ is usually quite sufficient for conventional ships.

8.3 Typical results

Fig. 8.7 shows the principal modes $w_r(x)$ $(r = 0, 1, 2, 3, 4)$ and curves of $M_r(x)$ and $V_r(x)$ $(r = 2, 3, 4)$ for the loaded tanker. These data appear in a much more condensed form in figs. 8.5(b) and 8.6(b). For the 50 sections (all of the same length) into which the hull is divided, the two-dimensional added mass $m(x, \omega_e)$ and fluid damping $N(x, \omega_e)$ can be calculated for a range of encounter frequency using a Lewis-form fit at each section. Hence the coefficients A_{rs}, B_{rs} and C_{rs} may be calculated.

A solution for the rth generalised coordinate p_r in equation (8.8) depends on a knowledge of the amplitude Ξ_s of the generalised wave force $(s = 0, 1, 2, 3, 4)$ as defined in equation (7.19). The complex function $Z(x)$ may be obtained from the strip theory discussed in section 7.1 and so the complex quantity $w_s(x)Z(x)$ may be determined. The moduli $|w_s(x)Z(x)|$ from which $|\Xi_s|$ may be determined for the loaded tanker travelling at 9 m/s in head seas $(\chi = 180°)$ may now be found, see equation (7.19). They are shown in fig. 8.8(a)–(e) for values of l/λ equal to 1.0, 8.8 and 15.0. The amplitudes $|\Xi_s|$ naturally vary according to the mode shapes $w_s(x)$, with dominant contributions coming from the $s = 0, 1$ modes at the smaller values of l/λ.

Table 8.3. *Hulls used in computation*

Parameter	Destroyer	Tanker in ballast	Tanker loaded
Length overall (m)	111	348	348
Total weight of dry hull (MN)	25	764	2792

All the necessary information is now to hand for estimating responses. Four types of response will be referred to:

the responses at the principal coordinates, $p_r(t)$,
displacement, $w(x, t)$,
bending moment, $M(x, t)$,
shearing force, $V(x, t)$.

The first of these quantities is found by means of equation (8.8) and the remaining three are obtained from the series (8.9).

Fig. 8.7. Results for the dry hull of the loaded tanker: (a) shows the principal modes $w_r(x)$ of lowest order ($r = 0, 1, 2, 3, 4$), (b) gives the corresponding modal bending moments $M_r(x)$ and (c) shows the shearing force $V_r(x)$. Notice that $M_0(x) = 0 = M_1(x) = V_0(x) = V_1(x)$.

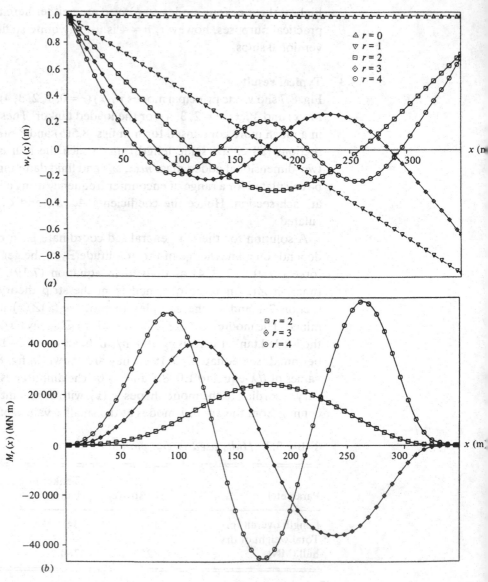

Fig. 8.7. (Continued)

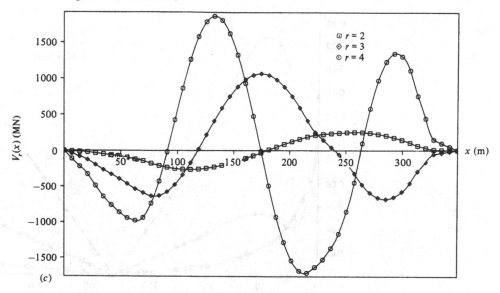

(c)

Fig. 8.8. The quantities $|w_s(x)Z(x)|$ from which the amplitudes of the generalised wave forces, $|\Xi_s|$, may be calculated. The graphs (a)–(e) refer to the modal indices $s = 0, 1, 2, 3, 4$ respectively. The results are found using Theory B for the fluid actions and refer to the loaded tanker travelling at 9 m/s in head seas.

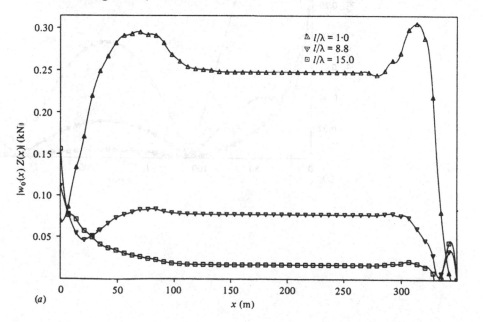

(a)

Fig. 8.8. (Continued)

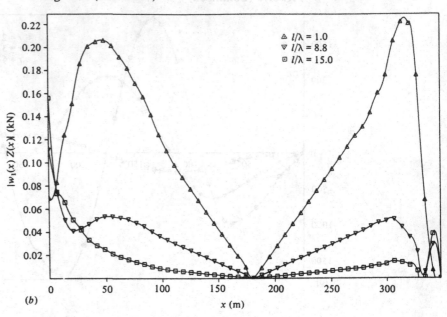

(b)

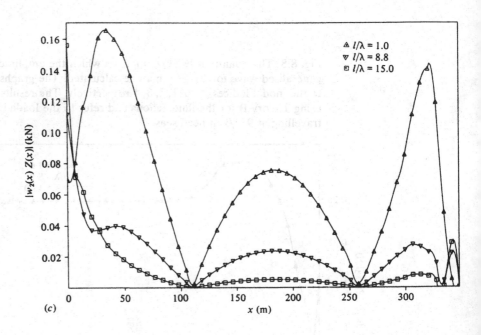

(c)

Fig. 8.8. (Continued)

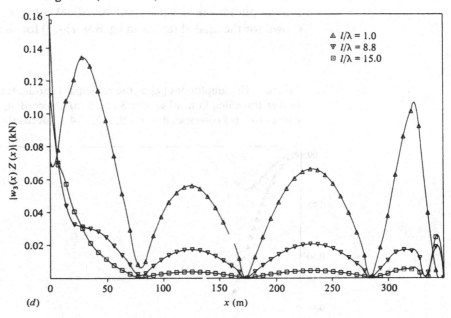

(d)

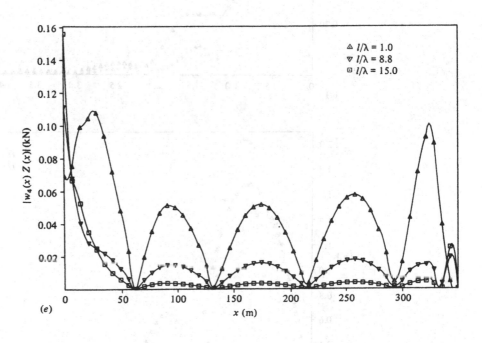

(e)

8.3.1   *Principal coordinates*

The amplitudes of the principal coordinates $|p_r|$ ($r = 0, 1, 2, 3, 4$) are shown for the loaded tanker in fig. 8.9(a)–(e) for ship speeds of 3 and

Fig. 8.9. The amplitudes $|p_r|$ of the principal coordinates for the loaded tanker travelling in head seas at 3 and 9 m/s, according to Theory B. The curves (a)–(e) correspond to $r = 0, 1, 2, 3, 4$ respectively.

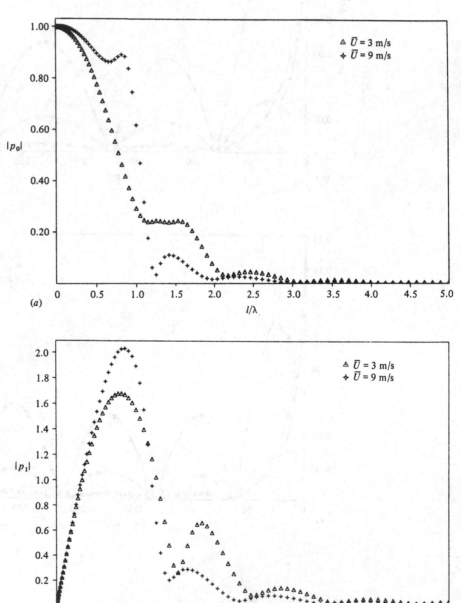

Fig. 8.9. (Continued)

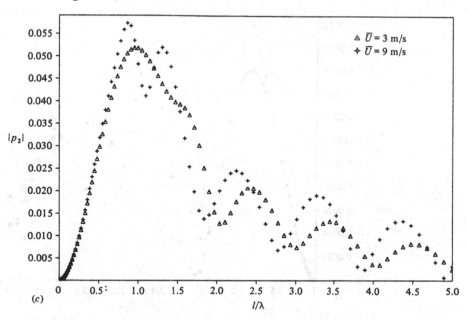

(c)

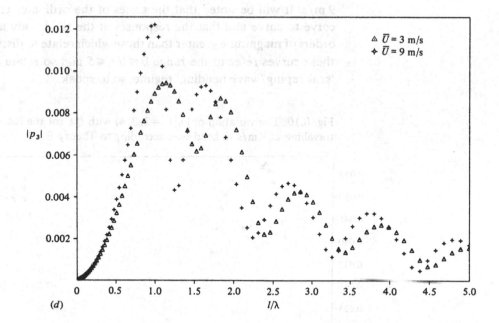

(d)

Fig. 8.9. (Continued)

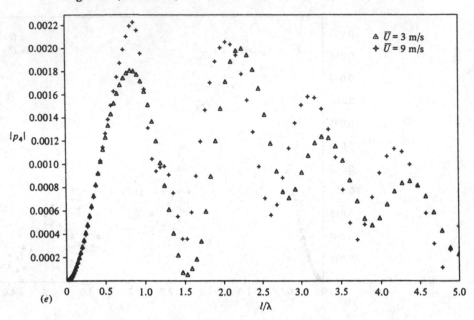

(e)

9 m/s. It will be noted that the scales of the ordinates change from curve to curve and that the responses at the rigid body motions are orders of magnitude greater than those which relate to distortions. All these curves refer to the range $0 \leqslant l/\lambda \leqslant 5$ and so relate only to the 'seakeeping/wave bending' regime, so to speak.

Fig. 8.10. The variation of $|p_r|$ $(r = 2, 3, 4)$ with l/λ for the loaded tanker travelling at 9 m/s in head seas according to Theory B.

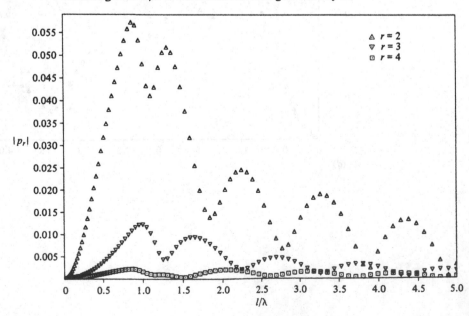

Speed has a comparatively marked influence on the quantity $|p_2|$ since the shape of the first peak alters (though there is no major variation of magnitude). The same behaviour is seen in mode 3, though it is less pronounced, and the changes appear largely to have disappeared in mode 4.

Fig. 8.10 shows the relative magnitudes of the amplitude $|p_r|$ ($r = 2, 3, 4$) when the ship speed is $\bar{U} = 9$ m/s. The appearance of the double peak in $|p_2|$ is very marked. The range of l/λ is still that of seakeeping/wave bending as in fig. 8.9. If that range is increased to encompass higher values of l/λ, the curves of fig. 8.11 are found. As one would expect, each mode now dominates one peak, the resonances falling at $l/\lambda = 8.8$, 26.2 and 49.5 as suggested in Table 8.1.

Instead of plotting $|p_r|$ against l/λ we can just as well use ω_e as the independent variable. Figs. 8.12(a)–(d) show $|p_0|$, $|p_1|$, $|p_2|$, $|p_3|$ plotted against ω_e; they relate to the destroyer travelling at 10 m/s with a heading angle $\chi = 135°$. The curves have peaks at $\omega_e = 1.5$, 8.9 and 18.3 rad/s approximately, as Table 8.1 suggests they should.

8.3.2 *Displacements*
The first of equations (8.9) shows that the amplitude of displacement at any section x is

$$w(x) = \sum_{r=0}^{n} p_r w_r(x).$$

That is to say, the amplitude is obtained from the principal modes $w_r(x)$ by adding them together with certain weighting functions p_r;

Fig. 8.11. Extension of fig. 8.10 to higher values of l/λ. It will be seen that the peaks fall at the values of l/λ that appear in Table 8.1.

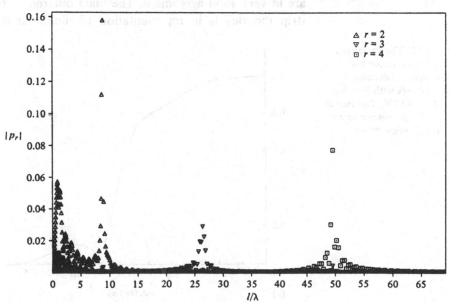

indeed at the stern, where $x = 0$, all the modal functions $w_r(x)$ are of unit value and the quantities p_r have merely to be added together. Notice, particularly, that the p_r are in general complex while the $w_r(x)$ are all real so that the extraction of the amplitude of displacement involves first addition in the complex plane and then finding the modulus of the resultant.

Plainly the values of the modal functions $w_r(x)$ are all important in the calculation of displacement. Since $w_0(x) = 1$, for all x, it is to be expected that $w_0 p_0$ makes a large contribution to the total displacement at any point of the hull unless p_0 is very small. By contrast, the displacement at $x = \bar{x}$ contains no contribution from p_1 since $w_1(\bar{x}) = 0$. Again, the deflection at $x = l/4$ is likely to be more sensitively dependent on p_0 and p_1 than it is on p_2, by reason of the various modal shapes. And so one may go on – different distortions $w_r(x) p_r(t)$ place emphasis at different points.

Fig. 8.13(a)–(d) show the amplitudes of displacement of the loaded tanker at two positions on the hull, two forward speeds and two heading angles. The two forms of strip theory employed are those of Theories A and B. It is seen that the different representations of hydrodynamic action make little difference to the calculated values of the amplitudes $|p_r|$, but that differences do tend to show up a little more in the displacement calculations for increasing speeds. Figs. 8.14(a) and (b) show the amplitude of displacement amidships in the destroyer travelling at 6 m/s and (with a more open scale of l/λ) 14 m/s in head seas, respectively. The greatest variation between the theories employed occurs in the range of low encounter frequencies (corresponding to low values of l/λ) and it increases with forward speed. At higher values of l/λ the results predicted by the two strip theories are in very good agreement. The main differences between the two strip theories is in representation of 'fluid damping' and as l/λ

Fig. 8.12. The variations of $|p_r|$ with ω_e for the destroyer proceeding at $\bar{U} = 10$ m/s with heading angle $\chi = 135°$. The curves (a)–(d) correspond to $r = 0, 1, 2, 3$ respectively.

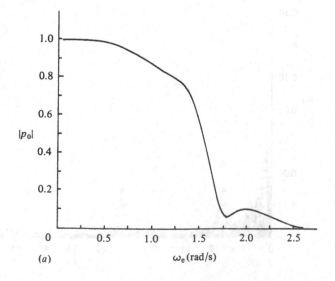

(a)

Fig. 8.12. (Continued)

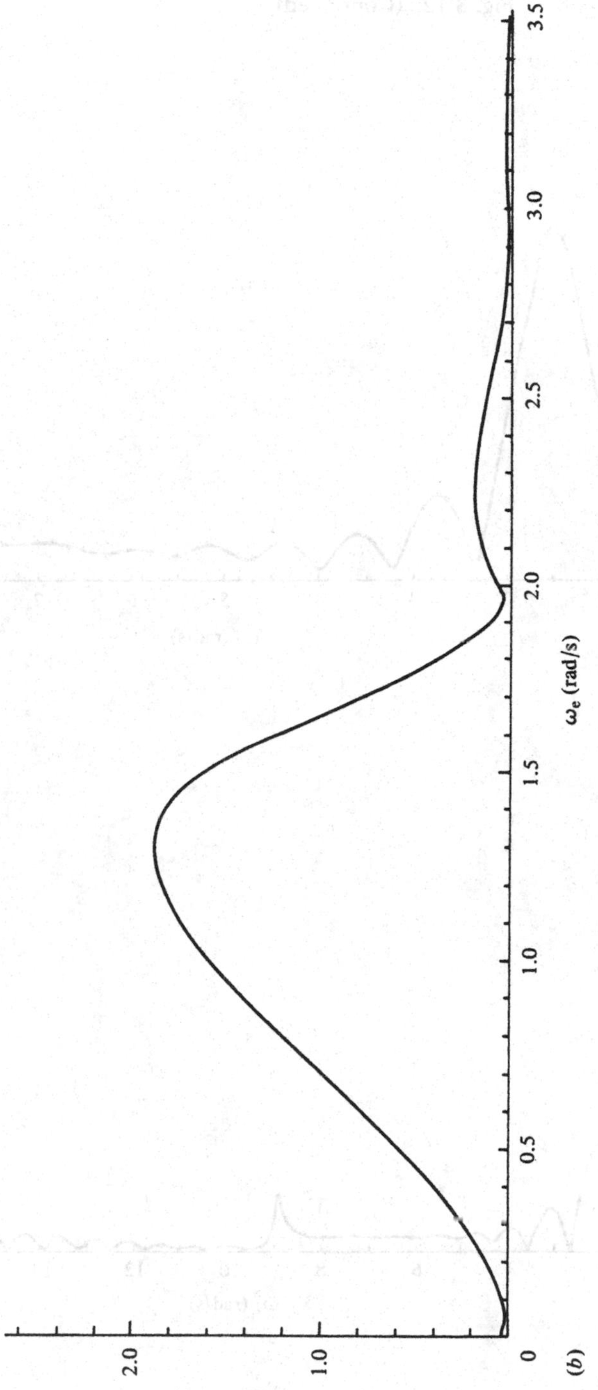

(b)

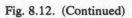

Fig. 8.12. (Continued)

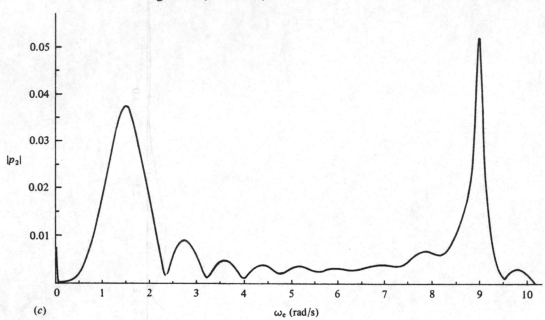

(c)

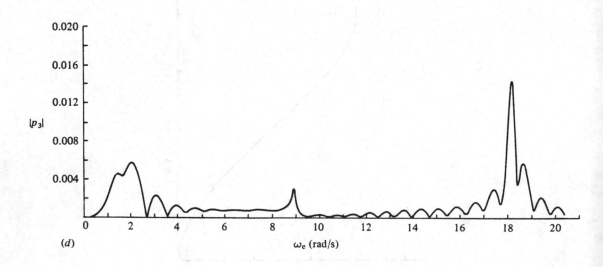

(d)

Fig. 8.13. Amplitude of displacement of loaded tanker: (a) $x = l/2$. (Note that there is no peak at $l/\lambda = 26.2$ (cf. Table 8.1) by reason of modal shape.) (b) $x = l/2$; comparable 'blow up' for small l/λ, with lower speed. (c) $x = l/4$. (Note the appearance of the peak corresponding to a dominant $r = 3$ mode at $l/\lambda = 34.4$.) (d) $x = l/4$; comparable 'blow up' for small l/λ, with lower speed.

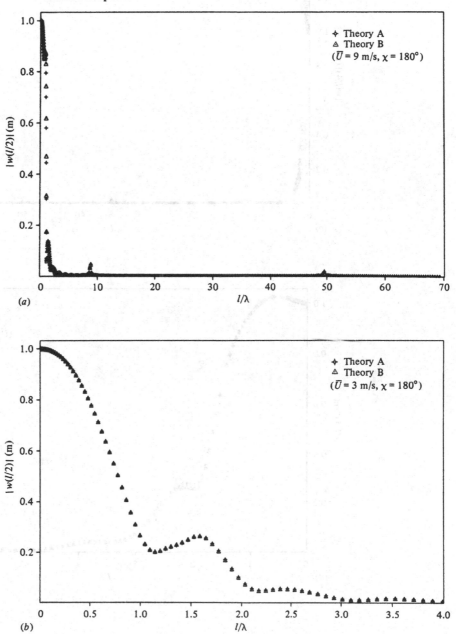

Fig. 8.13. (Continued)

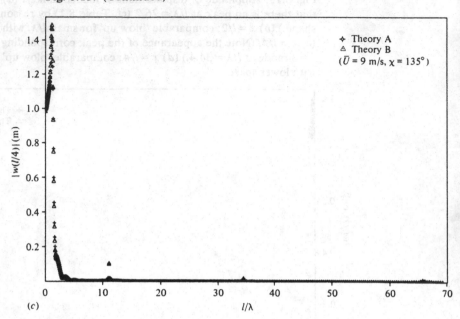

(c)

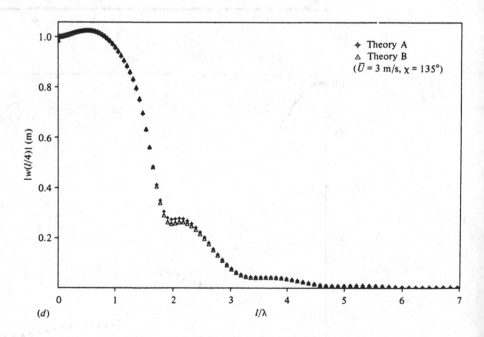

(d)

Fig. 8.14. Amplitude of displacement amidships of destroyer in head seas travelling (*a*) at 6 m/s and (*b*) at 14 m/s using a more open scale of l/λ.

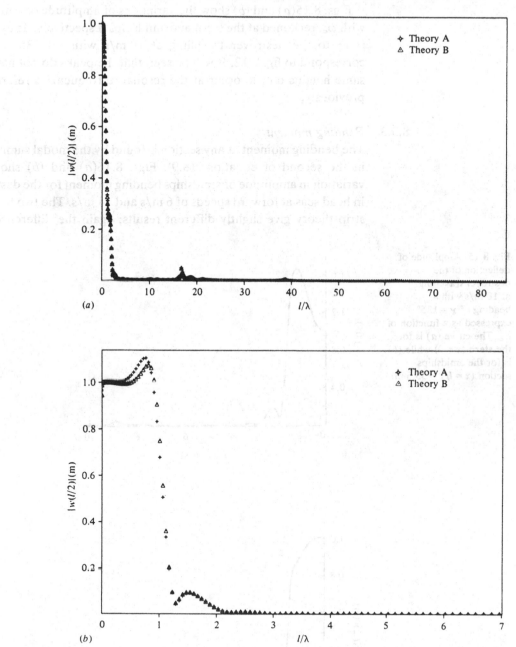

(*a*)

(*b*)

increases the effect of fluid damping decreases so that predictions on the basis of Theories A and B converge.

Figs. 8.15(a) and (b) show the variations of amplitude of deflection with ω_e, reckoned at the stern and amidships respectively. They again refer to the destroyer travelling at 10 m/s with $\chi = 135°$ and so correspond to fig. 8.12. It will be seen that the peaks do not have the same heights but do occur at the resonance frequencies referred to previously.

8.3.3 *Bending moment*
The bending moment at any section is found by the modal summation in the second of equations (8.9). Figs. 8.16(a) and (b) show the variation in amplitude of amidships bending moment for the destroyer in head seas at forward speeds of 6 m/s and 14 m/s. The two forms of strip theory give slightly different results; again the differences are

Fig. 8.15. Amplitude of deflection of the destroyer travelling at 10 m/s with a heading of $\chi = 135°$ expressed as a function of ω_e. The curve (a) is for the stern ($x = 0$) while (b) is for the amidships section ($x = l/2 = 55.5$ m).

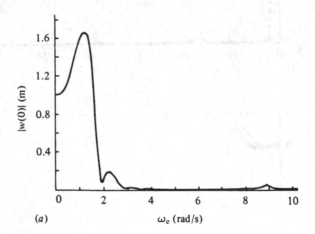

(a)

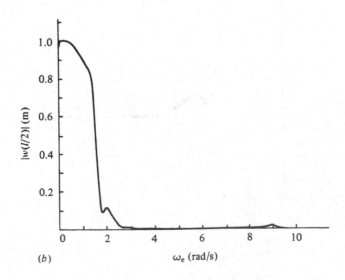

(b)

Fig. 8.16. Amplitude of bending moment amidships in the destroyer in head seas travelling (a) at 6 m/s and (b) at 14 m/s using a more open scale of l/λ.

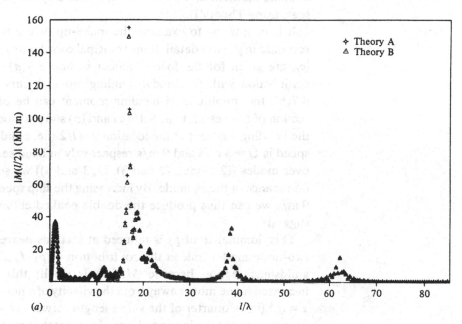

(a)

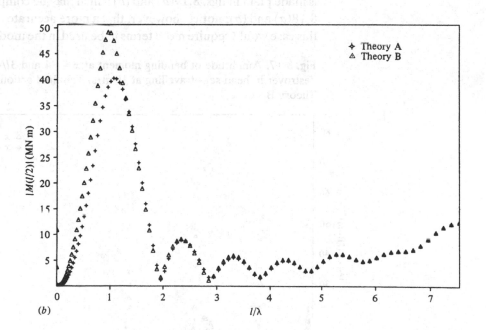

(b)

more pronounced at small values of l/λ and again they are found to increase with speed. Fig. 8.17 shows the variation in amplitude of bending moment at $x = l/4$ and $3l/4$ with a speed of 10 m/s in head seas, using Theory B.

It is of interest to examine the make-up of the bending moment response in greater detail. If the principal coordinates p_r whose moduli $|p_r|$ are given for the loaded tanker in figs. 8.9(a)–(e) are used in conjunction with the modal bending moment curves $M_r(x)$ in fig. 8.7(b), the amplitude of bending moment can be obtained for any section of the vessel. Figs. 8.18(a) and (b) show a modal synthesis for the bending moment at the location $x = l/2$ (i.e. amidships) when the speed is $\bar{U} = 3$ m/s and 9 m/s respectively in head seas. By summing over modes ((2 alone), (2 and 3), (2, 3 and 4)) we see the complete dominance of the p_2 mode. By increasing the ship speed from 3 m/s to 9 m/s we can thus produce the double peaked curve that fig. 8.9(c) suggests.

This dominance of p_2 is reduced at sections nearer a node of the two-node mode. Indeed the contribution of $p_3 M_3(x)$ is small at the amidships section because $M_3(l/2)$ is small; this contribution is increased if one moves away from the vicinity of a node. At the section $x = l/4$ (i.e. a quarter of the ship's length forward from the stern) the contributions of the lower order modes are rather more balanced. This is made plain in figs. 8.19(a) and (b) which may be compared with figs. 8.18(a) and (b); notice, however, that a more accurate specification in this case would require more terms to be used in the modal summation.

Fig. 8.17. Amplitude of bending moment at $x = l/4$ and $3l/4$ for the destroyer in head seas travelling at 10 m/s. The fluid actions are those of Theory B.

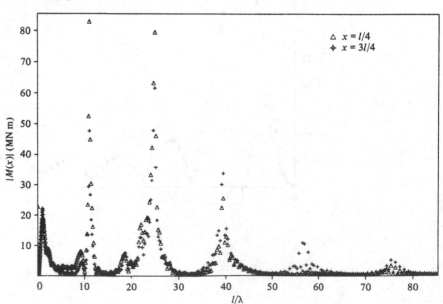

Fig. 8.18. Amplitude of bending moment amidships of the loaded tanker in head seas at (*a*) 3 m/s and (*b*) 9 m/s. By taking the highest order of mode used in the synthesis as $n = 2, 3, 4$ we see the complete dominance of the two-node dry mode in the 'wave bending' (low encounter frequency) range. Notice that increasing the ship speed from 3 m/s to 9 m/s accentuates the double peak of $|p_2|$ (see fig. 8.9(*c*)).

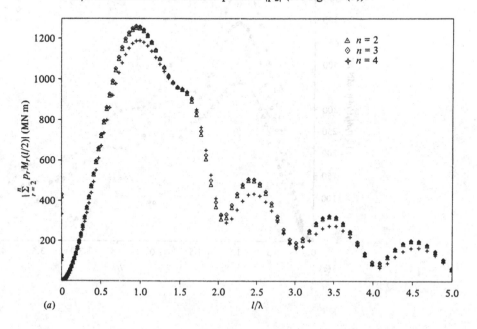

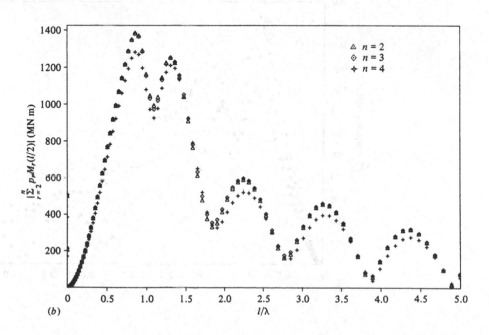

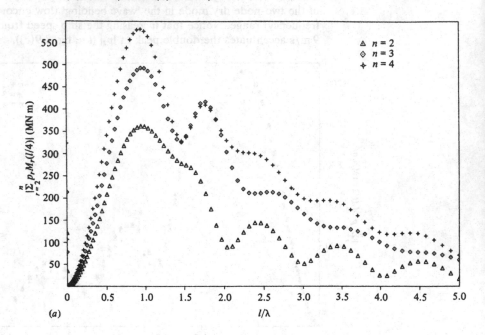

Fig. 8.19. Estimates, made by taking $n = 2, 3, 4$, of the amplitude of bending moment at the section $x = l/4$ of the loaded tanker in head seas (a) at 3 m/s and (b) at 9 m/s.

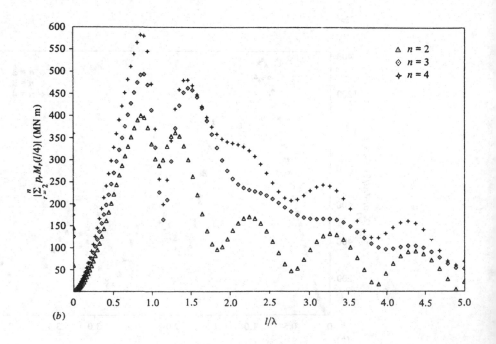

8.3.4 Shearing force

The shearing force is given by the third of equations (8.9). The results for shearing force are similar to those of bending moment. Fig. 8.20 illustrates the variation of the shearing force amplitude in the tanker in

Fig. 8.20. Amplitude of shearing force at certain sections of the tanker in ballast using Theory B, for headings $\chi = 180°$ and $135°$. (a) $x = l/2$, $\bar{U} = 3$ m/s, (b) $x = l/2$, $\bar{U} = 9$ m/s, (c) $x = 3l/4$, $\bar{U} = 3$ m/s, (d) $x = l/4$, $\bar{U} = 9$ m/s.

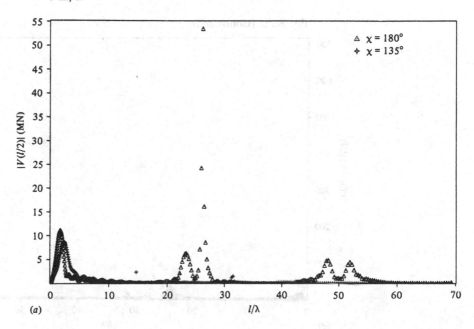

(a)

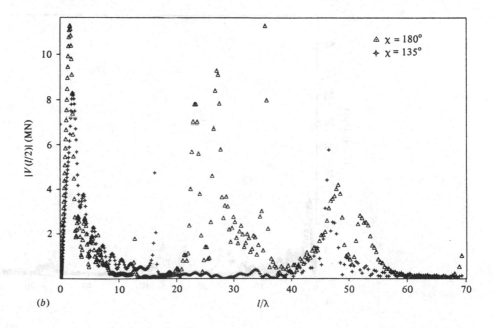

(b)

ballast for headings of 135° and 180°, using Theory B;

(a) gives amplitude at $x = l/2$ for 3 m/s,
(b) gives amplitude at $x = l/2$ for 9 m/s,
(c) gives amplitude at $x = 3l/4$ for 3 m/s,
(d) gives amplitude at $x = l/4$ for 9 m/s.

Fig. 8.21 shows the amplitude of shearing force amidships in the loaded tanker when it is found by summing the modes (2 alone), (2 and

Fig. 8.20. (Continued)

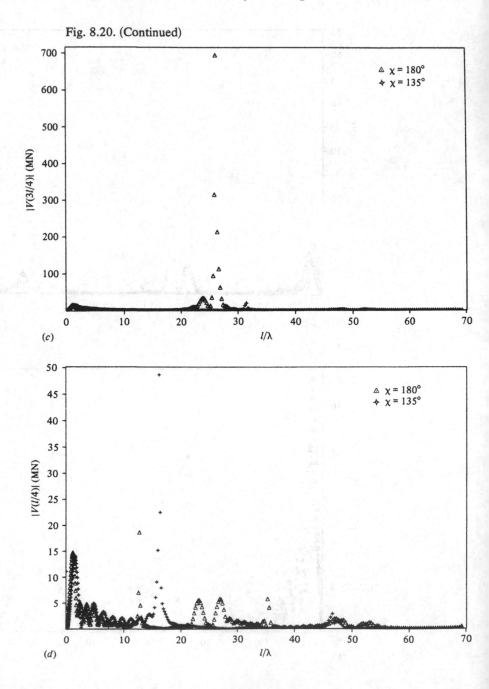

3) and (2, 3 and 4). The ship is assumed to encounter head seas and to travel (*a*) at 3 m/s and (*b*) at 9 m/s; then the curves of figs. 8.9(*a*)–(*e*) and 8.7(*c*) are relevant. Fig. 8.22 shows comparable results for the section $x = l/4$. It is seen that, whereas mode 2 plays a fairly modest

Fig. 8.21. Amplitude of shearing force amidships in the loaded tanker travelling at (*a*) 3 m/s and (*b*) 9 m/s in head seas (using Theory B). The summation is made over the modes (2 alone), (2 and 3), (2, 3 and 4); it will be seen that in neither case does a contribution $p_4 V_4(l/2)$ make any discernible difference to the result.

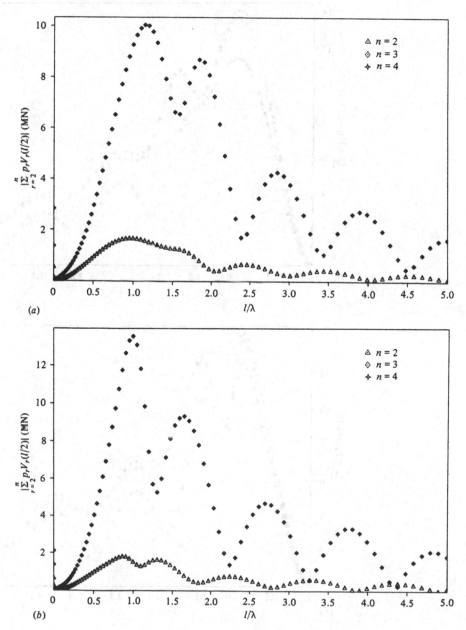

part in determining the shearing force amidships at either speed, the contribution of that mode to the shearing force at $x = l/4$ is dominant at both speeds.

8.4 Particular features of ship response

The possession of a general theory of ship response tempts one to examine a number of matters by way of 'parametric studies'. Such is

Fig. 8.22. Results comparable with those of fig. 8.21, but relating to the section $x = l/4$.

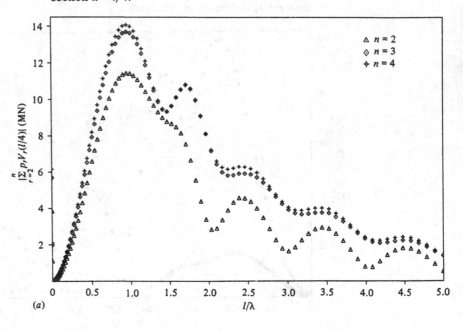

(a)

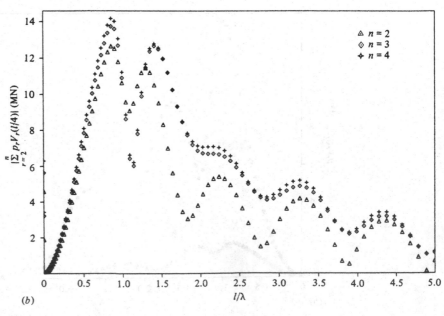

(b)

the nature of this field, however, such studies can only be made in a numerical form.

8.4.1 *Structural damping*

On the face of it, structural damping fills a very important role in ship dynamics, yet our understanding of it is very sketchy, as we noted in chapter 4. It is therefore natural to enquire how serious this state of affairs is.

Fig. 8.23. Bending moment amidships in head seas according to Theory B assuming the structural damping to be ν times the value quoted in Table 4.11 for (a) the destroyer at 10 m/s and (b) the loaded tanker at 6 m/s.

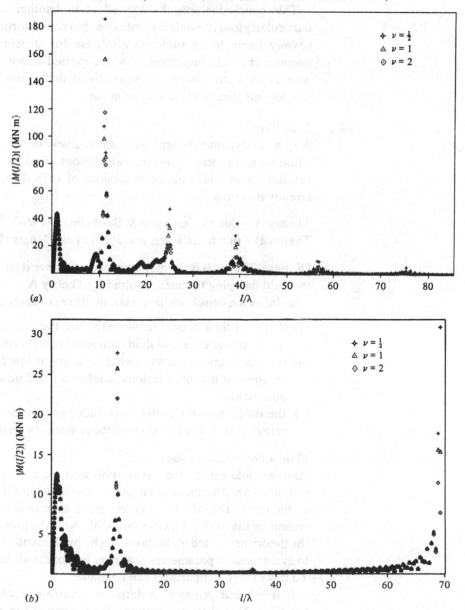

The values of structural damping used in the calculations may be varied and the resulting values of, for example, bending moment amidships compared. The results shown in figs. 8.23(a) and (b) are for the destroyer and loaded tanker respectively, in head seas. It will be seen that, for small values of l/λ, the bending moment is hardly changed, no matter what level of structural damping was chosen. For larger values of l/λ the magnitudes of the response at resonance are affected, though the corresponding shifting of the peaks along the l/λ axis is negligible. From these findings it appears that fluid damping is dominant when l/λ is small but structural damping becomes more and more significant with increasing l/λ.

This conclusion may be looked at in another way. Only in a particularly long, flexible and massive ship are distortions in a realistic seaway likely to be sufficiently intense for structural damping to become crucially important in wave-excited distortion. Usually, it would appear, the dominant frequencies of distortions in waves are too low for this parameter to matter much.

8.4.2   *Strip theories*

Any hydrodynamic theory – even a non-linear one if it were desired – could be incorporated into the calculations. Two have actually been catered for in this book, both versions of strip theory. They are, as already mentioned,

Theory A – due to Gerritsma & Beukelman (1964)
Theory B – due to Salvesen et al. (1970) and Vugts (1971).

Of these, Theory B is the more complicated since it adds extra terms in the fluid damping to those required by Theory A.

So far as the vessels we have examined are concerned, it appears that

(a) there is little to choose between the two theories for high values of l/λ since the effects of fluid damping are then small.
(b) their predictions are somewhat different at low l/λ, where wave response is the more serious problem since fluid damping is then dominant.
(c) the differences are in the magnitudes of resonant peaks of the p_r rather than in the locations of those peaks in terms of l/λ.

8.4.3   *Conventional calculations*

As one would expect, there is an enormous literature on the subject of ship strength. The theories employed are dealt with in books on naval architecture and are of a much more empirical nature than the technique that is developed in this book. Now the programs with which the theory presented in this book can be implemented can also be used to evaluate such parameters as appear in more traditional approaches. To this extent comparisons can be made.

It is not our purpose to derive the more conventional theories, however. Nor shall we examine their accuracy in the light of the dynamics approach used in this book, though this topic is touched upon to some extent in chapter 11. The interested reader will find such a comparison elsewhere (Bishop, Price & Tam, 1977).

8.4.4 Total hydrodynamic action

In chapter 7 we showed that, excluding the hydrostatic contribution, the total fluid action per unit length exerted by the sea on a ship hull may be expressed as the sum of two components:

$$F(x, t) = -H[w(x, t)] + Z[\zeta(x, t)]. \tag{8.12}$$

The quantity H is determined by the motion of the ship while Z depends on the incident waves. This total upward force per unit length is complex and depends on the wave system and the operating conditions.

Fig. 8.24 shows curves of $|w_0(x)F(x)|$ plotted for positions along the hull of the loaded tanker which is travelling at 9 m/s in head seas. The distribution naturally varies with l/λ and it is shown for $l/\lambda = 1.0$, 8.8 and 15.0. It will be seen that the fluctuation of $|w_0(x)F(x)|$ for $l/\lambda = 8.8$ is large; this value of l/λ corresponds to the resonant encounter frequency $\omega_e = 2.67$ rad/s which appears in Table 8.1.

8.4.5 Improvement of hull section representation

It has been mentioned that the results presented in this chapter have been computed for Lewis-form fits of the destroyer and tanker hulls. In chapter 7, however, we showed that more realistic representations of

Fig. 8.24. The total upward fluid force per unit length distributed along the hull of the loaded tanker travelling at 9 m/s in head seas, reckoned for three values of l/λ. Notice that the curve for $l/\lambda = 8.8$ corresponds to a condition of resonant encounter..

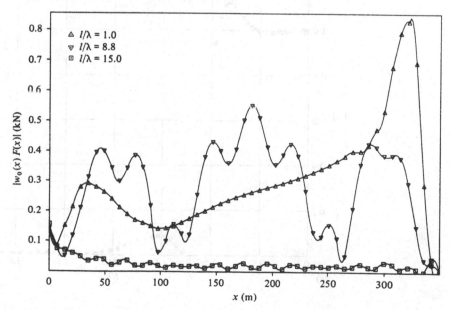

hull sections may be employed using a close-fit conformal mapping. It is natural to enquire to what extent use of the more advanced technique alters the predicted ship responses.

Using the destroyer data, and assuming that the ship proceeds at 14 m/s in a head sea of 1 m amplitude, we shall compare the predictions arrived at by the use of Lewis forms and the close-fit technique. In the latter, the number of parameters used varied from section to section but was usually six. Such were the lines of the destroyer, the Lewis-form fit was usually good while the six-parameter fit was extremely accurate. This comparative study therefore provides a searching test of the usefulness of close-fit mapping as it is to be expected that failure to use a close-fit method when hull sections are less well-conditioned could lead to much more serious errors.

Figs. 8.25(a)–(e) show the variation of $|p_r|$ with l/λ (with $r = 0, 1, 2, 3, 4$ respectively). Inspection reveals that in only one respect is there any serious discrepancy between predictions of the Lewis-form and close-fit predictions. There is an error of approximately 25% in the height of the highest peak in the curve of $|p_2|$. Apart from this the destroyer gives no justification for use of the more complicated conformal mapping in the computation of the $|p_r|$. (Note that this conclusion is not necessarily true for all types of hulls as discussed by Hoffman & van Hooff (1976).)

The amplitude of deflection amidships $|w(l/2)|$ is shown plotted against l/λ in fig. 8.26. It will be seen that the Lewis-form fit is perfectly adequate for this particular case.

Fig. 8.25. Comparison of predictions made with Lewis-form fits and with close-fit conformal mapping of the quantities $|p_r|$ for the destroyer travelling at 14 m/s in head seas. (a) $|p_0|$, (b) $|p_1|$, (c) $|p_2|$, (d) $|p_3|$ and (e) $|p_4|$.

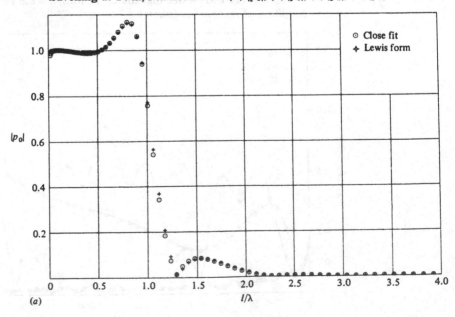

Fig. 8.25. (Continued)

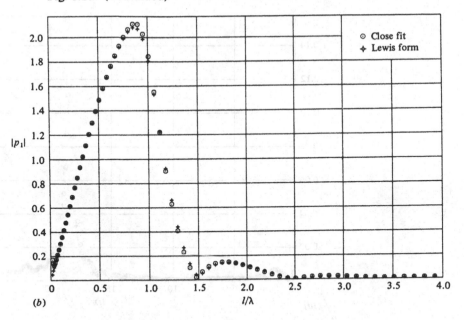

(b)

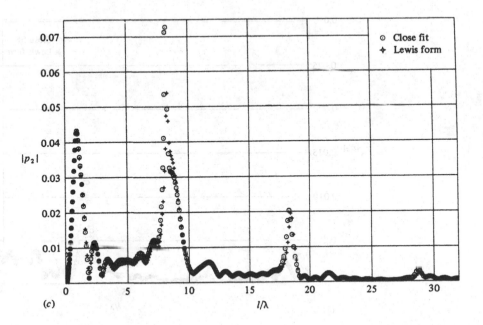

(c)

Fig. 8.25. (Continued)

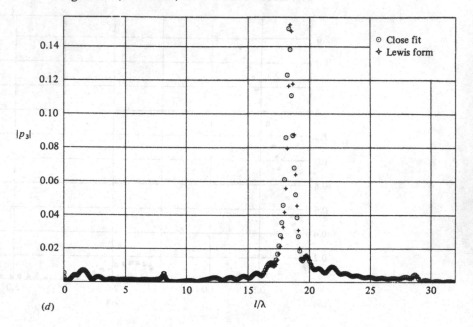

(d)

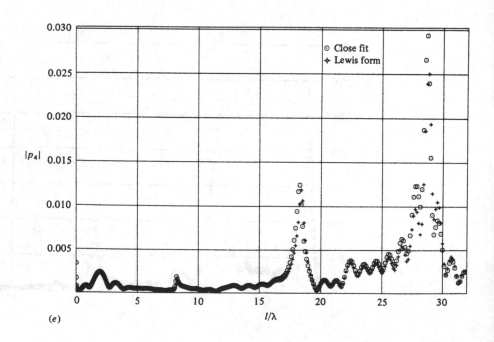

(e)

As one would expect, the discrepancy that we found in the p_2 peak is reflected in the amplitude of bending moment amidships. As fig. 8.27 shows, there is a disparity of about 25% in the height of the peak in the vicinity of $l/\lambda = 8$.

The shearing force amidships does not depend significantly on the deflection in the second dry mode since $V_2(l/2)$ is small. Accordingly the Lewis form gives as good a prediction as the close-fit conformal transformation for this particular section. This is shown in fig. 8.28. At other positions on the hull where $V_2(x)$ is not small (e.g. in the region $x = l/4$ or $3l/4$) discrepancies that are similar to those found in the bending moment at $x = l/2$ again appear.

To give some idea of the differences of representation that have produced the results shown in figs. 8.25–8.28, fig. 8.29 shows results for one particular section of the destroyer. The section is at station 4 of the hull (for which the distance from the stern is $x = 27.58$ m). The six-parameter mapping gives a significantly better fit to the hull offsets than the Lewis form does, as fig 8.29(a) shows. The corresponding differences of predicted added mass and fluid damping are shown in figs. 8.29(b) and (c) respectively; in both, the discrepancy is significant when $\omega^2 B/2g$ is equal to 14.0.

Fig. 8.30 shows the distributions of added mass $m(x)$ for infinite encounter frequency corresponding to the Lewis-form and the close-fit mapping techniques. It will be seen that the discrepancy in this case is fairly small.

Fig. 8.26. Amplitude of deflection amidships as a function of l/λ for the destroyer at 14 m/s in head seas. The close-fit technique and Lewis forms give almost identical results.

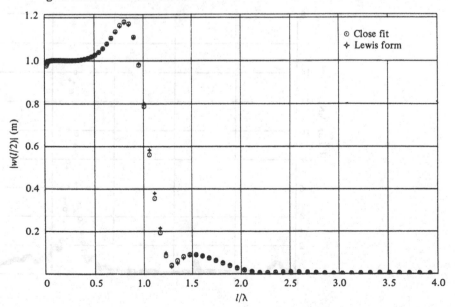

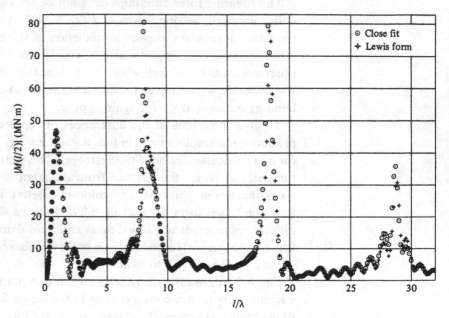

Fig. 8.27. Amplitude of bending moment amidships as a function of l/λ for the destroyer at 14 m/s in head seas. The discrepancy between the predictions made with the close-fit technique and with Lewis forms in the quantity $|p_2|$, shown in fig. 8.25(c), are reflected in a fairly significant discrepancy in the predictions of $|M(l/2)|$ at $l/\lambda \doteq 8$.

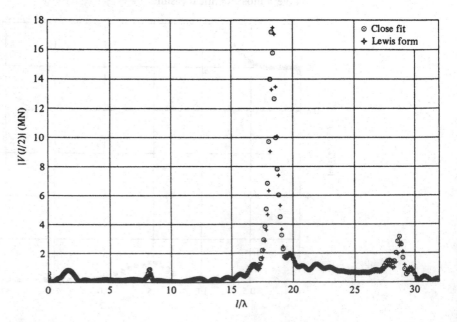

Fig. 8.28. Amplitude of shearing force amidships as a function of l/λ for the destroyer at 14 m/s in head seas. Here the Lewis-form and close-fit techniques give almost the same predictions.

Fig. 8.29. The section at station 4 of the destroyer (where $x = 27.58$ m).
(a) The offsets are represented by spots, the full line is for a 6-parameter
conformal mapping and the broken line refers to Lewis-form fits. (b)
Corresponding curves of non-dimensional added mass. (c) Corresponding
curves of the square of the 'amplitude ratio', $\bar{A}^2$; this is a non-dimensional
measure of the fluid damping.

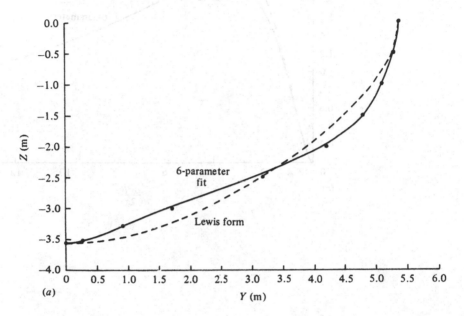

(a)

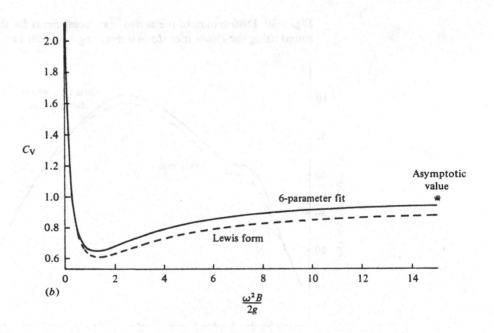

(b)

Fig. 8.29. (Continued)

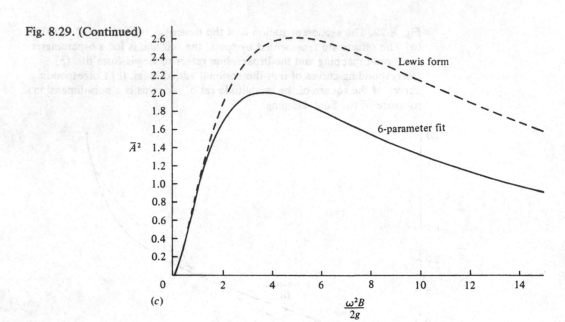

$\overline{A}^2$

Lewis form

6-parameter fit

$\dfrac{\omega^2 B}{2g}$

(c)

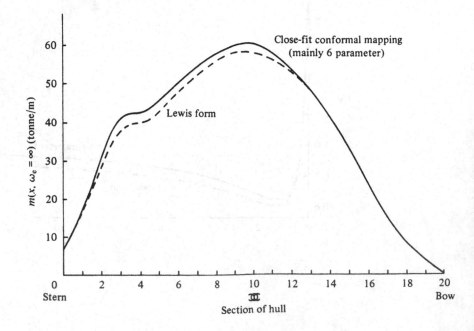

Fig. 8.30. Distribution of the added mass coefficients for the destroyer found using the close-fit conformal mapping and with Lewis forms.

Close-fit conformal mapping
(mainly 6 parameter)

Lewis form

$m(x, \omega_e = \infty)$ (tonne/m)

Stern

Ⅲ

Bow

Section of hull

8.5 Polar plotting of ship response

In several respects it is more illuminating to plot p_r on an Argand diagram than it is to show $|p_r|$ plotted against l/λ or ω_e, as we did in section 8.3.1. A form of resonance curve is used, to which values of ω_e may be attached at convenient equal intervals. This approach is widely employed in mechanical and aeronautical engineering since the properties of the polar plots not only give a clear picture of where resonances occur but also provide useful means of identifying principal modes (even when the system has close natural frequencies), giving estimates of modal damping factors and accurately identifying resonance frequencies.

A convenient starting point is to be found in the linear theory of non-conservative systems developed by Fawzy & Bishop (1976). Consider a system having n degrees of freedom, whose motion is governed by the equation

$$\mathbf{A\ddot{q} + B\dot{q} + Cq = \Phi}\, e^{i\omega t}, \tag{8.13}$$

where the $n \times n$ system matrices $\mathbf{A}, \mathbf{B}, \mathbf{C}$ are just arrays of real constants possessing no particular properties of symmetry or positive definiteness. We first note that the homogeneous equation

$$\mathbf{A\ddot{q} + B\dot{q} + Cq = 0}$$

has eigenvalues

$$\lambda_1, \bar{\lambda}_1 = \mu_1 \pm i\nu_1,$$
$$\lambda_2, \bar{\lambda}_2 = \mu_2 \pm i\nu_2,$$
$$\cdots \cdots \cdots$$
$$\lambda_n, \bar{\lambda}_n = \mu_n \pm i\nu_n,$$

each root having associated eigenvectors. The pair of eigenvalues $\lambda_r, \bar{\lambda}_r$ define a complex square matrix $\mathbf{W}^{(r)} = \mathbf{U}^{(r)} + i\mathbf{V}^{(r)}$, where $\mathbf{U}^{(r)}, \mathbf{V}^{(r)}$ are real and are determined by $\mathbf{A}, \mathbf{B}, \mathbf{C}$.

It may be shown from the theory referred to that the steady-state solution of equation (8.13) is capable of being written in the form

$$\mathbf{q} = \left\{ \sum_{r=1}^{n} \frac{\left[\mathbf{R}^{(r)} + i\dfrac{\omega}{\Omega_r}\mathbf{S}^{(r)} \right] e^{-i\theta_r}}{\left[\left(1 - \dfrac{\omega^2}{\Omega_r^2}\right)^2 + \dfrac{4\zeta_r^2(\omega^2)}{\Omega_r^2} \right]^{\frac{1}{2}}} \right\} \mathbf{\Phi}\, e^{i\omega t}. \tag{8.14}$$

In this result

$$\left. \begin{aligned} \Omega_r^2 &= \mu_r^2 + \nu_r^2 = |\lambda_r|^2 = |\bar{\lambda}_r|^2, \\ \zeta_r &= -\frac{\mu_r}{\Omega_r}, \\ \theta_r &= \tan^{-1}\left[\frac{2\zeta_r \omega/\Omega_r}{1 - (\omega/\Omega_r)^2} \right], \end{aligned} \right\} \tag{8.15}$$

while $\mathbf{R}^{(r)}, \mathbf{S}^{(r)}$ are constant $n \times n$ real matrices defined by $\mathbf{U}^{(r)}, \mathbf{V}^{(r)}$ and $\lambda_r, \bar{\lambda}_r$.

The contents of the curly brackets in equation (8.14) represent a complex square receptance matrix α, each element being the sum of n terms. In the construction of α, the matrix

$$\mathbf{R}^{(r)} + i\frac{\omega}{\Omega_r}\mathbf{S}^{(r)}$$

displays no particularly unusual features as ω/Ω_r passes through the value unity. By contrast the multiplier

$$G_r = \frac{e^{-i\theta_r}}{\left[\left(1 - \frac{\omega^2}{\Omega_r^2}\right)^2 + \frac{4\zeta_r^2\omega^2}{\Omega_r^2}\right]^{\frac{1}{2}}} \tag{8.16}$$

is very well known and provides all the familiar features of resonance. It is this multiplier which ensures that 'resonance loops' appear in a polar plot of any given receptance as ω passes through $\Omega_1, \Omega_2, \ldots, \Omega_n$, provided the system is stable (so that all the μ_r are negative and the ζ_r positive). It is this conclusion which, couched in rather different terms and arrived at in a different way, is reached by Fawzy & Bishop (1977).

Briefly, and roughly, the behaviour of G_r is such that if this quantity is plotted on an Argand diagram its modulus remains small until $\omega \to \Omega_r$. As this resonance condition is approached the modulus increases greatly while $\arg(G_r)$ passes the value $3\pi/2$. The locus near resonance is nearly circular and the frequency spacing opens out to a maximum at Ω_r. If such a locus is obtained, the rth damping factor ζ_r may be ascertained by noting ω_a and ω_b, the values of driving frequency for which $\arg(G_r)$ is $7\pi/4$ and $5\pi/4$ respectively; then

$$\frac{2\zeta_r\omega_a}{\Omega_r} = 1 - \left(\frac{\omega_a}{\Omega_r}\right)^2,$$

$$\frac{2\zeta_r\omega_b}{\Omega_r} = -\left[1 - \left(\frac{\omega_b}{\Omega_r}\right)^2\right],$$

whence

$$\zeta_r = \frac{\omega_b - \omega_a}{2\Omega_r}.$$

Thus in the sketch (fig. 8.31) the damping factor indicated is of the order

$$\frac{(1.2 - 0.8)\Omega_r}{2\Omega_r} \quad \text{or } 0.2.$$

A ship under way has an equation of motion of the type (8.13). The system matrices, denoted by $\mathcal{A}, \mathcal{B}, \mathcal{C}$, are all sums of two matrices, one of structural origin and one of hydrodynamic origin. The matrix $\mathbf{q}$ is identified with the vector of principal coordinates of the dry hull, the symbol $\mathbf{p}$ being employed. The driving frequency ω is identified with the encounter frequency ω_e and the excitation is written in the form

$\Xi\,e^{-i\omega_e t}$ where the vector Ξ is dependent on ω_e. The comparable equation for a ship is therefore

$$\mathscr{A}(\omega_e)\ddot{\mathbf{p}}+\mathscr{B}(\omega_e)\dot{\mathbf{p}}+\mathscr{C}\mathbf{p}=\Xi(\omega_e)\,e^{-i\omega_e t}, \tag{8.17}$$

where, in general, the real system matrices $\mathscr{A}$ and $\mathscr{B}$ are ω_e-dependent for a given speed $\bar{U}$ and heading χ in a long-crested sinusoidal sea; the complex vector Ξ, too, is dependent on ω_e, $\bar{U}$ and χ.

We may now proceed to solve equation (8.17) by analogy with the previous simpler case. If the equation of motion is written in the form

$$\mathbf{Dp}=\Xi\,e^{-i\omega_e t}$$

then

$$\mathbf{p}=\mathbf{D}^{-1}\Xi\,e^{-i\omega_e t}\equiv\alpha\Xi\,e^{-i\omega_e t}.$$

Evidently if we contemplate one fixed set of operating conditions $\bar{U}$, χ and if we admit only heave, pitch and $n-1$ symmetric distortion modes, we shall find

$$\mathbf{p}=\left\{\sum_{r=0}^{n}\frac{\left[\mathscr{R}^{(r)}(\omega_e)-i\dfrac{\omega_e}{\Omega_r}\mathscr{S}^{(r)}(\omega_e)\right]e^{i\theta_r}}{\left[\left(1-\dfrac{\omega_e^2}{\Omega_r^2}\right)^2+\dfrac{4\zeta_r^2\omega_e^2}{\Omega_r^2}\right]^{\frac{1}{2}}}\right\}\Xi(\omega_e)\,e^{-i\omega_e t}, \tag{8.18}$$

where

$$\theta_r=\tan^{-1}\left[\frac{2\zeta_r\omega_e/\Omega_r}{1-(\omega_e/\Omega_r)^2}\right]. \tag{8.19}$$

So familiar is polar plotting for simple passive systems in which **A, B, C** of equation (8.13) are positive definite, it is unnecessary to do more than present results for a particular ship and comment on any special feature that they display, wherever appropriate. The vessel is the destroyer and we shall again assume that it proceeds at 14 m/s in a

Fig. 8.31. A sketch showing how the complex multiplier G_r might be expected to vary as ω, the driving frequency, passes through the resonance frequency Ω_r.

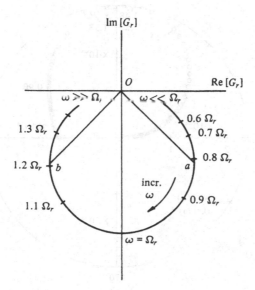

sinusoidal head sea whose amplitude of elevation is 1 m. The results are based on the close-fit hull representation referred to in section 8.4.5, using strip theory B; they were initially published by Bishop, Price, Tam & Temarel (1978).

To avoid congestion of the polar plots (which is very necessary in practice) each polar plot is given with three ranges of encounter frequency. In each range, equal increments of 0.02 rad/s in the encounter frequency are used, and in all the computations the modes of order 0, 1, 2, 3 only will be admitted – almost certainly all that matter in a practical analysis of the ship in question.

8.5.1   *Direct receptances*
The receptance matrix α is of the form

$$\alpha(\omega_e) = \begin{bmatrix} \alpha_{00} & \alpha_{01} & \alpha_{02} & \alpha_{03} \\ \alpha_{10} & \alpha_{11} & \alpha_{12} & \alpha_{13} \\ \alpha_{20} & \alpha_{21} & \alpha_{22} & \alpha_{23} \\ \alpha_{30} & \alpha_{31} & \alpha_{32} & \alpha_{33} \end{bmatrix}, \tag{8.20}$$

each element being the sum of four modal contributions. The elements in the leading diagonal are the 'direct receptances'. It is instructive to examine the polar plots of these quantities.

The curves for $\alpha_{00}(\omega_e)$ are shown in fig. 8.32. Except at very small values of ω_e, the loops are very nearly circular with well-defined resonances occurring at

$$\omega_e = 1.45, \ 8.88 \text{ and } 18.20 \text{ rad/s}.$$

There are thus three distinct resonances even though there are four coordinates used and the range of ω_e is sufficient to encompass the

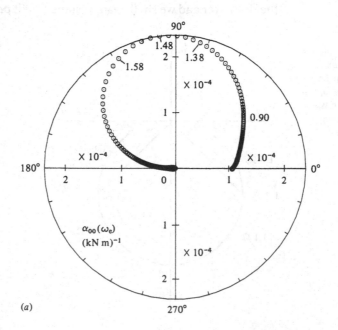

Fig. 8.32. Polar plots of $\alpha_{00}(\omega_e)$ for the destroyer travelling at 14 m/s in head seas. (*a*) the encounter frequency ω_e increases from 0.04 in steps of 0.02 to 7.04 rad/s; i.e. $\omega_e =$ 0.04 (0.02) 7.04 rad/s. There is a resonance at about $\omega_e = 1.45$ rad/s. (*b*) $\omega_e = 7.20$ (0.02) 14.02 rad/s. There is a resonance at about $\omega_e =$ 8.88 rad/s. (*c*) $\omega_e = 14.20$ (0.02) 21.02 rad/s. There is a resonance at about $\omega_e = 18.20$ rad/s.

(*a*)

Fig. 8.32. (Continued)

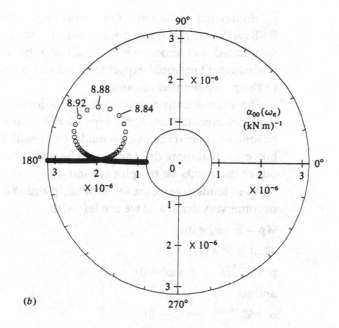

(b)

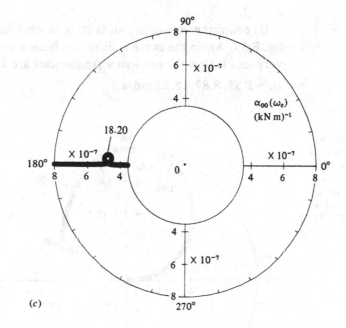

(c)

p_3-dominated resonance. On consulting Table 8.1 we notice that 8.88 rad/s corresponds approximately to resonance of the p_2-dominated wet mode while 18.20 can be identified with the p_3 resonance. One would expect the 1.45 rad/s resonance to correspond to the p_1-dominated response.

There is evidently no well-defined p_0-dominated resonance and this is not unexpected since the dependence of the inertia matrix $\mathscr{A}(\omega_e)$ becomes very marked as ω_e is made very small. It is of interest to note, however, that there does appear to be a slight resonance-like opening out of the points on the plot around $\omega_e = 0.90$ rad/s.

As ω_e tends to zero, the terms $\mathscr{A}(\omega_e)\ddot{\mathbf{p}}$ and $\mathscr{B}(\omega_e)\dot{\mathbf{p}}$ in equation (8.17) become very small and we are left with

$$\mathscr{C}\mathbf{p} = \Xi \quad (\omega_e = 0).$$

That is to say

$$\mathbf{p} = \mathscr{C}^{-1}\Xi \quad \text{(as } \omega_e \to 0)$$

and so

$$\alpha = \mathscr{C}^{-1} \quad \text{(as } \omega_e \to 0).$$

It follows that, as ω_e becomes very small, the receptances assume real and constant values. We notice that α_{00}, in particular, approaches a value of approximately 10^{-4} (kN m)$^{-1}$ which is the order of magnitude of $\mathscr{C}_{00}^{-1}$.

By contrast with α_{00}, α_{11} starts at the origin when $\omega_e = 0$, as shown in fig. 8.33. Again the curve is distorted from a circular arc when ω_e is very small. Now the resonance frequencies are at

$$\omega_e = 1.55, \ 8.87, \ 18.22 \ \text{rad/s}.$$

Fig. 8.33. Polar plot of $\alpha_{11}(\omega_e)$ for the destroyer travelling at 14 m/s in head seas. (a) $\omega_e = 0.04$ (0.02) 7.04 rad/s, (b) $\omega_e =$ 7.20 (0.02) 14.02 rad/s, (c) $\omega_e = 14.20$ (0.02) 21.02 rad/s.

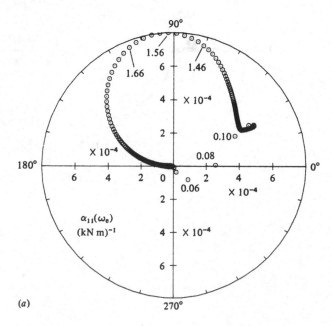

$\alpha_{11}(\omega_e)$
(kN m)$^{-1}$

(a)

Fig. 8.33. (Continued)

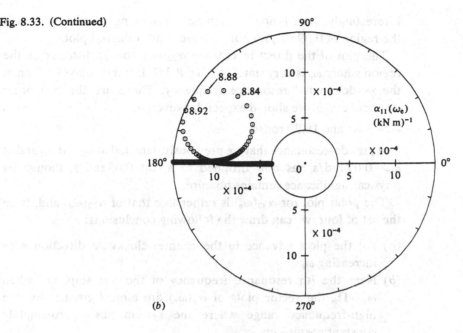

(b)

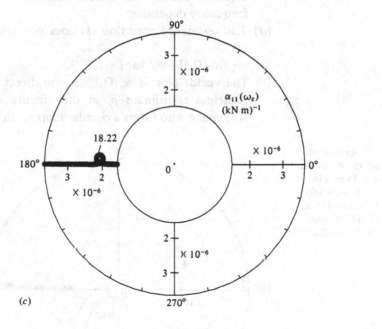

(c)

Interestingly, there is now a much clearer opening out of the points in the region of 0.09 rad/s – not 0.90 rad/s as in the α_{00} plot.

The plot of the direct receptance $\alpha_{22}(\omega_e)$, too, is distorted in the region where ω_e is very small (see fig. 8.34). But it displays no loop at the p_1-dominated resonance frequency. There are the two other resonances that we should expect, however, at

$$\omega_e = 8.87 \text{ and } 18.20 \text{ rad/s.}$$

The pseudo-resonance that we previously detected at $\omega_e = 0.90$ and at $\omega_e = 0.09$ rad/s has now dropped to about 0.05 rad/s, though its physical significance remains obscure.

The polar plot for $\alpha_{33}(\omega_e)$ is rather like that of $\alpha_{22}(\omega_e)$ and, from the set of four, we can draw the following conclusions:

(a) All the plots advance in the counter-clockwise direction with increasing ω_e.

(b) Near the ith resonance frequency of the wet ship, i.e. when $\omega_e \to \Omega_i$, the vector plots of $\alpha_{ii}(\omega_e)$ are almost circular for the high-frequency range where the system has approximately constant coefficients.

(c) When the encounter frequency is low, the resonance circle is distorted because the system matrices $\mathscr{A}$ and $\mathscr{B}$ are encounter frequency dependent.

(d) The resonance condition Ω_0 does not arise in the conventional way.

(e) $\arg[\alpha_{ii}(\Omega_i)] \doteqdot 90°$ for $i = 1, 2, 3$.

(f) The vector plot of $\alpha_{ii}(\Omega_j)$, i.e. the direct receptance at the ith principal coordinate p_i in the vicinity of the jth resonance frequency, also forms a circular loop when $j = 1, 2, 3$. This loop is

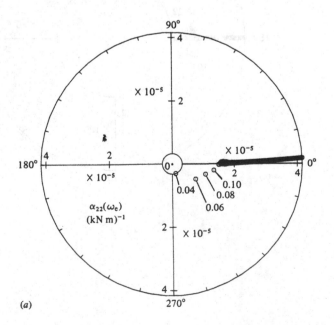

Fig. 8.34. Polar plot of $\alpha_{22}(\omega_e)$ for the destroyer travelling at 14 m/s in head seas. (a) $\omega_e = 0.04$ (0.02) 7.04 rad/s, (b) $\omega_e = 7.20$ (0.02) 14.02 rad/s, (c) $\omega_e = 14.20$ (0.02) 21.02 rad/s.

(a)

Fig. 8.34. (Continued)

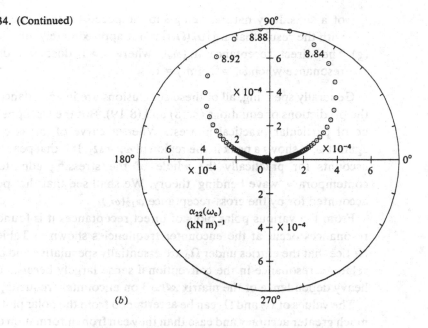

(b)

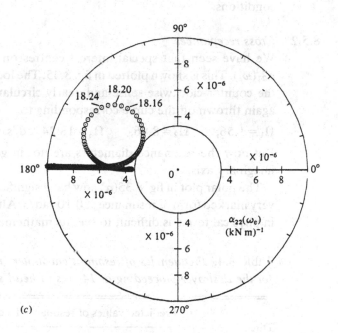

(c)

of a subsidiary nature, being superimposed on the main curve, with the result that $\arg[\alpha_{ii}(\Omega_j)]$ is not approximately 90°.

(g) The direct receptance $\alpha_{ii}(\omega_e)$, where $i > 1$ does not display resonance when $\omega_e = \Omega_j$ for $j = 1$.

Generally speaking, all of these conclusions are in accordance with the predictions of equations (8.18) and (8.19). But the last appears to be of particular practical interest. When a curve of $|p_2|$ is plotted against ω_e it shows a peak in the region of $\omega_e = \Omega_1$. It is that peak which accounts for practically the whole of the stressing admitted by contemporary 'wave bending' theory. We shall see that that peak is accounted for by the cross-receptance $\alpha_{12}(\omega_e)$.

From the various polar plots of direct receptances it is found that resonances occur at the encounter frequencies shown in Table 8.4. Notice that the entries under Ω_0 are essentially speculative and do not relate to resonance in the conventional sense largely because of the heavy dependence of the matrix $\mathcal{A}(\omega_e)$ on encounter frequency.

The values of Ω_2 and Ω_3 can be ascertained from the polar plots with much greater accuracy and ease than they can from information on the moduli of the receptances. They correspond very closely to the values quoted in Table 8.1 and they are virtually independent of operating conditions.

8.5.2 Cross receptances

We have seen that special interest centres on the cross-receptance $\alpha_{21}(\omega_e)$. This is shown plotted in fig. 8.35. The locus always proceeds in the counter-clockwise sense and nearly circular resonance loops are again thrown off the curve corresponding to

$$\Omega_1 = 1.56, \qquad \Omega_2 = 8.86, \qquad \Omega_3 = 18.24 \text{ rad/s}.$$

But now the resonance diameters are not in general parallel to the imaginary axis.

The polar plot in fig. 8.35(a) now has a significant feature. There is a very marked form of resonance at 0.10 rad/s. Although its explanation in physical terms is difficult to see, its mathematical interpretation is

Table 8.4. *Frequencies of resonant encounter (rad/s) for the destroyer proceeding at 14 m/s in head seas*

Direct receptance	Predicted values of resonance frequency			
	Ω_0?	Ω_1	Ω_2	Ω_3
α_{00}	0.90	1.45	8.88	18.20
α_{11}	0.09	1.54	8.87	18.22
α_{22}	0.05	–	8.87	18.20
α_{33}	0.05	–	8.87	18.21

Fig. 8.35. Polar plot of the cross receptance $\alpha_{21}(\omega_e)$ for the destroyer travelling at 14 m/s in head seas. (a) $\omega_e = 0.04$ (0.02) 7.04 rad/s, (b) $\omega_e = 7.00$ (0.02) 14.00 rad/s, (c) $\omega_e = 14.02$ (0.02) 21.02 rad/s.

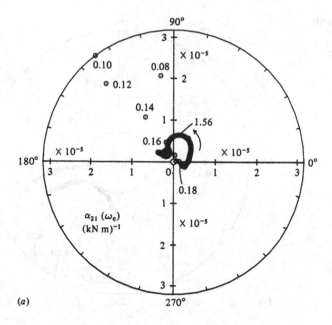

(a)

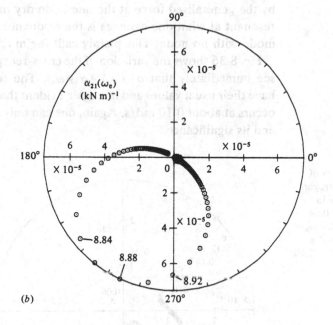

(b)

Fig. 8.35. (Continued)

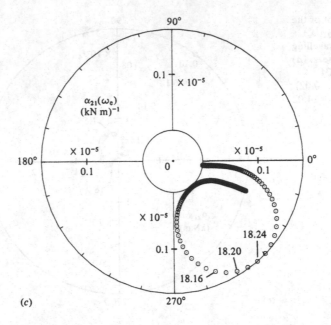

(c)

straightforward. It is that deflection in the two-node dry mode caused by the generalised force at the one-node dry mode becomes sharply resonant at what one assumes is the resonance frequency of the wet mode with no node. This plainly calls for much more investigation.

Fig. 8.36 shows the variation of the cross-receptance $\alpha_{12}(\omega_e)$ and we see immediately that $\alpha_{21}(\omega_e) \neq \alpha_{12}(\omega_e)$. The resonance frequencies have their usual values and it is again evident that a massive resonance occurs at about 0.10 rad/s. Again, one can only speculate on its cause and its significance.

Fig. 8.36. Polar plots of $\alpha_{12}(\omega_e)$ for the destroyer travelling at 14 m/s in head seas. (a) $\omega_e = 0.04$ (0.02) 7.04 rad/s, (b) $\omega_e =$ 7.02 (0.02) 14.02 rad/s, (c) $\omega_e = 14.00$ (0.02) 21.02 rad/s.

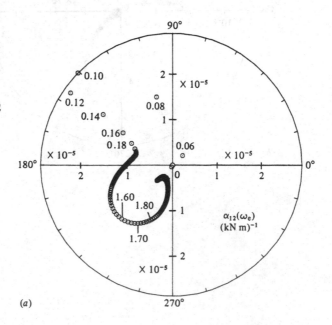

(a)

Fig. 8.36. (Continued)

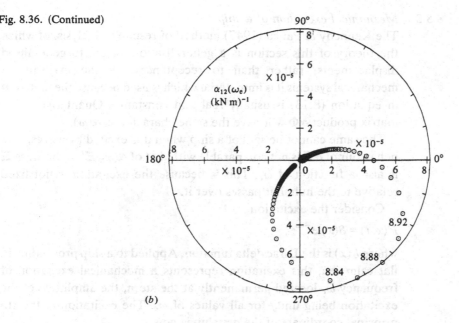

(b)

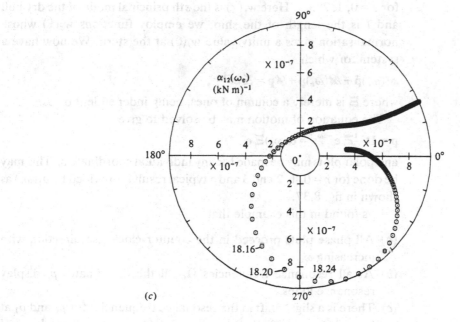

(c)

8.5.3   *Mechanical excitation of a ship*

The Kennedy & Pancu (1947) method of response analysis, of which the theory of this section is a generalisation, refers to generalised displacements, rather than to receptances. In the majority of mechanical systems it is immaterial which is used because the matrix $\mathbf{\Phi}$ in equation (8.13) is usually real and constant.† Qualitatively, the matrix product $\alpha\mathbf{\Phi}$ will have the same character as $\alpha(\omega)$.

The same cannot be said of a ship when it is excited by waves. The behaviour of $\alpha\mathbf{\Xi}$ is not comparable with that of $\alpha(\omega_e)\mathbf{\Xi}(\omega_e)$ because $\mathbf{\Xi}$ is also a function of ω_e. This is because the excitation is not fixed relative to the hull, but passes over it.

Consider the excitation

$$F(x, t) = \delta(x)\, e^{-i\omega_e t},$$

where $\delta(x)$ is the Dirac delta function. Applied to a ship proceeding in flat calm sea, this excitation represents a mechanical excitation of frequency ω_e located permanently at the stern, the amplitude of the excitation being unity for all values of ω_e. The excitation at the sth principal coordinate of the dry ship is now

$$\Xi_s(x, t) = \int_0^l F(x, t)w_s(x)\, \mathrm{d}x = w_s(0)\, e^{-i\omega_e t} = e^{-i\omega_e t},$$

for $s = 0, 1, 2, \ldots$. Here $w_s(x)$ is the sth principal mode of the dry hull and l is the length of the ship; we employ functions $w_s(x)$ whose normalisation gives a unity value $w_s(0)$ at the stern. We now have a system for which

$$\mathscr{A}(\omega_e)\ddot{\mathbf{p}} + \mathscr{B}(\omega_e)\dot{\mathbf{p}} + \mathscr{C}\mathbf{p} = \mathbf{\Xi}\, e^{-i\omega_e t},$$

where $\mathbf{\Xi}$ is merely a column of ones, being independent of ω_e.

The equation of motion may be solved to give

$$\mathbf{p} = \mathbf{D}^{-1}\mathbf{\Xi}\, e^{-i\omega_e t} = \alpha(\omega_e)\mathbf{\Xi}\, e^{-i\omega_e t}$$

and polar plots may be made for any individual coordinate p_r. This may be done for $r = 0, 1, 2$ and 3 and a typical result is provided by $p_1(\omega_e)$ as shown in fig. 8.37.

It is found in this example that:

(a) All phase plots proceed in the counter-clockwise direction with increasing ω_e.

(b) At all resonance frequencies Ω_i, all the coordinates p_r display resonance loops.

(c) There is a slight shift in the resonance frequencies for p_0 and p_1 at Ω_2 and Ω_3, possibly as a consequence of frequency-dependence of the system matrices.

(d) All resonance loops are nearly circular, except those for p_0 and p_1.

(e) $\arg[p_i(\Omega_i)] \doteq 90°$, for $i = 1, 2, 3$.

---

† Phase differences can arise, as with the torsional vibration of engine crankshafts.

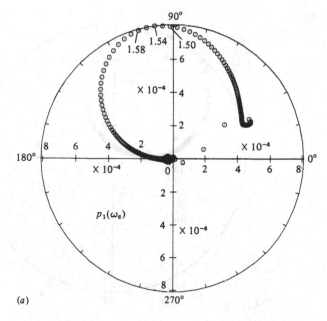

Fig. 8.37. Variation of $p_1(\omega_e)$ due to unit mechanical excitation at the stern of the destroyer when travelling at 14 m/s in calm water in head seas. (a) $\omega_e = 0.44$ (0.02) 7.44 rad/s, (b) $\omega_e = 7.20$ (0.02) 14.02 rad/s, (c) $\omega_e = 14.20$ (0.02) 21.02 rad/s.

(a)

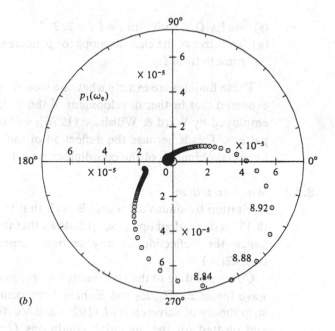

(b)

Fig. 8.37. (Continued)

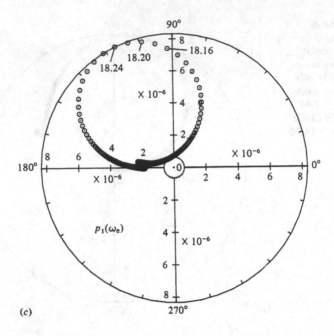

(c)

(f) $\arg[p_i(\Omega_j)] \neq 90°$, for $i \neq j$, $j = 1, 2, 3$.

(g) The sizes of the circular loops for p_i decrease as ω_e becomes more remote from Ω_i.

These findings are exactly what one would expect and it is also to be expected that further development of the technique, which has been employed by Ward & Willshare (1976), will conform to this general pattern. This is because the deflection of the ship at any section is a linear combination of the coordinates $p_r(\omega_e)$.

8.5.4 Wave excitation

Excitation by sinusoidal waves is such that the matrix Ξ in equation (8.17) is dependent upon $i\omega_e$. It follows that the responses $p_r(\omega_e)$, and hence the deflection at any section, depend upon the product $\alpha(\omega_e)\Xi(\omega_e)$.

Corresponding to the coordinates p_0, p_1, p_2 and p_3, the generalised wave forces Ξ_0, Ξ_1, Ξ_2 and Ξ_3 have been computed according to the strip theory of Salvesen et al. (1970) and Vugts (1971), i.e. Theory B, and plotted for the operating conditions $\bar{U} = 14$ m/s, $\chi = 180°$ and $a = 1$ m. The result for $\Xi_1(\omega_e)$ is shown in fig. 8.38. It is found that in every case the locus of $\Xi_s(\omega_e)$ moves in the clockwise direction with increasing ω_e.

8.5.5 Wave response

In view of the forms of the elements of the matrices $\alpha(\omega_e)$ and $\Xi(\omega_e)$ it is not to be expected that elementary analysis of wave response is

Fig. 8.38. The generalised wave force $\Xi_1(\omega_e)$ kN m applied to the destroyer when travelling at 14 m/s in a head sea, the amplitude of whose elevation is 1 m. (a) $\omega_e = 0.44$ (0.02) 7.44 rad/s, (b) $\omega_e = 7.20$ (0.02) 14.02 rad/s, (c) $\omega_e = 14.20$ (0.02) 21.02 rad/s.

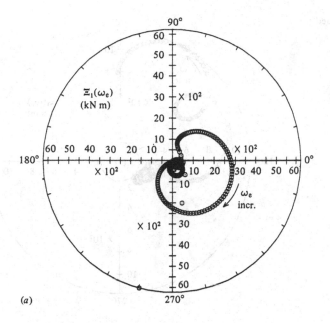

(a)

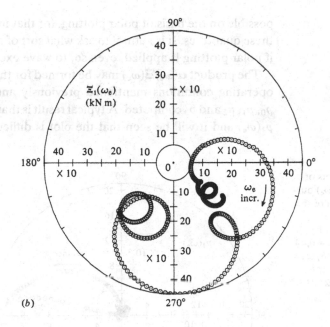

(b)

Fig. 8.38. (Continued)

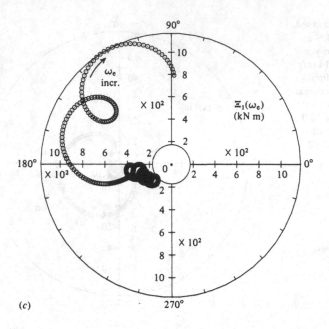

(c)

possible on the basis of polar plotting, for that involves the product of
these quantities. It is natural to ask what sort of result can be expected
if polar plotting is applied, even so, to wave excitation.

The product $\alpha(\omega_e)\Xi(\omega_e)$ may be formed for the destroyer under the
operating conditions mentioned previously and the polar plots for
p_0, p_1, p_2 and p_3 computed. A typical result is that shown in fig. 8.39 for
$p_1(\omega_e)$ and it will be seen that the plot is difficult to interpret in the

Fig. 8.39. Polar plots of
the coordinate $p_1(\omega_e)$ during wave excitation of the
destroyer under the
conditions to which fig.
8.38 relates. (a) $\omega_e = 0.44$
(0.02) 7.44 rad/s, (b) $\omega_e =$
7.20 (0.02) 14.02 rad/s,
(c) $\omega_e = 14.20$ (0.02) 21.02
rad/s.

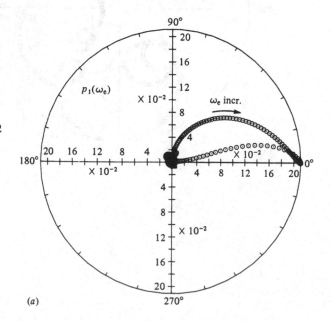

(a)

Fig. 8.39. (Continued)

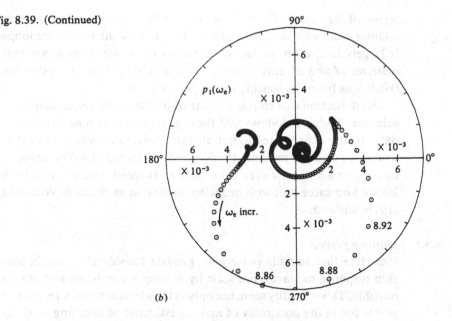

(b)

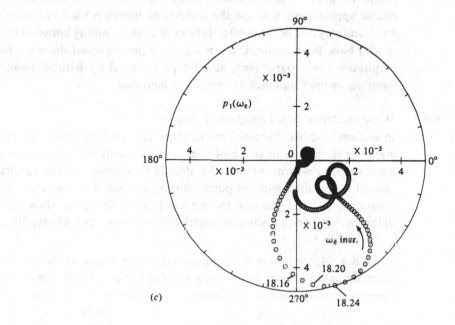

(c)

terms of fig. 8.31. There is no invariable direction, clockwise or counter-clockwise, as ω_e increases. Circularity of the resonance loops is largely lost, even for the large values of ω_e which give constant matrices $\mathcal{A}$ and $\mathcal{B}$. It is, however, still possible to identify resonance conditions from the spacing of the plotted points.

The deflection of a ship is a linear combination of deflections in its principal modes. It follows that these general conclusions obtain, not only for the coordinates p_r but also for deflections $w(x, t)$. In other words, an attempt to apply the Kennedy & Pancu (1947) technique directly to a ship in waves is unlikely to succeed whereas it is both logical and successful with propeller excitation as Ward & Willshare (1976) showed.

8.5.6 *Damping levels*
It is clear that suitable polar plots provide considerable insight into ship response to waves, not least by making accurate measurements possible. This is readily seen, not only in the identification of resonance points but in the possibility of making estimates of damping levels as suggested in section 8.5.

By constructing the resonance circles at $\alpha_{11}(\Omega_1)$, $\alpha_{22}(\Omega_2)$ and $\alpha_{33}(\Omega_3)$ and noting the frequencies ω_a and ω_b on each, the results found in Table 8.5 are obtained. These measures of damping are, of course approximate because the underlying theory is used in a semi-empirical way and because all questions of cross-damping between the modes have been ignored. Even so, the figures quoted do seem to emphasise the progressively smaller part played by hydrodynamic damping as the frequency of encounter increases.

8.5.7 *Wave excitation based on close-fit mapping*
In section 8.4.5 we discussed the effect of using a close-fit conformal mapping technique in place of the more conventional, though less accurate, Lewis-form method. As already mentioned, all the results quoted in this discussion of polar plotting are based on the close-fit approach. It is of interest to employ polar plotting to show the differences between results obtained with the two types of mapping.

Table 8.5. *Approximate damping factors of wet modes of the destroyer proceeding at 14 m/s in a head sea of amplitude 1 m*

Modal index	Estimated damping factor from plots	Structural damping assumed in calculations	Damping factor attributable to fluid action
1	0.231	0	0.231
2	0.010	0.006	0.004
3	0.009	0.009	0

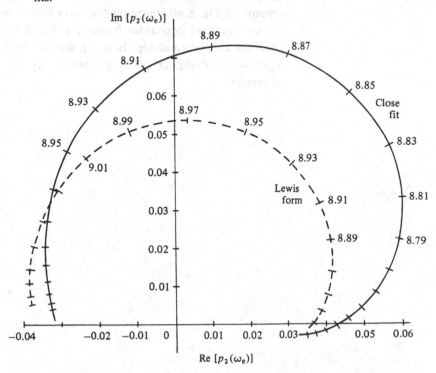

Fig. 8.40. Polar plot of p_2 in the region of $\omega_e = \Omega_2$, for the destroyer travelling at 14 m/s in head seas. The full line gives the result found with the close-fit method while the broken line relates to the use of Lewis-form fits.

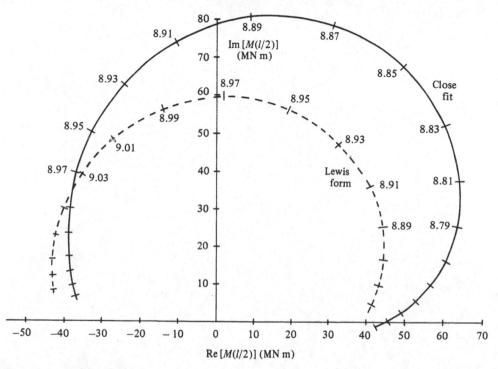

Fig. 8.41. Polar plot of bending moment amidships for the destroyer using the close-fit and Lewis-form techniques.

Once again we shall refer to the destroyer travelling at 14 m/s in head seas of 1 m amplitude, using Theory B for the estimation of fluid actions. The form of close fit described in section 8.4.5 has been employed. Fig. 8.40 compares the variations of p_2 in the region of the second resonant encounter frequency. Fig. 8.41 shows the variation obtained for the amidships bending moment in the same range. Once again we see that polar plotting substantially clarifies the presentation of results.

9 Transient loading

Oh, what is the matter, and what is the clatter?
He's glowering at her, and threatens a blow!
Oh, why does he batter the girl he did flatter?
And why does the latter recoil from him so?
The Sorcerer

9.1 Some comments on slamming

The phenomenon of slamming was discussed briefly in section 1.1.3. It will be remembered that under certain conditions the sea imparts a severe transient loading to the hull, owing to rapid and deep immersion of the bow when there is a pronounced flare or as a consequence of local emergence of the hull either near the stern or at the forefoot. While bow emergence is the commonest cause of slamming, we shall use the word 'slam' in a general sense when dealing with transient loading.

Bottom slamming forward has been the most widely studied of these phenomena and both theoretical and experimental investigations have been carried out. While some writers have investigated local plate forces and localised damage, the majority have examined the overall response of the hull. Tick (1958) and Ochi (1964) conclude that the main prerequisite for this type of slamming to occur is the emergence of the bow, although there are other necessary conditions to be met as regards crossing a threshold of relative velocity, unfavourable phase differences between bow and waves and the angle between the wave at the impact point and keel line (which must be small). Much effort has been spent determining the appropriate slam force.

One of the earliest theoretical investigations into the magnitude of slam loading was not performed for ships but rather for the landing of seaplanes (e.g. see Lewison, 1970a,b). Although the effects of water compressibility and air cushioning are ignored, the theory of Wagner (1932) has been the basis of many subsequent theoretical and experimental studies. The reader may wish to consult the work of, for instance, Bisplinghoff & Doherty (1952), Szebehely (1956), Borg (1957), Ochi & Bledsoe (1960), Chu & Abramson (1961), Ferdinande (1966), Foxwell & Madden (1969), Chuang (1970), Madden (1970), Chuang & Milne (1971) and of Hagiwara & Yuhara (1974).

In this chapter, the actual magnitude of the slam is described in two distinct ways. The first method (which we shall call that of 'impact slamming') attempts to evaluate the forces at the instant when the hull strikes the free surface of the waves. This type of slam is of short duration and in it the effects of spray, the compressibility of the water and the influence of air cushioning are probably significant; moreover the impact pressure depends critically on the relative velocity at re-entry. The second approach (that of 'momentum slamming') describes the effect of pressure variations around the hull surface as it penetrates the moving fluid after the initial entry. In fact, it appears to be arguable that slamming is really a combination of these events (i.e. of the initial impingement and the subsequent hull re-entry), whilst flare slamming is adequately described by the second approach only. Both approaches are based on empiricism and both will be explained briefly.

If slamming occurs, the problem is strictly one of non-linear dynamics. It is not practical to tackle it as such, however, and one must seek simplifications of one sort or another. In this chapter we shall outline a linearised theory of slamming dynamics.

Linear analysis admits the possibility of adding solutions of the appropriate equations to give other solutions. If the principle of superposition is applied to slamming, one is led to discuss two distinct problems and to add the two relevant results. In the first place the ship responds to a sinusoidal sea in the manner discussed in the previous chapter. From considerations of the sinusoidal sea one seeks to estimate a transient input (a 'slam') using one or both of the theories already mentioned. The response of the ship to this latter excitation is then estimated as if the vessel suffers this transient loading while it is under way in a flat calm sea. This chapter is on the transient problem.

Now it is known from full-scale experiments (e.g. see Aertssen, 1963, 1966, 1968; and Aertssen & van Sluijs, 1972) that when slamming occurs the hull of a ship vibrates at a much higher frequency than that of pitching in the one-node wet mode. The vibration persists for a large number of cycles, the rate of decay being small. It follows that complete oscillations of the hull take place while the bows are deeply immersed in the sea and also possibly during periods when the forefoot is clear of the water. As we have said, the problem is not linear and one is led to enquire how accurate a linear approximation based on the above argument is likely to be. The position is made even more obscure by uncertainty about the effects of pronounced bow flare in extreme motions. In the end, these are questions that can only be answered by experiment; our purpose in this book is merely to show that a linear theory can be devised and that it appears to give sensible answers.

The structural response of the dry hull during slamming almost certainly conforms to linear equations of motion since amplitudes of deflection are far smaller than the overall dimensions of the hull. It is on the hydrodynamics that suspicion falls. We shall work in terms of the principal coordinates of the dry hull and it therefore seems safe to say that non-linearity will not render the linear technique useless in a qualitative sense but, rather, may affect responses in a quantitative way. What is almost certainly true is that a linear theory is an essential prerequisite of a believable non-linear one and could well prove to be all that is of practical value to naval architects.

That part of the slamming theory which relates to an impact imparted to the hull while it is moving over a flat calm sea – the second of the two problems mentioned – raises other complications. The constants A_{rs}, B_{rs} and C_{rs} are, in general, frequency dependent and if the excitation is impulsive it is necessary to enquire what value of frequency should be used when defining them. The response turns out to be a slowly decaying one of fairly high frequency; although it is natural, then, to employ the infinite-frequency values of the fluid coefficients under the assumption that the motion is sinusoidal, there does remain an element of uncertainty about the matter.

The precise nature of slamming remains a subject of speculation. While the time variation, location and distribution of the fluid action during a slam are all matters that have received a good deal of attention they cannot yet be described as adequately understood. Nor has the sensitivity of slamming responses on these parameters been adequately investigated. It is not known if these features of slamming play a decisive role in determining responses – any more than it is known that linear theory really must be discarded in favour of non-linear.

There is one final matter that must be mentioned and it goes back to the basic question of why one wishes to analyse slamming responses. It is known that slamming can cause local fatigue in hull plating; this may well be associated with high levels of transient vibration in individual plates and sub-structures. This possibility of local vibration is merely a special case of a very wide subject indeed.

9.1.1 *Intuitive assessment*

Suppose that a concentrated unit impulse is applied at some section distant x from the stern of a hull travelling through *calm* water, at time $t = 0$. It will throw the hull into vibration and one would expect the response at the various principal coordinates of the dry hull to be in the nature of 'decaying sinusoidal' fluctuations. One would expect the 'frequencies' of these fluctuations to be approximately equal to the corresponding frequencies of resonant encounter in the sinusoidal sea.

The *intensity* of the vibration in any one dry mode would depend on where the unit impulse is applied. If it were applied at a node of the two-node dry mode, for instance, one would expect there to be no vibration at the appropriate coordinate $p_2(t)$, save only as a result of coupling.

The rates of decay of the responses at the coordinates $p_0, p_1, p_2, \ldots$ will not be the same. As discussed in section 7.3, those of $p_0(t)$ and $p_1(t)$ will be dominated by fluid damping whereas those of the oscillatory modes – i.e. the rates at which $p_2(t), p_3(t), \ldots$ decay – will be determined almost solely by structural damping; since the modal damping then rises with the order of the mode, the rates of decay will progressively increase.

All these are intuitive conclusions based on the supposition that a ship under way is a linear passive system. In fact, as we have seen, a ship is not passive; it is decidedly non-conservative, so we should be prepared for some areas of obscurity.

In this chapter we shall first demonstrate the essential correctness of all these predictions. The actual displacements at any section, caused by a unit impulse, are found from linear combinations of the responses at the coordinates $p_r(t)$ ($r = 0, 1, 2, \ldots$). These forms of response, too, will be examined.

Having found responses to unit impulses we shall 'build up' responses for more representative slamming using the technique of convolution. We shall refer exclusively to the dynamics of the destroyer that is discussed in chapter 4.

9.2   Response to a unit impulse

Consider the ship proceeding at constant speed in flat calm water when, at the instant $t = 0$, a unit impulse $\delta(t)$ of unit magnitude (e.g. of 1 kN s) is applied to it at some section distant $x = x'$ forward of the stern, i.e. at a location given by $\delta(x - x')$. The excitation at the sth coordinate is

$$\Xi_s(t) = \int_0^l \delta(x - x')\delta(t)w_s(x)\,\mathrm{d}x = w_s(x')\,\delta(t),$$

for $s = 0, 1, 2, \ldots$. If we employ n coordinates, then, we may write

$$\Xi(t) = \mathbf{w}(x')\delta(t), \tag{9.1}$$

where

$$\Xi(t) = \begin{bmatrix} \Xi_0(t) \\ \Xi_1(t) \\ \cdots\cdots \\ \Xi_{n-1}(t) \end{bmatrix}; \ \mathbf{w}(x') = \begin{bmatrix} w_0(x') \\ w_1(x') \\ \cdots\cdots \\ w_{n-1}(x') \end{bmatrix}.$$

Suppose that the generalised input at every coordinate is a unit impulse. The responses would then be given by

$$\mathbf{p}(t) = \mathbf{h}(t)\mathbf{1}, \tag{9.2}$$

where $\mathbf{h}(t)$ is the $(n \times n)$ 'impulse response matrix' (which is sometimes referred to as the 'impulsive receptance matrix' and denoted by $\boldsymbol{\alpha}^\delta(t)$) and $\mathbf{1}$ is a column of ones. With the present loading then

$$\mathbf{p}(t) = \mathbf{h}(t)\mathbf{w}(x'). \tag{9.3}$$

The problem initially confronting us is that of determining $\mathbf{h}(t)$ and we shall now describe two methods of doing so.

9.2.1 Fourier transformation of the receptance matrix

The equation of motion of the ship may be written in the form

$$\mathscr{A}(\omega_e)\ddot{\mathbf{p}}(t) + \mathscr{B}(\omega_e)\dot{\mathbf{p}}(t) + \mathscr{C}\mathbf{p}(t) = \Xi\, e^{-i\omega_e t}, \tag{9.4}$$

where Ξ is an $(n \times 1)$ column of constants. We showed in chapter 8 that the solution can be found in the form

$$\mathbf{p}(t) = \mathbf{H}(\omega_e)\Xi\, e^{-i\omega_e t},$$

where

$$\mathbf{H}(\omega_e) = [-\omega_e^2\mathscr{A}(\omega_e) - i\omega_e\mathscr{B}(\omega_e) + \mathscr{C}]^{-1}.$$

According to the principle of convolution, $\mathbf{p}(t)$ may be expressed in the alternative Duhamel integral form

$$\mathbf{p}(t) = \int_{-\infty}^{\infty} \mathbf{h}(\tau)\Xi(t - \tau)\, d\tau.$$

It follows that

$$\mathbf{p}(t) = \int_{-\infty}^{\infty} \mathbf{h}(\tau)\Xi\, e^{-i\omega_e(t-\tau)}\, d\tau$$

$$= \Xi\, e^{-i\omega_e t} \int_{-\infty}^{\infty} \mathbf{h}(\tau)\, e^{i\omega_e \tau}\, d\tau.$$

If we now adopt a new variable $\omega = -\omega_e$, we find that $\mathbf{h}(\tau)$ and $\mathbf{H}(\omega)$ form a Fourier transform pair, i.e.

$$\mathbf{H}(\omega) = \int_{-\infty}^{\infty} \mathbf{h}(\tau)\, e^{-i\omega\tau}\, d\tau \tag{9.5}$$

whence

$$\mathbf{h}(t) = \frac{1}{2\pi} \int_{-\infty}^{\infty} \mathbf{H}(\omega)\, e^{i\omega t}\, d\omega. \tag{9.6}$$

It is clear that, far from restricting attention to sinusoidal loading and responses, results of the type discussed in chapter 8 can be used in the solution of transient oscillation problems. Unfortunately, however, this approach is not convenient as it stands for use in practical computation. This is because a full knowledge of the hydrodynamic characteristics $\mathscr{A}(\omega_e)$ and $\mathscr{B}(\omega_e)$ is required over the entire frequency range so that a satisfactory integration may be performed.

9.2.2 The Hamiltonian method

Consider a set of $2n$ variables

$$\left.\begin{array}{ll}
y_0(t) = \dot{p}_0(t), & y_n(t) = p_0(t), \\
y_1(t) = \dot{p}_1(t), & y_{n+1}(t) = p_1(t), \\
\cdots\cdots\cdots\cdots & \cdots\cdots\cdots\cdots \\
y_{n-1}(t) = \dot{p}_{n-1}(t), & y_{2n-1}(t) = p_{n-1}(t).
\end{array}\right\} \tag{9.7}$$

With these, the matrix equation (9.4) reduces to

$$\mathbf{E}\dot{\mathbf{y}}(t) + \mathbf{F}\mathbf{y}(t) = \left[\frac{\mathbf{0}}{\Xi(t)}\right], \tag{9.8}$$

where the dashed line partitions the $(2n \times 1)$ column matrix into two matrices of order $(n \times 1)$ and where

$$\mathbf{y}(t) = \{y_0(t), y_1(t), \dots, y_{2n-1}(t)\}.$$

The matrices $\mathbf{E}$ and $\mathbf{F}$ are both of order $(2n \times 2n)$, being given by

$$\mathbf{E} = \left[\begin{array}{c|c} 0 & \mathscr{A} \\ \hline \mathscr{A} & \mathscr{B} \end{array}\right], \quad \mathbf{F} = \left[\begin{array}{c|c} -\mathscr{A} & 0 \\ \hline 0 & \mathscr{C} \end{array}\right].$$

The equations of motion can now be written in the alternative form

$$\mathbf{I}\dot{\mathbf{y}}(t) + \mathbf{E}^{-1}\mathbf{F}\mathbf{y}(t) = \mathbf{E}^{-1}\left[\frac{\mathbf{0}}{\Xi(t)}\right].$$

Provided the inverse of $\mathscr{A}$ exists, the inverse of matrix $\mathbf{E}$ is

$$\mathbf{E}^{-1} = \left[\begin{array}{c|c} -\mathscr{A}^{-1}\mathscr{B}\mathscr{A}^{-1} & \mathscr{A}^{-1} \\ \hline \mathscr{A}^{-1} & 0 \end{array}\right]$$

and matrix multiplication now shows that

$$\mathbf{E}^{-1}\mathbf{F} = \left[\begin{array}{c|c} \mathscr{A}^{-1}\mathscr{B} & \mathscr{A}^{-1}\mathscr{C} \\ \hline -\mathbf{I} & 0 \end{array}\right] = \mathbf{G}$$

while

$$\mathbf{E}^{-1}\left[\frac{\mathbf{0}}{\Xi(t)}\right] = \left[\frac{\mathscr{A}^{-1}\Xi(t)}{0}\right] = \mathbf{H}(t),$$

where $\mathbf{G}$ is of order $(2n \times 2n)$ and $H(t)$ is a column of order $(2n \times 1)$. Thus the initial matrix equation (9.4) may be thrown into the form

$$\mathbf{I}\dot{\mathbf{y}}(t) + \mathbf{G}\mathbf{y}(t) = \mathbf{H}(t). \tag{9.9}$$

Whereas equation (9.4) is sometimes referred to as being of the 'Lagrangean' form, equation (9.9) is said to be of the 'Hamiltonian' form. Whereas the former is the more familiar, the latter version (being of lower order) is sometimes very useful. The relationships that exist between the two methods have been investigated by Fawzy & Bishop (1976).

The general solution of equation (9.9) is the sum of the complementary function and a particular integral. The former is the general solution of the homogeneous equation

$$\mathbf{I}\dot{\mathbf{y}}(t) + \mathbf{G}\mathbf{y}(t) = \mathbf{0}$$

for which a suitable trial solution is

$$\mathbf{y}(t) = \mathbf{Y}\,e^{-\lambda t},$$

whence

$$(\mathbf{G} - \lambda\mathbf{I})\mathbf{Y} = \mathbf{0}.$$

The trial solution is non-trivial provided

$$\det(\mathbf{G} - \lambda\mathbf{I}) = 0,$$

and so λ may take eigenvalues $\lambda_0, \lambda_1, \ldots, \lambda_{2n-1}$, identified with the $(2n \times 1)$ column eigenvectors $\mathbf{Y}^{(0)}, \mathbf{Y}^{(1)}, \ldots, \mathbf{Y}^{(2n-1)}$ respectively. The complementary function is thus

$$\mathbf{y}_{CF}(t) = C_0 \mathbf{Y}^{(0)} e^{-\lambda_0 t} + C_1 \mathbf{Y}^{(1)} e^{-\lambda_1 t} + \ldots + C_{2n-1} \mathbf{Y}^{(2n-1)} e^{-\lambda_{2n-1} t},$$

where $C_0, C_1, \ldots, C_{2n-1}$ are constants determined by initial conditions of this free motion.

It is convenient to form a $(2n \times 2n)$ matrix

$$\mathbf{Y} = [\mathbf{Y}^{(0)} | \mathbf{Y}^{(1)} | \ldots | \mathbf{Y}^{(2n-1)}]$$

and also the diagonal matrix

$$\mathbf{\Lambda}(t) = \begin{bmatrix} e^{-\lambda_0 t} & 0 & \ldots & 0 \\ 0 & e^{-\lambda_1 t} & \ldots & 0 \\ \cdots\cdots\cdots\cdots\cdots\cdots\cdots \\ 0 & 0 & \ldots & e^{-\lambda_{2n-1} t} \end{bmatrix}.$$

The complementary function can be written

$$\mathbf{y}_{CF}(t) = \mathbf{Y}\mathbf{\Lambda}(t) \begin{bmatrix} C_0 \\ C_1 \\ \vdots \\ C_{2n-1} \end{bmatrix} = \mathbf{Y}\mathbf{\Lambda}(t)\mathbf{C} \quad \text{(say)}$$

and an expression can now be written down for the column matrix $\mathbf{C}$. If $\mathbf{y}(0)$ is a $(2n \times 1)$ column matrix representing $\mathbf{y}_{CF}(t)$ at the instant $t = 0$ we may write $\mathbf{C} = \mathbf{Y}^{-1}\mathbf{y}(0)$; this gives

$$\mathbf{y}_{CF}(t) = \mathbf{Y}\mathbf{\Lambda}(t)\mathbf{Y}^{-1}\mathbf{y}(0) \qquad\qquad (9.10)$$

and we note that the product $\mathbf{Y}\mathbf{\Lambda}(t)\mathbf{Y}^{-1} = \mathbf{I}$ when $t = 0$, as it should.

Suppose that every element in the vector $\mathbf{H}(t)$ is a delta function so that

$$\int_{0-dt}^{0+dt} \mathbf{H}(t)\, dt = \{1, 1, \ldots, 1\}.$$

If this excitation is applied to the system at rest, integration of equation (9.9) shows that

$$\mathbf{I}\mathbf{y}(0) = \{1, 1, \ldots, 1\}.$$

Thus a free motion ensues in which

$$\mathbf{y}(0) = \begin{bmatrix} 1 \\ 1 \\ \vdots \\ 1 \end{bmatrix}$$

and

$$\mathbf{y}_{CF}(t) = \mathbf{Y}\mathbf{\Lambda}(t)\mathbf{Y}^{-1}\mathbf{y}(0).$$

The particular integral of equation (9.9) can now be expressed in terms of a convolution integral:

$$y_{PI}(t) = \int_0^t \mathbf{Y}\mathbf{\Lambda}(\tau)\mathbf{Y}^{-1}\mathbf{H}(t-\tau)\,d\tau, \tag{9.11}$$

assuming that $\mathbf{H}(t) = 0$ for $t \leqslant 0$.

The complete solution of equation (9.9) can now be assembled; it may be written

$$\mathbf{y}(t) = \mathbf{Y}\mathbf{\Lambda}(t)\mathbf{Y}^{-1}\mathbf{y}(0) + \int_0^t \mathbf{Y}\mathbf{\Lambda}(\tau)\mathbf{Y}^{-1}\left[\frac{\mathscr{A}^{-1}}{\mathbf{0}}\right]\mathbf{\Xi}(t-\tau)\,d\tau. \tag{9.12}$$

This form of the convolution integral shows that the $(2n \times n)$ matrix

$$\mathbf{Y}\mathbf{\Lambda}(\tau)\mathbf{Y}^{-1}\left[\frac{\mathscr{A}^{-1}}{\mathbf{0}}\right]$$

is the impulse response matrix for the system whose response matrix is $\mathbf{y}(t)$.

The definition of $\mathbf{y}(t)$ in equations (9.7) shows that only the last n terms are the generalised coordinates $\mathbf{p}(t)$. By retaining only the impulse response functions for these last n terms, we find that the impulse response functions for the generalised coordinates are given by the $(n \times n)$ matrix

$$\mathbf{h}(t) = \begin{cases} [\ \mathbf{0} \ \vdots \ \mathbf{I}\ ]\mathbf{Y}\mathbf{\Lambda}(t)\mathbf{Y}^{-1}\left[\dfrac{\mathscr{A}^{-1}}{\mathbf{0}}\right] & (t \geqslant 0), \\[2em] \mathbf{0} & (t < 0). \end{cases} \tag{9.13}$$

Although it is not often of advantage to change to the Hamiltonian form of the equations of motion, it is of interest to note a particular feature of the relevant eigenvectors. Let $\boldsymbol{\lambda}$ be the diagonal matrix

$$\boldsymbol{\lambda} = \begin{bmatrix} \lambda_0 & 0 & \cdots & 0 \\ 0 & \lambda_1 & \cdots & 0 \\ \cdots\cdots\cdots\cdots\cdots\cdots \\ 0 & 0 & \cdots & \lambda_{2n-1} \end{bmatrix}.$$

It has been shown that

$$\mathbf{G}\mathbf{Y}^{(r)} = \lambda_r \mathbf{I}\mathbf{Y}^{(r)}$$

so that

$$[\mathbf{G}\mathbf{Y}^{(0)} \vdots \mathbf{G}\mathbf{Y}^{(1)} \vdots \ldots \vdots \mathbf{G}\mathbf{Y}^{(2n-1)}] = [\lambda_0 \mathbf{Y}^{(0)} \vdots \lambda_1 \mathbf{Y}^{(1)} \vdots \ldots \vdots \lambda_{2n-1}\mathbf{Y}^{(2n-1)}],$$

whence

$$\mathbf{G}\mathbf{Y} = \mathbf{Y}\boldsymbol{\lambda}$$

and so

$$\boldsymbol{\lambda} = \mathbf{Y}^{-1}\mathbf{G}\mathbf{Y}. \tag{9.14}$$

A simple relationship thus exists between the eigenvalues, the eigenvectors and the matrix **G**. In practice, this method of determining the impulse response matrix **h** – i.e. by the use of equation (9.13) – is superior to that outlined in section 9.2.1.

9.2.3 *Response to a unit impulse*

Reverting now to the Lagrangean approach we can write down an expression for the transient response at the generalised coordinates. It is

$$\mathbf{p}(t) = \int_0^t \mathbf{h}(\tau)\mathbf{\Xi}(t-\tau)\,\mathrm{d}\tau, \tag{9.15}$$

where $\mathbf{h}(t)$ is given by equation (9.13). For the unit-impulsive loading administered at the section $x = x'$, this becomes

$$\mathbf{p}(t) = \int_{-\infty}^{\infty} \mathbf{h}(t)\mathbf{w}(x')\delta(t-\tau)\,\mathrm{d}\tau$$
$$= \mathbf{h}(t)\mathbf{w}(x'), \tag{9.16}$$

since $\mathbf{\Xi}(t-\tau) = 0$ for $\tau > t$. This result was arrived at in equation (9.3).

Having found an expression for the variation of the generalised coordinates following the impulsive loading, we can now find the deflection, bending moment and shearing forces. The deflection for instance, is given by

$$w(x, t) = \sum_{r=0}^{n-1} w_r(x)p_r(t).$$

If $\mathbf{w}(x) = \{w_0(x), w_1(x), \ldots, w_{n-1}(x)\}$ it follows that

$$w(x, t) = \mathbf{w}^T(x)\mathbf{p}(t) = \mathbf{w}^T(x)\mathbf{h}(t)\mathbf{w}(x'), \tag{9.17}$$

where the superscript T indicates the transposed matrix. Similarly

$$M(x, t) = \sum_{r=0}^{n-1} M_r(x)p_r(t) \equiv \mathbf{M}^T(x)\mathbf{p}(t) = \mathbf{M}^T(x)\mathbf{h}(t)\mathbf{w}(x') \tag{9.18}$$

and

$$V(x, t) = \sum_{r=0}^{n-1} V_r(x)p_r(t) \equiv \mathbf{V}^T(x)\mathbf{p}(t) = \mathbf{V}^T(x)\mathbf{h}(t)\mathbf{w}(x'). \tag{9.19}$$

9.2.4 *Unit impulsive loading of the destroyer*

For the sake of discussion, suppose that $x' = 0.8l$, where l is the length of the vessel. That is to say, we suppose that a unit upward impulse is applied at a particular section in the region of the fo'c'sle. We shall consider only the modes $r = 0, 1, 2, 3$ and 4; the vector $\mathbf{w}(x')$ is therefore of order (5×1) and is known once the modal vectors $w_r(x)$ have been found.

The matrix **G** may now be found and, hence the eigenvalues $\lambda_0, \lambda_1, \ldots, \lambda_9$ and their associated eigenvectors $\mathbf{Y}^{(0)}, \mathbf{Y}^{(1)}, \ldots, \mathbf{Y}^{(9)}$.

Thus the (5×5) impulse response matrix $\mathbf{h}(t)$ may be computed by the use of equation (9.13). This approach has been adopted in arriving at the results to be quoted in this chapter, using the infinite-frequency values of all the hydrodynamic coefficients.

Fig. 9.1 shows the variations of the five principal coordinates of lowest order, for a forward speed of 14 m/s. It will be seen that the variations are all 'decaying sinusoids' (though there is clear evidence of coupling) and that approximate 'frequencies' can be assigned to the variations of $p_2(t)$, $p_3(t)$ and $p_4(t)$; they are

1.41 Hz (8.84 rad/s),

2.91 Hz (18.3 rad/s),

4.41 Hz (27.8 rad/s).

It was pointed out in section 8.1.3 that a ship has 'resonant encounter frequencies'. For this ship, those corresponding to the wet modes dominated by p_2, p_3 and p_4 are

8.79 rad/s, 18.3 rad/s and 27.8 rad/s

respectively, as shown in Table 8.1. Note that in calculating the variations of the principal coordinates following an impulsive loading the generalised hydrodynamic coefficients are calculated for infinite frequency using the conformal transformation method, whereas for Table 8.1 they were calculated at the frequency of encounter of ship with wave and the 'Lewis-form' method was used.

Fig. 9.1. Variation of the coordinates $p_r(t)$, $r = 0, 1, 2, 3, 4$, following the application of a unit impulse at a distance $0.8l$ forward of the stern. The ship speed is 14 m/s.

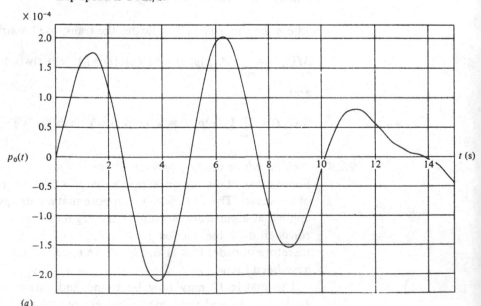

(a)

Fig. 9.1. (Continued)

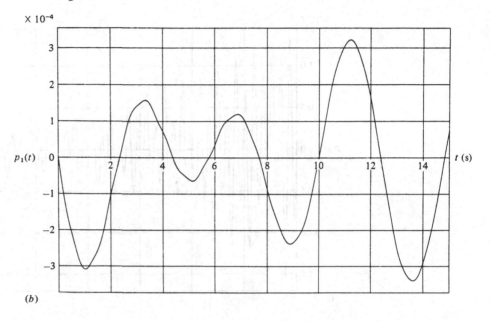

(b)

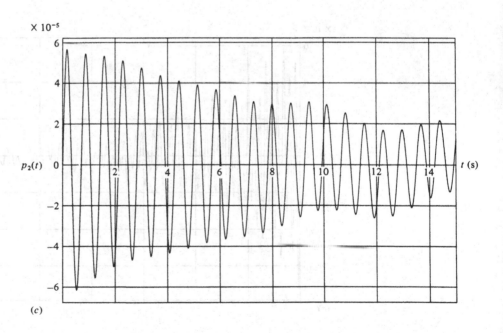

(c)

Fig. 9.1. (Continued)

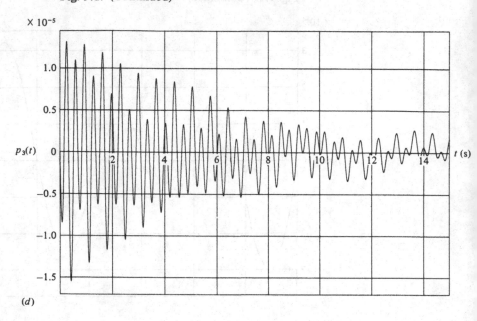

(d)

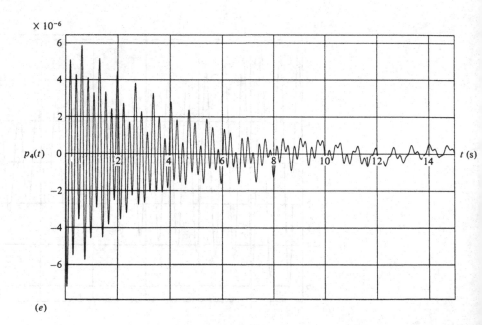

(e)

It was also noted in section 8.1.3 that these resonant encounter frequencies are probably independent of ship speed, although the theory does not appear to ensure it. If results like those of fig. 9.1 are calculated for a speed of 6 m/s (instead of 14 m/s) it is found that the curves are changed rather little and that the 'frequencies' of $p_2(t)$, $p_3(t)$ and $p_4(t)$ are

8.86 rad/s,   18.5 rad/s,   27.9 rad/s.

It seems safe to conclude that, although free vibrations in wet modes involve significant coupling of motions in the dry modes, there are discernible 'frequencies'. Moreover these correspond to resonant encounter frequencies and are more or less independent of speed. The position is a little more obscure for the 'heave' and 'pitch' modes, however, since 'periodic times' are not as well defined and one would wish to average results over a longer time. There is no difficulty in so doing, and some longer traces appear later, but it is clear that heave and pitch do represent a pair of special cases which need special study.

Having found the time history of the quantities $p_r(t)$ for $r = 0, 1, 2, 3, 4$ we are in a position to find the variation of displacement, bending moment and shearing force at any section x. The variation of displacement can be subdivided into two, namely that corresponding to motion of the ship as a 'rigid body' (i.e. in the $r = 0, 1$ modes only) and that due to distortion. Thus we can now identify:

$$
\left.
\begin{aligned}
w_{\text{rigid}}(x, t) &= p_0(t)w_0(x) + p_1(t)w_1(x), \\
w_{\text{dist}}(x, t) &= \sum_{r=2}^{4} p_r(t)w_r(x), \\
w_{\text{total}}(x, t) &= \sum_{r=0}^{4} p_r(t)w_r(x), \\
M(x, t) &= \sum_{r=2}^{4} p_r(t)M_r(x), \\
V(x, t) &= \sum_{r=2}^{4} p_r(t)V_r(x).
\end{aligned}
\right\}
\qquad (9.20)
$$

The variations of these parameters are shown in fig. 9.2 for $x = l/2$.

The curves shown in fig. 9.2 all show slowly decaying fluctuations as one would expect. Those of $w_{\text{dist}}(l/2, t)$ and $M(l/2, t)$ occur with a dominant 'frequency' of about 8.84 rad/s since they contain large contributions from $p_2(t)$ and small contributions from $p_3(t)$. By contrast $V(l/2, t)$ contains a distinct element of higher frequency as one would expect from the fact that $V_2(l/2)$ is small.

It can readily be demonstrated that these varying degrees of dominance of the modes result from the choice made in selecting the section for which responses are to be estimated. Fig. 9.3 shows responses for the section $x = 3l/4$ and it will be seen that they display

Fig. 9.2. Variation of (a) displacement as a rigid body, (b) displacement due to distortion, (c) total displacement, (d) bending moment and (e) shearing force caused by a unit impulse applied at the section $x = 0.8l$. All responses are calculated for the section $x = l/2$. The ship speed is 14 m/s.

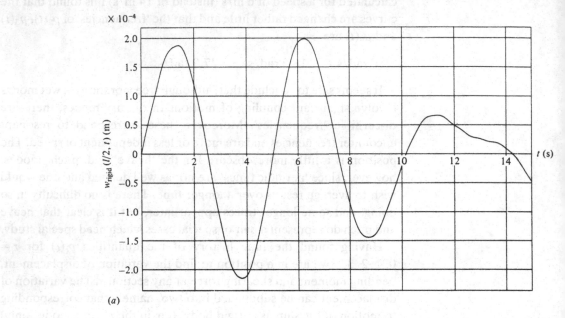

(a)

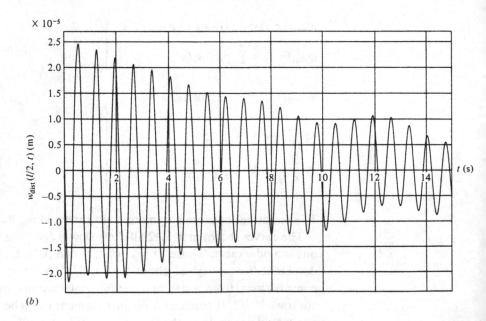

(b)

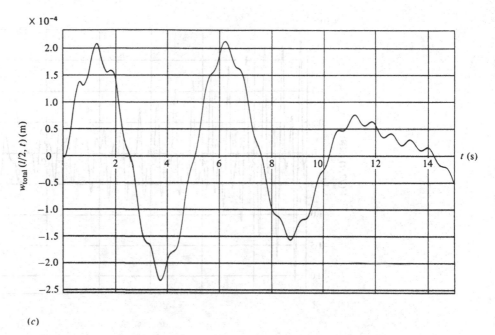

Fig. 9.2. (Continued)

(c)

(d)

Fig. 9.2. (Continued)

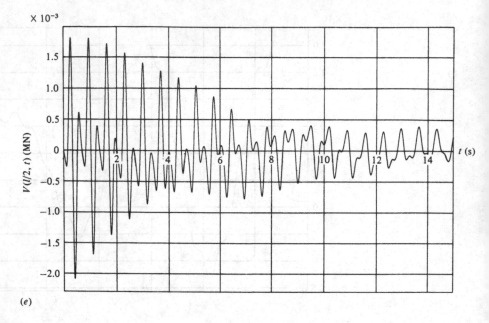

(e)

Fig. 9.3. Results corresponding to those of fig. 9.2 but relating to the section $x = 3l/4$ instead of $l/2$.

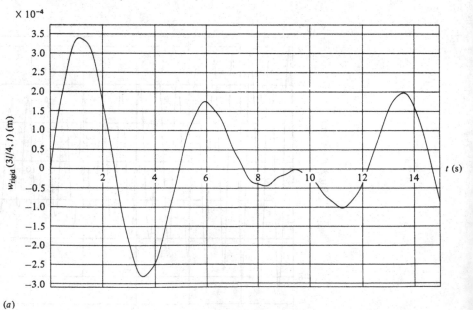

(a)

Fig. 9.3. (Continued)

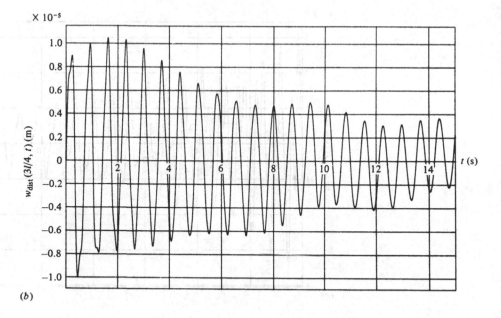

(b)

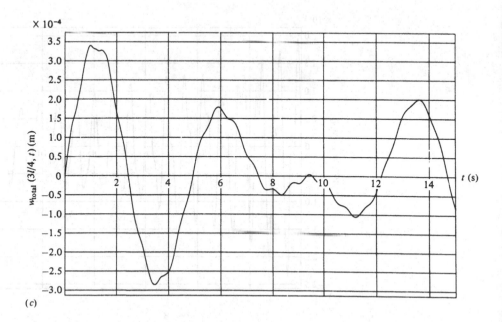

(c)

Fig. 9.3. (Continued)

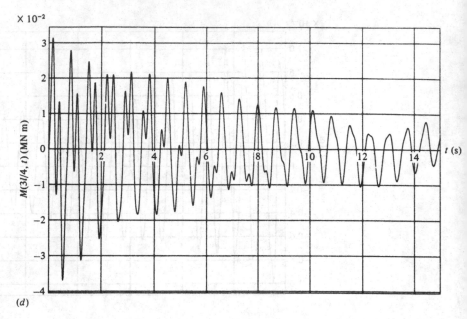

(d)

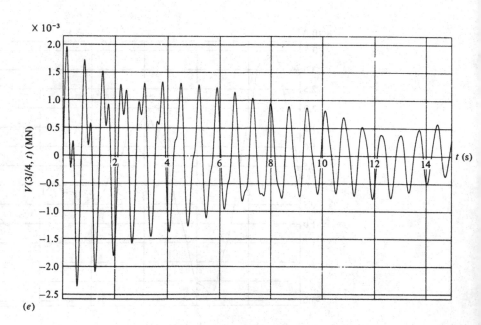

(e)

significantly different frequency contents. Notice that components of higher frequency die away more rapidly than those of lower frequency; this is particularly plain in fig. 9.3(e).

9.3 The occurrence of slamming in regular waves

The generalised force applied at the sth coordinate is

$$\Xi_s(t) = \int_0^l F(x, t) w_s(x) \, dx,$$

where $F(x, t)$ represents the transient force per unit length acting on the hull. For example if $x = x'$ is the aftermost point on the hull at which bow emergence occurs then,

$$F(x, t) \geq 0 \qquad (x \geq x'),$$
$$= 0 \qquad (x < x').$$

For a given ship speed $\bar{U}$, heading χ, and for sinusoidal waves, the time history of the relative displacement, velocity and acceleration, as between any point of the hull and the wave, may be determined. Thus in the notation employed previously the relative displacement at a section x on the hull is

$$w_{rel}(x, t) = w(x, t) - \zeta(x, t), \tag{9.21}$$

where $w(x, t)$ is the hull displacement and $\zeta(x, t)$ is the wave displacement. The relative velocity and acceleration are $Dw_{rel}(x, t)/Dt$, $D^2 w_{rel}(x, t)/Dt^2$ respectively, where the total time derivative

$$\frac{D}{Dt} = \frac{\partial}{\partial t} - \bar{U} \frac{\partial}{\partial x}.$$

On the assumption that a linear response may be found, a wave amplitude may be determined which is such that a specified station on the hull will emerge from the water;† i.e.

$$|w_{rel}(x, t)| \geq T(x),$$

where $T(x)$ is the local draught, while the instantaneous immersion of a section is

$$I(x, t) = T(x) - w_{rel}(x, t).$$

---

† The computer program written to determine the slam forces, UCLSLAM, is written in such a way that the wave amplitude is increased automatically if bow emergence is not predicted at the station where the slam is assumed to occur. On the basis of regular wave tests, Ochi (1958) concluded that the minimum wave height which causes slamming is approximately 0.02 of the wavelength but there is no hard and fast rule concerning this value. In fact, under the conditions imposed in formulating the problem, calculations suggested that a much higher value of wave height would be required before slamming occurred (see section 9.4.1).

In the computations to which we shall refer, the aftermost position at which emergence was assumed to occur was taken as $x = x' = 0.75l$.

Fig. 9.4 is a sketch showing time history of relative displacement at station x in which:

(a) $t_s(x)$ is the time when the slam occurs. This is the time elapsed from the instant when there existed a wave crest at the stern, and
(b) $t_d(x)$ is the time it takes for the section to submerge to its still water draught, at which juncture the slam is usually assumed to cease (though it might equally be argued presumably that the transient loading persists until the relative velocity is zero).

If the relative displacement is defined as

$$w_{rel}(x, t) = w_{rel}(x) \cos (\omega_e t - \varepsilon_{rel}),$$

where ω_e is the frequency of wave encounter, $w_{rel}(x)$ is the local amplitude of the relative displacement and ε_{rel} is the phase angle between the relative displacement at position x and the wave crest at the stern, then it follows that at time $t = t_s(x)$,

$$w_{rel}(x) \cos [\omega_e t_s(x) - \varepsilon_{rel}] = T(x), \tag{9.22}$$

whence

$$\omega_e t_s(x) = \varepsilon_{rel} + \cos^{-1}\left[\frac{T(x)}{|w_{rel}(x)|}\right] \quad \text{if } \varepsilon_{rel} + \cos^{-1}\left[\frac{T(x)}{|w_{rel}(x)|}\right] \geq 0,$$

$$= 2\pi + \varepsilon_{rel} + \cos^{-1}\left[\frac{T(x)}{|w_{rel}(x)|}\right] \quad \text{if } \varepsilon_{rel} + \cos^{-1}\left[\frac{T(x)}{|w_{rel}(x)|}\right] < 0. \tag{9.23}$$

Similarly, by letting $t = -t_d(x)$ measured from the instant at which the slam is completed, we find that

$$\omega_e t_d(x) = \frac{\pi}{2} - \cos^{-1}\left[\frac{T(x)}{|w_{rel}(x)|}\right]. \tag{9.24}$$

Therefore, for a given frequency of encounter, the quantities $t_s(x)$ $t_d(x)$ may be determined as well as the time lag between slams at different positions $x_1, x_2, \ldots$, as illustrated in fig. 9.5. The actual magnitude of

Fig. 9.4. The period occupied by a slam.

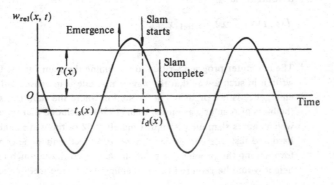

the transient loading at each position may be obtained by using one of two approaches – i.e. by employing the concepts of 'impact slamming' or 'momentum slamming'.

The results quoted so far refer to a single impulse concentrated at one section of the ship.† Now a representative slam can be synthesised by spreading appropriately scaled impulses out, both in space along the hull and in time, and the total response is found by adding the responses to the various scaled impulses. The technique of convolution may be employed for this purpose.

9.4   The application of 'impact slamming' theory

Although the transient forces are continuously distributed over a length of the hull which was initially clear of the water near the forefoot, the distributed force must be 'discretised' for the purposes of calculation. This can conveniently be accomplished by assuming the transient to be stepwise distributed over, say, 5 of the 20 slices into which the hull is imagined cut for the purposes of stuctural or hydrodynamic calculations, the centres of the slices being located at the sections given by $x/l = 0.775, 0.825, 0.875, 0.925, 0.975$ (see fig. 4.1). For each of these slices, the relative motion between the ship and water in sinusoidal waves may be calculated, and also the time elapsed before impact (measured from an instant at which a wave crest is located at the stern). Then the time history of the force at each section may be estimated by the technique proposed by Ochi & Motter (1973) and Kawakami, Michimoto & Kobayashi (1977).

The transient force and consequently the slam pressure at any slice is assumed to be approximately proportional to the square of the magnitude of the relative vertical velocity between the hull and wave at the instant of impact. That is to say,

$$p(x, t) = \tfrac{1}{2}\rho k_1 |\mathrm{D} w_{\mathrm{rel}}(x, t)/\mathrm{D} t|^2, \tag{9.25}$$

† Whether or not a slam can be adequately represented by a judiciously scaled and located impulse of this sort is certainly worth investigation, but is not our concern here.

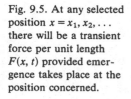

Fig. 9.5. At any selected position $x = x_1, x_2, \ldots$ there will be a transient force per unit length $F(x, t)$ provided emergence takes place at the position concerned.

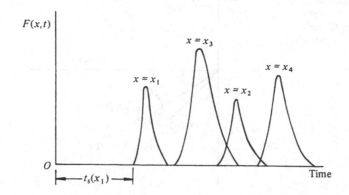

where ρ is the mass density of the water, and k_1 is the non-dimensional pressure factor which depends on the section shape and whose value may be determined from drop tests. Several different body shapes have been used in the determination of this coefficient; e.g. see Foxwell & Madden (1969), Chuang (1970), Madden (1970), Chuang & Milne (1971) and Hagiwara & Yuhara (1974). Lloyd (1976) has shown that, for a local sectional deadrise angle β greater than 24°, the value

$$k_1 = 1 + \left(\frac{\pi \cot \beta}{2}\right)^2 \tag{9.26}$$

proposed by Wagner (1932) for wedge impact over the range $0 \leqslant \beta \leqslant \pi/2$ seems to be a good approximation. It is evident from this expression that the theory predicts very large pressures for the small deadrise angles which are on the whole more typical of cargo ships than destroyers. For $\beta = 0$, the theory predicts an infinite pressure because no account is taken of the compressibility of the water and air.

Ochi & Motter (1973) proposed an empirical non-dimensional formula for k_1. It is a regression expression derived from the results of seakeeping tests and drop tests on ship models, see Ochi & Motter (1971):

$$k_1 = \exp\left(1.377 + 2.419a_1 - 0.873a_3 + 9.624a_5\right). \tag{9.27}$$

The constants a_1, a_3 and a_5 are the conformal transformation coefficients of the section's $0.1T(x)$ profile when a three-parameter transformation is employed as explained in section 7.1.3.

In this approach the loading is assumed to act instantaneously over the bottom one-tenth of the draught (i.e. over a depth of $0.1T$) but the pressure varies over this region, being assumed zero at $0.1T$ and a maximum, p_{max}, at the keel. Fig. 9.6(a) shows the profile of the region

Fig. 9.6. (a) The distribution of normal pressure p_n over the bottom one-tenth of the draught at any instant. It is assumed that $p_n(\phi) \simeq p_{max}(y(\phi)/d)$, where $d \equiv 0.1T(x)$. (b) Notation used in writing down that contribution made at the ith element of the ship's bottom to the total transient force per unit length.

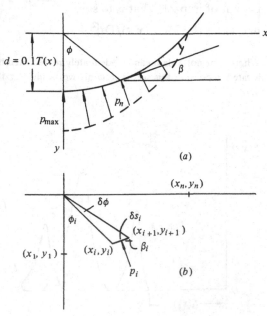

of interest. As a simplification, the normal pressure at any angle ϕ is assumed to be given by

$$p_n(\phi) = p_{max}\left(\frac{y(\phi)}{d}\right),$$

where $d = 0.1T(x)$ and p_{max} is deduced from equation (9.25). The net transient force at any section may be evaluated by integrating the vertical component around the section profile. The contribution to the net vertical transient force per unit length along the hull exerted at the ith element of the section is given by

$$p_i \cos \beta_i \, \delta s_i,$$

where β_i and δs_i are as shown in fig. 9.6(b). The total instantaneous vertical force per unit length on the section is thus

$$F(x) = 2 \sum_{i=1}^{n-1} p_i \cos \beta_i \, \delta s_i.$$

But since $y_1 = d$,

$$p_i = p_{max}\left(\frac{y_{i+1} + y_i}{2y_1}\right)$$

and

$$\cos \beta_i = \frac{x_{i+1} - x_i}{\delta s_i},$$

it follows that

$$F(x) = p_{max} G(x),$$

where $p = p_{max}$ is given by equation (9.25) and the shape factor

$$G(x) = \frac{1}{y_1} \sum_{i=1}^{n-1} (y_{i+1} + y_i)(x_{i+1} - x_i). \tag{9.28}$$

Turning now to the temporal variation, we employ an assumption due to Kawakami et al. (1977). It is that the transient loading is of the form

$$F(x, t) = \frac{p_{max}}{T_0} G(x)t \, e^{(1-t/T_0)}, \tag{9.29}$$

where T_0 is the time that elapses between the instant at which the bottom strikes the wave surface and the instant at which the loading reaches its maximum value. By employing Froude's law in conjunction with the data given by Kawakami et al., it is found that T_0 is about 0.01 s for the destroyer.

9.4.1 Impact slamming of the destroyer
Consider the destroyer travelling at 14 m/s in a sinusoidal head sea with

wave frequency $\omega = 0.76$ rad/s,

encounter frequency $\omega_e = 1.58$ rad/s,

wavelength $\lambda = 107.3$ m.

That is to say the wavelength is equal to the ship length. For these
conditions, calculations suggest that slamming will occur when the
amplitude of wave elevation a is 2 m; we shall therefore derive results
for that case.

Fig. 9.7 shows the results of impact slamming computations for these
conditions. It shows five peaks, one for each slice. The time elapsed
before impact happens to be greatest for the aftermost slice and the
intensity of the force per unit length is also the greatest for that slice.
This is thus a case where the slam 'runs back' from the bow. The force
per unit length $F(x, t)$ for any given slice is spread out in time in a
manner determined by the relative motion of ship and water.

The slam does not always run back from the bow and fig. 9.8 shows a
case in which the reverse is true. The transient excitation at the fifth
slice is the first to commence and end. It is interesting to note that the
magnitude of the slams are much larger in fig 9.7 than fig. 9.8.

Returning to our previous case, each pulse in fig. 9.7 is associated
with a particular slice of the ship. By multiplying the ordinate of a curve
for any instant by the thickness Δx of a slice, an instantaneous
equivalent force acting at the centre of the slice may be found. (Note
that, by decreasing Δx, a more exact description of the transient force
may be obtained.) By dividing the pulse into intervals of 0.002 s and
integrating for each strip, this time-dependent equivalent force can be

Fig. 9.7. Force/unit length $F(x, t)$ at each of 5 slices of the hull. On any
one slice $F(x, t)$ is assumed to be uniformly distributed in space but to
vary with time in the manner shown. The time variation for the section
centred on $x = 0.775l$ is that of the right-hand (the highest) peak while
that for the $x = 0.975l$ section is the smallest one on the left.

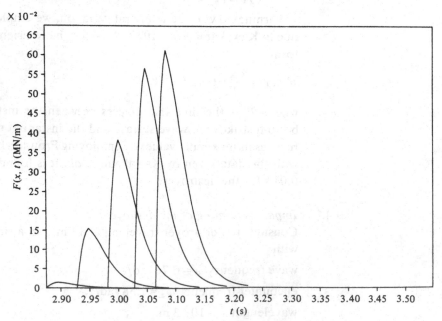

represented as a series of impulses. Thus the slamming problem can be broken down into a large number of problems like those we discussed in the last section. The total response is found by adding the various responses to the scaled impulses.

Fig. 9.9 shows the variations of the $p_r(t)$ ($r = 0, 1, 2, 3, 4$) for the excitation indicated in fig. 9.7. The effects of close coupling between $p_0(t)$ and $p_1(t)$ can readily be seen, as can the increase of frequency and rate of decay with order of mode.

Turning next to the responses of displacement, bending moment and shearing force, we find the results given in fig. 9.10 for the amidship section $x = l/2$.

9.5 The application of 'momentum slamming' theory

A quite different theory of slamming excitation has been proposed by Leibowitz (1962, 1963). In this approach, which is applicable to bottom slamming (see Mansour & d'Oliveira, 1975) and possibly flare slamming (see Church, 1962; and Kaplan & Sargent, 1972), the time history of the transient force is related to the rate of change of momentum of the surrounding fluid and the instantaneous buoyancy. That is to say, at any time t

$$F(x, t) = -\left\{ \frac{D}{Dt} \left[m(x, t) \frac{D}{Dt} w_{\text{rel}}(x, t) \right] - \rho g S(x, t) \right\}, \qquad (9.30)$$

Fig. 9.8. Variation of force/unit length at each of 5 slices of the hull. The operating conditions were: forward speed = 14 m/s in head seas, $\omega = 0.54$ rad/s, $\omega_e = 0.95$ rad/s, $\lambda/l = 2$. In this case, by contrast to that shown in fig. 9.7, the value of $F(x, t)$ for $x = 0.775l$ is that of the left-hand peak while that for $x = 0.975l$ is that of the right-hand peak.

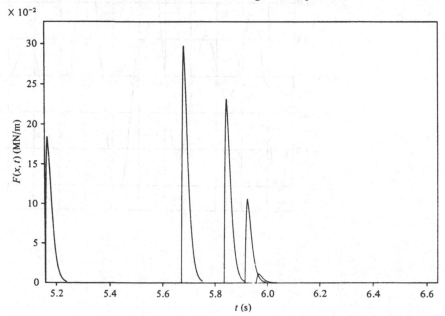

Fig. 9.9. Slamming responses at the coordinates $p_r(t)$, $r = 0, 1, 2, 3, 4$, for the destroyer travelling at 14 m/s in head seas when the excitation is that of fig. 9.7.

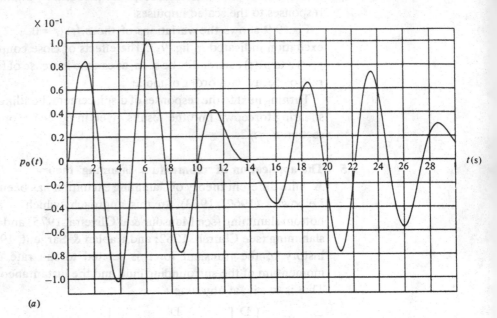

(a)

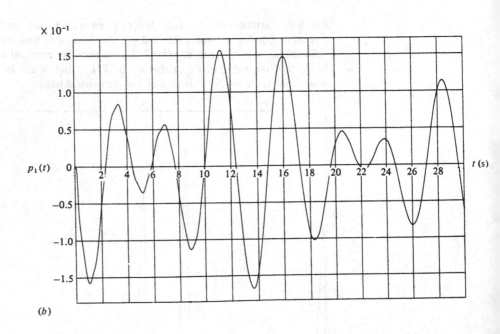

(b)

Fig. 9.9. (Continued)

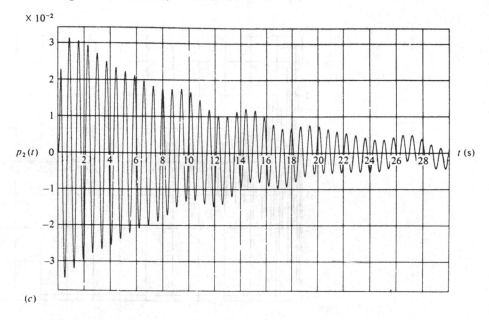

(c)

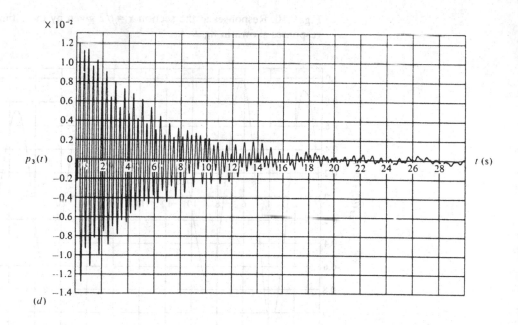

(d)

Fig. 9.9. (Continued)

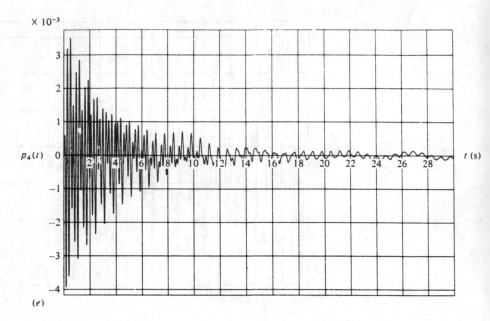

(e)

Fig. 9.10. Responses at the section $x = l/2$ given by the estimated responses shown in fig. 9.9.

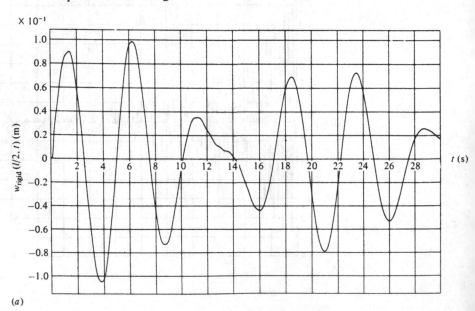

(a)

Fig. 9.10. (Continued)

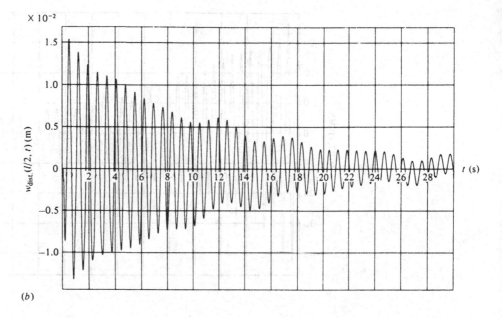

(b)

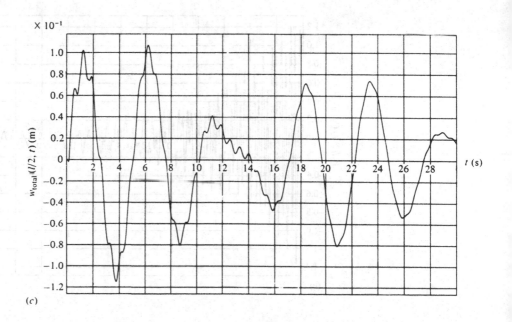

(c)

Fig. 9.10. (Continued)

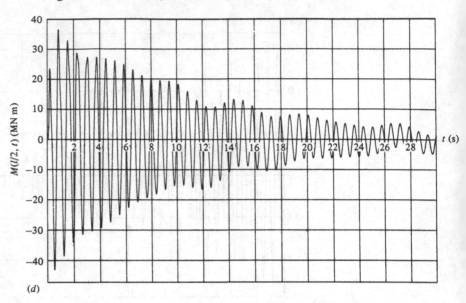

(d)

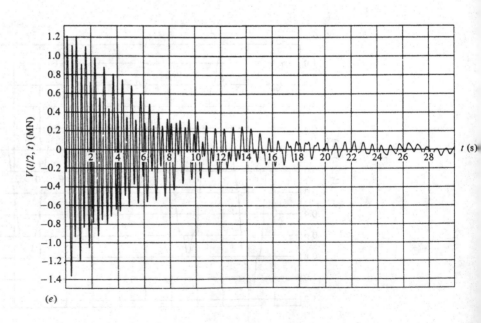

(e)

where $m(x, t)$ and $S(x, t)$ are the instantaneous added mass and submerged area of the hull section as it re-enters the water, i.e. when the instantaneous immersion is $I(x, t)$. These parameters are calculated at successive instants of time as the hull section penetrates the wave, using a multiparameter conformal mapping of the type discussed in section 7.1.3 and the asymptotic value of the added mass at infinite frequency.

This latter restriction may be discarded and the added mass of the instantaneously immersed hull section may be calculated at the frequency of wave encounter. If this is the case, a wave damping term may also be included in the expression for $F(x, t)$ as discussed in chapters 7 and 8. The disadvantage of such refinement is the increased burden of computing.

The duration of the slam is now not so well defined. For the purposes of calculation it may be taken as the interval between impingement of the hull section on the wave and the instant at which the section is immersed to its still water draught. This, however, is by no means a well-founded rule and another possibility is the interval between impingement and the instant of zero relative velocity between hull and wave (i.e. maximum overshoot).

9.5.1 Momentum slamming of the destroyer
Consider the same operating conditions as before. Fig. 9.11 shows five curves of $F(x, t)$, one for each of the slices which suffer slamming. The

Fig. 9.11. Force–time history of a slam at each of the five forward sections of the hull. On any particular slice, $F(x, t)$ is assumed to be uniformly distributed in space but to vary with time in the manner shown. The curves are those according to the theory of Leibowitz (1962, 1963).

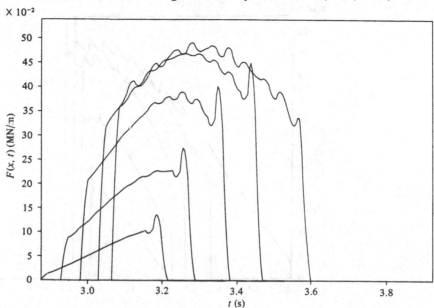

transient load has been calculated in a stepwise manner for intervals of 0.01 s. The smallest curve corresponds to the conditions at the slice whose centre is at $x = 0.975l$ while the largest relates to the aftermost slice whose centre is at $x = 0.775l$. It will be seen that the various pulses begin at the same instants as those indicated in fig. 9.7 but that the time histories at the five slices are now quite different. The force–time history which is equivalent to that shown in fig. 9.8 is illustrated in fig. 9.12. (Again note the difference in actual magnitudes.)

The total responses at the $p_r(t)$ are shown in fig. 9.13. The responses at $x = l/2$ can now be found as before and are shown in fig. 9.14. It will be seen that while the forms of these results differ only slightly from those of fig. 9.10, the differences of magnitude are substantial.

9.6　Concluding remarks

In this chapter the phenomenon of slamming has been analysed as what it undoubtedly is – i.e. as a problem of dynamics. Linear theory has been employed and, although the results are believable, their main practical value may well be as a guide in formulating simpler theory. There is obviously very great scope for useful parametric studies of slamming, though the need for experimental validation is all too plain.

Existing theories have been employed in modelling the slamming and it is clear that much more attention will have to be paid to this phenomenon since the two theories that have been used give very different answers. There appears to be strong physical evidence to support the view that a better representation of a slam would be

Fig. 9.12. Force–time history of a slam according to the Leibowitz theory (1962, 1963) but corresponding to the conditions referred to in fig. 9.8.

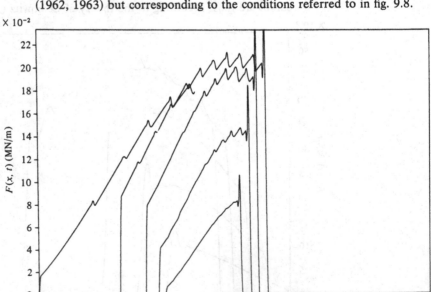

Fig. 9.13. Slamming responses at the coordinates $p_r(t)$, $r = 0, 1, 2, 3, 4$, for the destroyer travelling at 14 m/s in head seas when the excitation is that shown in fig. 9.11.

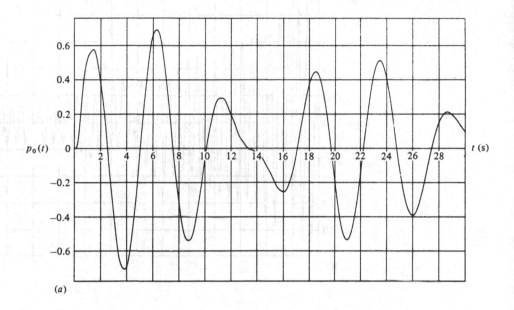

(a)

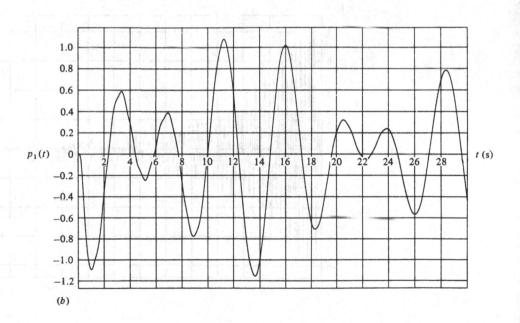

(b)

Fig. 9.13. (Continued)

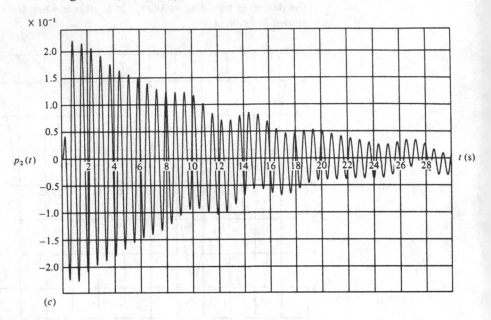

(c)

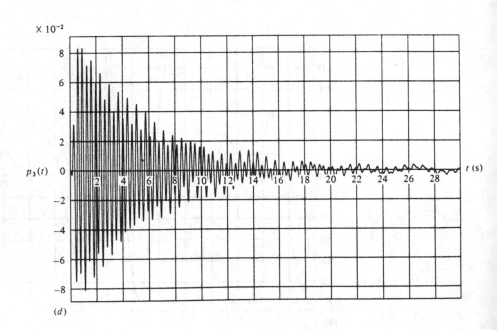

(d)

Fig. 9.13. (Continued)

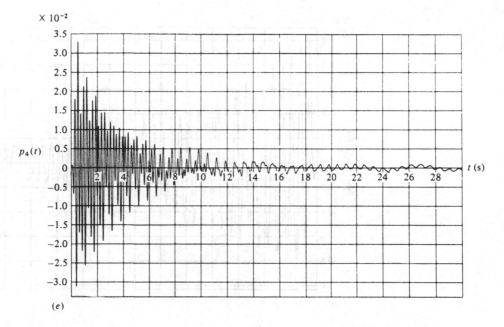

(e)

Fig. 9.14. Responses at the section $x = l/2$ given by the estimated responses shown in fig. 9.13.

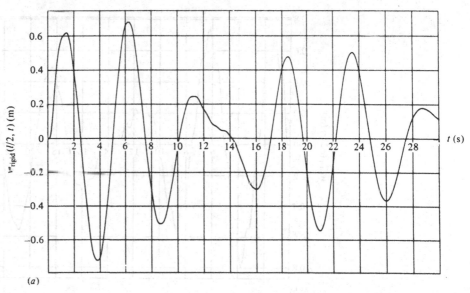

(a)

Fig. 9.14. (Continued)

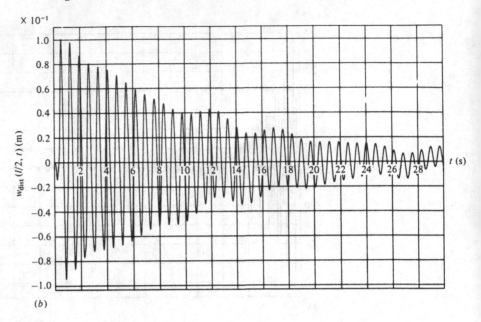

(b)

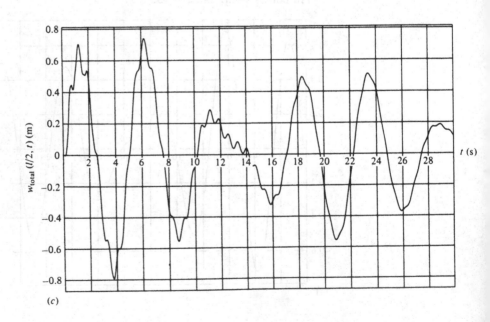

(c)

Fig. 9.14. (Continued)

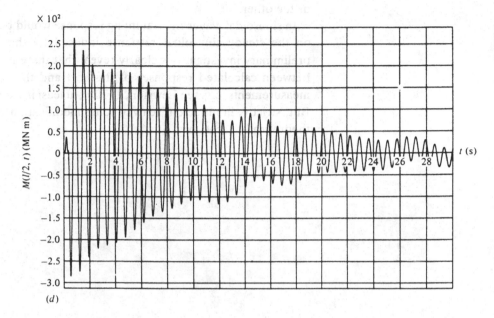

(d)

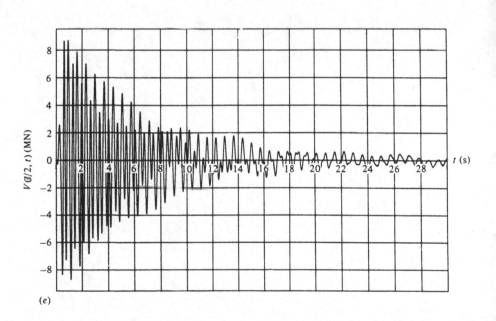

(e)

obtained from both of the theories used together rather than from one or the other.

In sinusoidal waves the slamming response would be superimposed on the steady sinusoidal response induced by the waves. Indeed preliminary investigations clearly reveal that there is great similarity between calculated responses of this sort and the results of actual measurements (e.g. see Lewison, 1970b), at least in a qualitative sense. But, strictly, in an actual seaway some modification of the results is theoretically necessary.

10 Antisymmetric response to wave excitation

There is beauty in the bellow of the blast,
There is grandeur in the growling of the gale,
There is eloquent outpouring
When the lion is a-roaring,
And the tiger is a-lashing of his tail!
The Mikado

10.1 Antisymmetric wave forces

Antisymmetric wave forces (see fig. 10.1) cause a ship to move bodily in sway, roll and yaw and also to distort by lateral bending and by twisting. The response may be examined in modal form, the modes being of the types discussed in chapters 2 and 3. It is probably true to

Fig. 10.1. The wave loading applied to the frigate is causing the vessel to roll. The hull is also twisting and suffering horizontal bending.
[Photograph by courtesy of the Royal Navy.]

say that the behaviour of most immediate interest is likely to be that of roll, though the twisting of a ship with large deck openings also raises some serious questions (see e.g. fig. 2.8).

The calculations that we shall now discuss are much more complicated than those of symmetric response. Even so, the general approach is the same as that which we have already discussed. Although numerous fresh matters of detail crop up (some of which expose our lack of essential information on actual ships) there are no fresh matters of principle. We can therefore complete our discussion of antisymmetric motions in rather less detail. The relevant structural theory having been introduced in chapters 2 and 3, we shall first discuss generalised antisymmetric fluid forces and then present results for a particular ship.

10.2 Relative motions of water and ship in oblique waves

The potential function for a train of sinusoidal waves moving in the fixed AX direction in fig. 6.5 was shown in equation (6.11) to be

$$\Phi = -\frac{ag}{\omega}\frac{\cosh\left[k(Z+d)\right]}{\cosh kd}\sin(kX-\omega t).$$

Now

$$\frac{\cosh\left[k(Z+d)\right]}{\cosh kd}=\frac{e^{k(Z+d)}+e^{-k(Z+d)}}{e^{kd}+e^{-kd}}$$

and for deep water $kd\to\infty$ and so the terms $e^{-k(Z+d)}$, e^{-kd} may be ignored provided $|Z|\ll d$. The velocity potential becomes

$$\Phi = -\frac{ag}{\omega}e^{kZ}\sin(kX-\omega t), \tag{10.1}$$

where, of course $Z\le 0$.

Referring to the equilibrium axes $Oxyz$ in fig. 6.5, we have

$$X = X_0\cos\chi+Y_0\sin\chi = (\bar{U}t+x)\cos\chi+y\sin\chi,$$
$$Z = z,$$

so that

$$\Phi = -\frac{ag}{\omega}e^{kz}\sin(kx\cos\chi+ky\sin\chi-\omega_e t). \tag{10.2}$$

The corresponding surface elevation was found in equation (6.18) to be

$$\zeta(x,y,t)=a\cos(kx\cos\chi+ky\sin\chi-\omega_e t). \tag{10.3}$$

According to the analysis by which the Smith correction is applied, a modified amplitude $a\,e^{-k\bar{T}}$ of surface elevation is assumed. This is held to account for all variations with respect to z, being a weighted average over the draught of the hull. Thus we shall assume that

$$\zeta(x,y,t)=a\,e^{-k\bar{T}}\cos(kx\cos\chi+ky\sin\chi-\omega_e t), \tag{10.4}$$

$$\Phi(x, y, t) = -\frac{ag}{\omega} e^{-k\bar{T}} \sin(kx \cos \chi + ky \sin \chi - \omega_e t). \tag{10.5}$$

During antisymmetric motion of the hull there is relative motion between the water and the hull. This takes two forms, which we now examine separately.

10.2.1 Relative motions of wave slope and hull twist

The component of wave slope in the athwartships direction will be denoted by γ. It is

$$\gamma = \frac{\partial \zeta}{\partial y} = -ak (\sin \chi) e^{-k\bar{T}} \sin(kx \cos \chi + ky \sin \chi - \omega_e t). \tag{10.6}$$

The rates of change of γ are

$$\frac{D\gamma}{Dt} = \left(\frac{\partial}{\partial t} - \bar{U}\frac{\partial}{\partial x}\right) \frac{\partial \zeta}{\partial y} = ak\omega (\sin \chi) e^{-k\bar{T}} \cos(kx \cos \chi + ky \sin \chi - \omega_e t) \tag{10.7}$$

and

$$\frac{D^2\gamma}{Dt^2} = ak\omega^2 (\sin \chi) e^{-k\bar{T}} \sin(kx \cos \chi + ky \sin \chi - \omega_e t). \tag{10.8}$$

The average component of wave slope in the athwartships direction at any given section is

$$\bar{\gamma}(x, t) = \frac{1}{B(x)} \int_{-0.5B(x)}^{0.5B(x)} \gamma(x, y, t) \, dy$$
$$= -ak (\sin \chi) \alpha(x) e^{-k\bar{T}} \sin(kx \cos \chi - \omega_e t), \tag{10.9}$$

where

$$\alpha(x) = \frac{\sin [0.5kB(x) \sin \chi]}{0.5kB(x) \sin \chi}. \tag{10.10}$$

This last quantity, $\alpha(x)$, was introduced by Kaplan (1969), Kaplan, Sargent & Raff (1969) and Bishop & Price (1977), both to account for the variation of wave parameters across the beam of the ship and as a means of allowing for the influence of short waves. The average rates of change of wave slope may be determined in a similar way; they are

$$\frac{\overline{D\gamma}}{Dt} = \frac{1}{B(x)} \int_{-0.5B(x)}^{0.5B(x)} \frac{D\gamma(x, y, t)}{Dt} \, dy$$
$$= ak\omega (\sin \chi) \alpha(x) e^{-k\bar{T}} \cos(kx \cos \chi - \omega_e t), \tag{10.11}$$

$$\frac{\overline{D^2\gamma}}{Dt^2} = ak\omega^2 (\sin \chi) \alpha(x) e^{-k\bar{T}} \sin(kx \cos \chi - \omega_e t). \tag{10.12}$$

The approximate relative angle of roll of the hull section with respect to the waves is

$$\bar{\phi}(x, t) = \phi(x, t) - \bar{\gamma}(x, t), \tag{10.13}$$

where $\phi(x, t)$ is the local angle of twist of the hull.

10.2.2 *Relative velocity in the sway direction*

It will be helpful to dispose of a notation problem right away. Following normal custom we have used the symbol v both for lateral deflection of the hull and for fluid velocity in the Oy-direction. We must now distinguish between the two very carefully. Accordingly, we shall employ the capital script $\mathscr{V}$ to represent the fluid velocity and retain v for lateral displacement in the Oy-direction. We shall only adopt this change of notation in the present chapter.

The component of the orbital velocity in the direction Oy is

$$\mathscr{V}(x, y, t) = -\frac{\partial \Phi}{\partial y} = \frac{agk \sin \chi}{\omega} e^{-k\bar{T}} \cos (kx \cos \chi + ky \sin \chi - \omega_e t).$$

The average value of this quantity across the section is

$$\bar{\mathscr{V}}(x, t) = \frac{1}{B(x)} \int_{-0.5B(x)}^{0.5B(x)} \mathscr{V}(x, y, t)\, dy$$

$$= \frac{agk \sin \chi}{\omega} \alpha(x)\, e^{-k\bar{T}} \cos (kx \cos \chi - \omega_e t).$$

For waves in deep water $k = \omega^2/g$ and so we may write

$$\bar{\mathscr{V}}(x, t) = a\omega\, (\sin \chi)\, \alpha(x)\, e^{-k\bar{T}} \cos (kx \cos \chi - \omega_e t). \tag{10.14}$$

The average component in the direction Oy of the fluid's orbital acceleration may be found in a similar way. The component is

$$\frac{D\mathscr{V}(x, y, t)}{Dt} = agk\, (\sin \chi)\, e^{-k\,T} \sin (kx \cos \chi + ky \sin \chi - \omega_e t)$$

and the required average is

$$\overline{\frac{D\mathscr{V}(x, y, t)}{Dt}} = \frac{1}{B(x)} \int_{-0.5B(x)}^{0.5B(x)} \frac{D\mathscr{V}(x, y, t)}{Dt}\, dy$$

$$= agk\, (\sin \chi)\, \alpha(x)\, e^{-k\bar{T}} \sin (kx \cos \chi - \omega_e t). \tag{10.15}$$

The relative velocity between the ship's section concerned and the fluid particles may be found by writing down both velocities with respect to the Ox-axis and subtracting. This can be done for a point in the plane of port and starboard symmetry located at any required depth above or below the plane Oxy. For reasons that we shall explain later, we shall write down the relative velocity for a point at the level of the still water surface. In fig. 2.9, that is, we make the plane OXY the still water plane so that OX intersects the section of interest at some point A and then we write the relative velocity at A. (Notice that we now revert to our previous labelling of the equilibrium axes as $OXYZ$ instead of $Oxyz$.)

The displacement at A of the ship section is

$$v_A(x, t) = v(x, t) - z_S(x)\phi(x, t),$$

$v(x, t)$ being the deflection at the shear centre S and $z_S(x)$ the distance by which S lies below A. The velocity of A is $Dv_A(x, t)/Dt$ and the required relative velocity is

$$\frac{D\bar{v}(x, t)}{Dt} = \frac{Dv_A(x, t)}{Dt} - \bar{\mathcal{V}}(x, t),$$ (10.16)

it being assumed that a positive relative velocity corresponds to the hull moving faster than the water. In the same way, the relative transverse acceleration is

$$\frac{D^2\bar{v}(x, t)}{Dt^2} = \frac{D^2v_A(x, t)}{Dt^2} - \frac{\overline{D\mathcal{V}}(x, t)}{Dt},$$ (10.17)

since $\overline{D\mathcal{V}}/Dt$ is the average of the fluid's orbital acceleration. Note that terms involving the wave components which are written in the form $D\bar{\mathcal{V}}/Dt, \partial\bar{\mathcal{V}}/\partial t, D\bar{\gamma}/Dt, \partial\bar{\gamma}/\partial x$, etc. imply $\overline{D\mathcal{V}}/Dt, \overline{\partial\mathcal{V}}/\partial t, \overline{D\gamma}/Dt, \overline{\partial\gamma}/\partial x$, etc. For example the rate of change of the average, $D\bar{\mathcal{V}}/Dt$ is not implied but it is the average of the rate of change, $\overline{D\mathcal{V}}/Dt$ which is to be assumed. The approximation of averaging in the athwartships direction is applied to the components of wave slope and fluid orbital motions after the necessary differentiation operation.

10.2.3 *Use of the complex exponential representation*

It will be useful, henceforth, to employ the complex exponential representation. If we agree to take the real part of the complex function, then the relevant results obtained are

$$\left.\begin{array}{l} \bar{\gamma}(x, t) = iak\,(\sin\chi)\,\alpha(x)\,e^{-k\bar{T}}e^{i(kx\cos\chi - \omega_e t)}, \\[2mm] \dfrac{\overline{D\gamma}(x, t)}{Dt} = ak\omega\,(\sin\chi)\,\alpha(x)\,e^{-k\bar{T}}e^{i(kx\cos\chi - \omega_e t)}, \\[2mm] \dfrac{\overline{D^2\gamma}(x, t)}{Dt^2} = -iak\omega^2\,(\sin\chi)\,\alpha(x)\,e^{-k\bar{T}}e^{i(kx\cos\chi - \omega_e t)}, \\[2mm] \bar{\mathcal{V}}(x, t) = a\omega\,(\sin\chi)\,\alpha(x)\,e^{-k\bar{T}}e^{i(kx\cos\chi - \omega_e t)}, \\[2mm] \dfrac{\overline{D\mathcal{V}}(x, t)}{Dt} = -iagk\,(\sin\chi)\,\alpha(x)\,e^{-k\bar{T}}e^{i(kx\cos\chi - \omega_e t)}, \end{array}\right\}$$ (10.18)

corresponding to

$$\left.\begin{array}{l} \Phi(x, y, t) = \dfrac{iag}{\omega}e^{-k\bar{T}}e^{i(kx\cos\chi + ky\sin\chi - \omega_e t)}, \\[2mm] \zeta(x, y, t) = a\,e^{-k\bar{T}}e^{i(kx\cos\chi + ky\sin\chi - \omega_e t)}. \end{array}\right\}$$ (10.19)

It will also be convenient at this point to present the relevant portion of fig. 2.9 again, showing the equilibrium axes and the point A in the still water surface of a typical slice of the hull. We shall henceforth use the information contained in fig. 10.2.

10.3   The fluid actions

The hydrodynamic action on an elemental section of the hull can be subdivided into a 'dynamic' contribution (denoted by the subscript D) and a 'Froude–Krylov' contribution (denoted by a subscript FK). Thus the transverse fluid force per unit length is given by

$$F_A(x, t) = F_D(x, t) + F_{FK}(x, t), \tag{10.20}$$

while the rolling moment per unit length about A, applied by the waves, is

$$K_A(x, t) = K_D(x, t) + K_{FK}(x, t). \tag{10.21}$$

In the equations of motion, (2.42) or (3.64), the fluid actions are defined with respect to the centre of mass C of the section. It follows that

$$\left.\begin{aligned}
Y(x, t) &= F_A(x, t) = F_D(x, t) + F_{FK}(x, t), \\
K(x, t) &= K_A(x, t) - z_C F_A(x, t) \\
&= K_D(x, t) - z_C F_D(x, t) + K_{FK}(x, t) - z_C F_{FK}(x, t) \\
&= K_D(x, t) - z_C F_D(x, t) + K_C(x, t),
\end{aligned}\right\} \tag{10.22}$$

where $K_C(x, t)$ is the Froude–Krylov contribution to the roll moment about the centre of mass C. It is assumed that the dynamic contribution consists of an in phase (or 'added mass') term and a quadrature (or 'fluid damping') term.

10.3.1 Dynamic contributions to fluid actions

In the theories discussed by Salvesen, Tuck & Faltinsen (1970) and by Vugts (1971), the hydrodynamic actions are found from potential flow theory in which the fluid is assumed irrotational, incompressible and inviscid. The strip theory approximation is introduced in the latter stages of the analysis. Using such a theory, Wahab & Vink (1973) found the dynamic fluid force per unit length in the form

$$\begin{aligned}
F_D(x, t) = &-\frac{D}{Dt}\left\{\left[m_y(x) + \frac{i}{\omega_e} N_y(x)\right] \frac{D\bar{v}(x, t)}{Dt}\right\} \\
&-\frac{D}{Dt}\left\{\left[m_{y\phi}(x) + \frac{i}{\omega_e} N_{y\phi}(x)\right] \frac{D\bar{\phi}(x, t)}{Dt}\right\}
\end{aligned} \tag{10.23}$$

Fig. 10.2. The section of a ship deflected antisymmetrically with respect to equilibrium axes $OXYZ$.

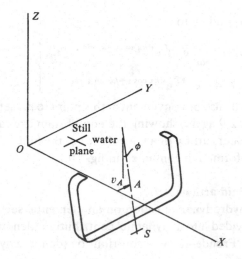

and the dynamic rolling moment per unit length in the form

$$K_D(x, t) = -\frac{D}{Dt}\left\{\left[m_{\phi y}(x) + \frac{i}{\omega_e}N_{\phi y}(x)\right]\frac{D\bar{v}(x, t)}{Dt}\right\}$$

$$-\frac{D}{Dt}\left\{\left[I_x(x) + \frac{i}{\omega_e}N_\phi(x)\right]\frac{D\bar{\phi}(x, t)}{Dt}\right\}, \qquad (10.24)$$

where $m_y(x), N_y(x)$ are the added mass and damping coefficients associated with sway, $m_{y\phi}(x), N_{y\phi}(x)$ are the cross-coupling sway added mass and damping coefficients due to roll, $I_x(x)$ and $N_\phi(x)$ are the roll added moment of inertia and damping coefficients due to pure roll, while $m_{\phi y}(x), N_{\phi y}(x)$ are the cross-coupling roll added moment of inertia and damping coefficients due to sway. This is 'Theory B'.

For ease of writing, the functions $m_y(x), N_{y\phi}(x)$, etc. have all been written as if they depend solely on position x. In fact they are also functions of the frequency of encounter ω_e as well, that is to say

$$m_y(x) \equiv m_y(x, \omega_e), \text{ etc.}$$

Other forms of strip theory have been proposed (e.g. see Raff, 1972; Flokstra, 1974). In particular, another suggested expression for the dynamic fluid actions per unit length is ('Theory A')

$$F_D(x, t) = -\frac{D}{Dt}\left[m_y(x)\frac{D\bar{v}(x, t)}{Dt}\right] - N_y(x)\frac{D\bar{v}}{Dt}$$

$$-\frac{D}{Dt}\left[m_{y\phi}(x)\frac{D\bar{\phi}(x, t)}{Dt}\right] - N_{y\phi}(x)\frac{D\bar{\phi}(x, t)}{Dt}, \qquad (10.25)$$

$$K_D(x, t) = -\frac{D}{Dt}\left[m_{\phi y}(x)\frac{D\bar{v}(x, t)}{Dt}\right] - N_{\phi y}(x)\frac{D\bar{v}}{Dt}$$

$$-\frac{D}{Dt}\left[I_x(x)\frac{D\bar{\phi}(x, t)}{Dt}\right] - N_\phi(x)\frac{D\bar{\phi}(x, t)}{Dt}. \qquad (10.26)$$

On comparison with the previous expressions for $F_D(x, t)$ and $K_D(x, t)$ it will be seen that the terms involving fluid damping differ. The differences apparently originate in the interpretation of the 'momentum of the fluid motion'. We shall employ the expressions (10.23) and (10.24) in the subsequent analysis; indeed the theory we shall adopt contains that represented by expressions of the type (10.25) and (10.26) and we shall underline the additional terms. It should be said that, in adapting these existing theories to our present needs, we take them to some extent out of their originally intended context. Strictly, then, they are open to further discussion.

10.3.2   *Froude–Krylov contributions to fluid actions*
Expressions may be written down for $F_{FK}(x, t)$ and $K_C(x, t)$. Suppose the average wave slope is $\bar{\gamma}$, as indicated in fig. 10.3. The Froude (1861) and Krylov (1896) assumption is that the fluid action is represented by a quasi-hydrostatic buoyancy force $F_B(x, t)$ per unit length whose magnitude is what it would be if the hull were replaced by water.

If $S(x)$ is the sectional submerged area

$$F_B = \rho g S(x).$$

It follows that

$$F_{FK}(x, t) = -\rho g S(x) \bar{\gamma}$$
$$= -iak\rho g S(x)\alpha(x) \sin \chi \; e^{-k\bar{T}} e^{i(kx \cos \chi - \omega_e t)}. \qquad (10.27)$$

The distance from B, the centroid of $S(x)$ to the local metacentre M is given by the well-known expression

$$BM = \frac{I}{\nabla} = \frac{I}{S(x) \, \Delta x},$$

where I is the appropriate second moment of area in the water plane and Δx is the thickness of the slice of the hull; i.e.

$$BM = \frac{[B(x)]^3}{12 S(x)}.$$

If the slice is rotated by the angle $\bar{\phi}$ relative to the average wave slope, the centre of buoyancy moves from B to B' and the moment of F_B about C is

$$K_C(x, t) \, \Delta x = -F_B \overline{CM} \bar{\phi}$$
$$= -\rho g S(x) \, \Delta x [\overline{BM} - \overline{BC}] \bar{\phi}$$
$$= -\rho g \, \Delta x \left\{ \frac{[B(x)]^3}{12} - \overline{BC} \, S(x) \right\} \bar{\phi}.$$

That is to say

$$K_C(x, t) = -\rho g \left\{ \frac{[B(x)]^3}{12} - \overline{BC} \, S(x) \right\} \bar{\phi}, \qquad (10.28)$$

which agrees with the form proposed by Conolly (1969).

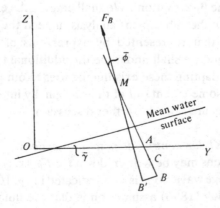

Fig. 10.3. If the hull slice has an angle of roll equal to the local mean slope $\bar{\gamma}$ of the water surface, so that $\bar{\phi} = 0$, the direction of the Froude–Krylov quasi-hydrostatic buoyancy force F_B is inclined at the angle $\bar{\gamma}$ to the vertical and acts through the centre of buoyancy B. If $\bar{\phi} \neq 0$, B moves to B' and the angle between F_B and the line $BCAM$ of port and star-board symmetry of the section is $\bar{\phi}$, M being the local metacentre.

10.4   Generalised fluid actions

Equations (2.42) or (3.64) show that the sth generalised fluid force may be written as

$$\int_0^l [(v_s - \bar{z}\phi_s)Y + \phi_s K]\,dx$$

$$= \int_0^l [(v_s - z_s\phi_s)F_D + \phi_s(K_D + K_C) + (v_s - \bar{z}\phi_s)F_{FK}]\,dx,$$

where $s = 0, 1, 2, 3, \ldots$, since $\bar{z} = z_S - z_C$. (Note that the subscript S of z_S refers to the shear centre and is not a modal index.) Now

$$\frac{D\bar{v}(x,t)}{Dt} = \left(\frac{\partial}{\partial t} - \bar{U}\frac{\partial}{\partial x}\right)(v - z_s\phi) - \bar{\mathscr{V}}$$

$$= \sum_{r=0}^{\infty} \dot{p}_r(v_r - z_s\phi_r) - \bar{U}\sum_{r=0}^{\infty} p_r(v_r - z_s\phi_r)' - \bar{\mathscr{V}}$$

and, taking by way of example the first term in the contribution of $F_D(x, t)$ in the form (10.23) to the sth generalised force, we have

$$-\int_0^l (v_s - z_s\phi_s)\frac{D}{Dt}\left[\left(m_y + \frac{i}{\omega_e}N_y\right)\frac{D\bar{v}(x,t)}{Dt}\right]dx$$

$$= -\sum_{r=0}^{\infty} \ddot{p}_r \int_0^l \left(m_y + \frac{iN_y}{\omega_e}\right)(v_s - z_s\phi_s)(v_r - z_s\phi_r)\,dx$$

$$+ \bar{U}\sum_{r=0}^{\infty} \dot{p}_r \int_0^l \left(m_y + \frac{iN_y}{\omega_e}\right)[(v_s - z_s\phi_s)(v_r - z_s\phi_r)'$$

$$- (v_r - z_s\phi_r)(v_s - z_s\phi_s)']\,dx$$

$$+ \bar{U}^2 \sum_{r=0}^{\infty} p_r \int_0^l \left(m_y + \frac{iN_y}{\omega_e}\right)(v_s - z_s\phi_s)'(v_r - z_s\phi_r)'\,dx$$

$$+ \bar{U}\sum_{r=0}^{\infty} \dot{p}_r \left[\left(m_y + \frac{iN_y}{\omega_e}\right)(v_s - z_s\phi_s)(v_r - z_s\phi_r)\right]_0^l$$

$$- \bar{U}^2 \sum_{r=0}^{\infty} p_r \left[\left(m_y + \frac{iN_y}{\omega_e}\right)(v_s - z_s\phi_s)(v_r - z_s\phi_r)'\right]_0^l$$

$$+ \int_0^l (v_s - z_s\phi_s)\left[\left(m_y + \frac{iN_y}{\omega_e}\right)\frac{D\bar{\mathscr{V}}}{Dt} - \bar{U}\left(m_y + \frac{iN_y}{\omega_e}\right)\bar{\mathscr{V}}\right]dx$$

$$(s = 0, 1, 2, 3, \ldots). \quad (10.29)$$

For a sinusoidal response $p_r(t)$ expressed in the form

$$p_r(t) = p_r\,e^{-i\omega_e t}, \qquad \dot{p}_r(t) = -i\omega_e p_r(t), \qquad \ddot{p}_r(t) = -\omega_e^2 p_r(t),$$

the sth generalised fluid force can be found by proceeding in this way. It is straightforward, if tedious, now to write out the complete expression for the required generalised fluid force. That part of the expression which contains $p_r(t)$ and its derivatives may be written

$$-\sum_{r=0}^{\infty} (\ddot{p}_r A_{rs} + \dot{p}_r B_{rs} + p_r C_{rs}).$$

Here, the coefficients A_{rs}, B_{rs} and C_{rs} are given by

$$
\begin{aligned}
A_{rs} = & \int_0^l (v_s - z_S\phi_s)[m_y(v_r - z_S\phi_r) + m_{y\phi}\phi_r]\,\mathrm{d}x \\
& + \int_0^l \phi_s[I_x\phi_r + m_{\phi y}(v_r - z_S\phi_r)]\,\mathrm{d}x \\
& + \frac{\bar{U}}{\omega_e^2}\int_0^l (v_s - z_S\phi_s)[N_y(v_r - z_S\phi_r)' + N_{y\phi}\phi_r']\,\mathrm{d}x \\
& + \frac{\bar{U}}{\omega_e^2}\int_0^l \phi_s[N_\phi\phi_r' + N_{\phi y}(v_r - z_S\phi_r)']\,\mathrm{d}x \\
& - \frac{\bar{U}}{\omega_e^2}\int_0^l (v_s - z_S\phi_s)'\underline{[N_y(v_r - z_S\phi_r) + N_{y\phi}\phi_r]}\,\mathrm{d}x \\
& - \frac{\bar{U}}{\omega_e^2}\int_0^l \phi_s'[N_\phi\phi_r + N_{\phi y}(v_r - z_S\phi_r)]\,\mathrm{d}x \\
& + \frac{\bar{U}^2}{\omega_e^2}\int_0^l (v_s - z_S\phi_s)'[m_y(v_r - z_S\phi_r)' + m_{y\phi}\phi_r']\,\mathrm{d}x \\
& + \frac{\bar{U}^2}{\omega_e^2}\int_0^l \phi_s'[I_x\phi_r' + m_{\phi y}(v_r - z_S\phi_r)']\,\mathrm{d}x \\
& + \frac{\bar{U}}{\omega_e^2}\Big[(v_s - z_S\phi_s)\underline{\{N_y(v_r - z_S\phi_r) + N_{y\phi}\phi_r\}} \\
& \qquad + \phi_s\underline{\{N_\phi\phi_r + N_{\phi y}(v_r - z_S\phi_r)\}}\Big]_0^l \\
& - \frac{\bar{U}^2}{\omega_e^2}\Big[(v_s - z_S\phi_s)\{m_y(v_r - z_S\phi_r)' + m_{y\phi}\phi_r'\}\Big]_0^l \\
& - \frac{\bar{U}^2}{\omega_e^2}\Big[\phi_s\{I_x\phi_r' + m_{\phi y}(v_r - z_S\phi_r)'\}\Big]_0^l,
\end{aligned}
\tag{10.30}
$$

$$
\begin{aligned}
B_{rs} = & \int_0^l (v_s - z_S\phi_s)[N_y(v_r - z_S\phi_r) + N_{y\phi}\phi_r]\,\mathrm{d}x \\
& + \int_0^l \phi_s[N_\phi\phi_r + N_{\phi y}(v_r - z_S\phi_r)]\,\mathrm{d}x \\
& - \bar{U}\int_0^l (v_s - z_S\phi_s)[m_y(v_r - z_S\phi_r)' + m_{y\phi}\phi_r']\,\mathrm{d}x \\
& - \bar{U}\int_0^l \phi_s[I_x\phi_r' + m_{\phi y}(v_r - z_S\phi_r)']\,\mathrm{d}x \\
& + \bar{U}\int_0^l (v_s - z_S\phi_s)'[m_y(v_r - z_S\phi_r) + m_{y\phi}\phi_r]\,\mathrm{d}x \\
& + \bar{U}\int_0^l \phi_s'[I_x\phi_r + m_{\phi y}(v_r - z_S\phi_r)]\,\mathrm{d}x \\
& + \frac{\bar{U}^2}{\omega_e^2}\int_0^l (v_s - z_S\phi_s)'\underline{[N_y(v_r - z_S\phi_r)' + N_{y\phi}\phi_r']}\,\mathrm{d}x
\end{aligned}
$$

$$+\frac{\bar{U}^2}{\omega_e^2}\int_0^l \phi_s'\left[N_\phi\phi_r' + N_{\phi y}(v_r - z_S\phi_r)'\right]\mathrm{d}x$$

$$-\bar{U}\left[(v_s - z_S\phi_s)\{m_y(v_r - z_S\phi_r) + m_{y\phi}\phi_r\}\right.$$

$$\left. +\phi_s\{I_x\phi_r + m_{\phi y}(v_r - z_S\phi_r)\}\right]_0^l$$

$$-\frac{\bar{U}^2}{\omega_e^2}\left[(v_s - z_S\phi_s)\{N_y(v_r - z_S\phi_r)' + N_{y\phi}\phi_r'\}\right]_0^l$$

$$-\frac{\bar{U}^2}{\omega_e^2}\left[\phi_s\{N_\phi\phi_r' + N_{\phi y}(v_r - z_S\phi_r)'\}\right]_0^l \qquad (10.31)$$

and

$$C_{rs} = \int_0^l \rho g \phi_r \phi_s\left[\frac{B^3}{12} - \overline{BC}\,S(x)\right]\mathrm{d}x. \qquad (10.32)$$

We turn next to those terms in the expressions for the generalised fluid force which do not depend on the $p_r(t)$. The complete expression may be written

$$\Xi_s(t) = \int_0^l (v_s - z_S\phi_s)$$

$$\times\left\{\frac{\mathrm{D}}{\mathrm{D}t}\left[\left(m_y + \frac{\mathrm{i}}{\omega_e}N_y\right)\bar{V}\right] + \frac{\mathrm{D}}{\mathrm{D}t}\left[\left(m_{y\phi} + \frac{\mathrm{i}}{\omega_e}N_{y\phi}\right)\frac{\mathrm{D}\bar{\gamma}}{\mathrm{D}t}\right]\right\}\mathrm{d}x$$

$$+\int_0^l \phi_s\left\{\frac{\mathrm{D}}{\mathrm{D}t}\left[\left(m_{\phi y} + \frac{\mathrm{i}}{\omega_e}N_{\phi y}\right)\bar{V}\right] + \frac{\mathrm{D}}{\mathrm{D}t}\left[\left(I_x + \frac{\mathrm{i}}{\omega_e}N_\phi\right)\frac{\mathrm{D}\bar{\gamma}}{\mathrm{D}t}\right]\right\}\mathrm{d}x$$

$$-\int_0^l (v_s - \bar{z}\phi_s)\rho g S(x)\bar{\gamma}\,\mathrm{d}x$$

$$+\int_0^l \phi_s\rho g\left[\frac{B^3}{12} - \overline{BC}\cdot S(x)\right]\bar{\gamma}\,\mathrm{d}x \qquad (s = 0, 1, 2, \ldots). \quad (10.33)$$

It follows that

$$\Xi_s(t) = \int_0^l (v_s - z_S\phi_s)\left[\frac{\mathrm{D}}{\mathrm{D}t}(m_y\bar{V}) + \frac{\mathrm{i}N_y}{\omega_e}\frac{\partial\bar{V}}{\partial t} - \frac{\mathrm{i}\bar{U}}{\omega_e}\frac{\partial(N_y\bar{V})}{\partial x}\right.$$

$$\left. +\frac{\mathrm{D}}{\mathrm{D}t}\left(m_{y\psi}\frac{\mathrm{D}\bar{\gamma}}{\mathrm{D}t}\right) + \frac{\mathrm{i}}{\omega_e}N_{y\phi}\frac{\partial}{\partial t}\left(\frac{\mathrm{D}\bar{\gamma}}{\mathrm{D}t}\right) - \frac{\mathrm{i}\bar{U}}{\omega_e}\frac{\partial}{\partial x}\left(N_{y\phi}\frac{\mathrm{D}\bar{\gamma}}{\mathrm{D}t}\right)\right]\mathrm{d}x$$

$$-\int_0^l (v_s - \bar{z}\phi_s)\rho g S(x)\bar{\gamma}\,\mathrm{d}x + \int_0^l \phi_s\left\{\frac{\mathrm{D}}{\mathrm{D}t}(m_{\phi y}\bar{V}) + \frac{\mathrm{i}}{\omega_e}N_{\phi y}\frac{\partial}{\partial t}\bar{V}\right.$$

$$-\frac{\mathrm{i}\bar{U}}{\omega_e\partial x}(N_{\phi y}\bar{V}) + \frac{\mathrm{D}}{\mathrm{D}t}\left(I_x\frac{\mathrm{D}\bar{\gamma}}{\mathrm{D}t}\right) + \frac{\mathrm{i}}{\omega_e}N_\phi\frac{\partial}{\partial t}\left(\frac{\mathrm{D}\bar{\gamma}}{\mathrm{D}t}\right)$$

$$\left. -\frac{\mathrm{i}\bar{U}\partial}{\omega_e\partial x}\left(N_\phi\frac{\mathrm{D}\bar{\gamma}}{\mathrm{D}t}\right) + \rho g\left[\frac{B^3}{12} - \overline{BC}\,S(x)\right]\bar{\gamma}\right\}\mathrm{d}x$$

$$(\text{for } s = 0, 1, 2, \ldots). \quad (10.34)$$

For deep water, we may use the values (10.18) for $\bar{\gamma}$ and $\mathrm{D}\bar{\gamma}/\mathrm{D}t$; notice that, for example, $\mathrm{D}\bar{\gamma}/\mathrm{D}t \Rightarrow \overline{\mathrm{D}\gamma/\mathrm{D}t}$, etc. The generalised fluid force reduces to

$$\Xi_s(t) = e^{-i\omega_e t} \int_0^l [(v_s - \bar{z}\phi_s)T_1 + (v_s - z_s\phi_s)(T_2 - iT_3) + \phi_s(T_4 - iT_5)]\, dx$$

$$= \Xi_s e^{-i\omega_e t} \quad (s = 0, 1, 2, \ldots), \tag{10.35}$$

where

$$
\left.
\begin{aligned}
T_1 &= -i\rho\omega^2 S(x)T_0, \\[4pt]
T_2 &= \omega\left\{\left[\left(1 + \underline{\frac{\bar{U}k\cos\chi}{\omega_e}}\right)N_y - \bar{U}m_y'\right]\right. \\
&\quad \left. + k\left[\left(1 + \underline{\frac{\bar{U}k\cos\chi}{\omega_e}}\right)N_{y\phi} - \bar{U}m_{y\phi}'\right]\right\}T_0, \\[4pt]
T_3 &= \omega^2\left[m_y + \underline{\frac{\bar{U}N_y'}{\omega\omega_e}} + k\left(m_{y\phi} + \underline{\frac{\bar{U}N_{y\phi}'}{\omega\omega_e}}\right)\right]T_0, \\[4pt]
T_4 &= \omega\left\{\left[\left(1 + \underline{\frac{\bar{U}k\cos\chi}{\omega_e}}\right)N_{\phi y} - \bar{U}m_{\phi y}'\right]\right. \\
&\quad \left. + k\left[\left(1 + \underline{\frac{\bar{U}k\cos\chi}{\omega_e}}\right)N_\phi - \bar{U}I_x'\right]\right\}T_0, \\[4pt]
T_5 &= \omega^2\left\{m_{\phi y} + \underline{\frac{\bar{U}N_{\phi y}'}{\omega\omega_e}} + k\left(I_x + \underline{\frac{\bar{U}N_\phi'}{\omega\omega_e}}\right)\right. \\
&\quad \left. - \rho\left[\frac{B^3}{12} - \overline{BC}\, S(x)\right]\right\}T_0,
\end{aligned}
\right\} \tag{10.36}
$$

the quantity T_0 being given by

$$T_0 = a\alpha(x)\sin\chi\, e^{-k\bar{T}}\, e^{ikx\cos\chi}. \tag{10.37}$$

It will be noticed that if the theory upon which equations (10.25) and (10.26) are based is used, the expressions are simplified by omission of the underlined terms. If the theory of equations (10.23) and (10.24) is employed, on the other hand, another form of simplification is valid; thus, in the expressions for T_2 and T_4, equation (10.13) shows that

$$1 + \frac{\bar{U}k\cos\chi}{\omega_e} = \frac{\omega}{\omega_e}.$$

In the remainder of this chapter the underlining of terms to indicate the difference between the two representations of the fluid actions is discontinued.

The complete set of equations can now be expressed in the form

$$\sum_{r=0}^{\infty} [a_{rs}\delta_{rs}\ddot{p}_r + \omega_r^2 a_{rs}\delta_{rs}p_r + (\alpha_{rs} + \beta_{rs} + \Gamma_{rs})\dot{p}_r]$$

$$+ \sum_{r=0}^{\infty} (\ddot{p}_r A_{rs} + \dot{p}_r B_{rs} + p_r C_{rs}) = \Xi_s\, e^{-i\omega_e t} \quad (s = 0, 1, 2, 3, \ldots). \tag{10.38}$$

This familiar set of equations readily lends itself to a matrix formulation. Notice that, in it, the $p_r(t)$ are associated with distortions (of frequency ω_e) at the centre of shear S.

10.5   Special cases

It will be convenient to examine two special cases of the general equations (10.38), not merely by way of illustration, but because they are of importance in their own right.

10.5.1   *The rigid body modes*

Consider the equations governing the rigid body modes ($r, s = 0, 1, 2$), as in the 'seakeeping' analysis of sway ($r = 0$), yaw ($r = 1$) and roll ($r = 2$) motions. The appropriate components of fluid action in the first three of equations (2.42) or (3.64) are given by

$$A_{00} = \int_0^l m_y \, dx + \frac{\bar{U}}{\omega_e^2} [N_y]_0^l,$$

$$B_{00} = \int_0^l N_y \, dx - \bar{U}[m_y]_0^l,$$

$$A_{10} = \int_0^l v_1 m_y \, dx + \frac{\bar{U}}{\omega_e^2} \int_0^l v_1' N_y \, dx + \frac{\bar{U}}{\omega_e^2} [v_1 N_y]_0^l - \frac{\bar{U}^2}{\omega_e^2} [v_1' m_y]_0^l,$$

$$B_{10} = \int_0^l v_1 N_y \, dx - \bar{U} \int_0^l v_1' m_y \, dx - \bar{U} [v_1 m_y]_0^l - \frac{\bar{U}^2}{\omega_e^2} [v_1' N_y]_0^l,$$

$$A_{20} = \int_0^l m_{y\phi} \, dx + \frac{\bar{U}}{\omega_e^2} [N_{y\phi}]_0^l,$$

$$B_{20} = \int_0^l N_{y\phi} \, dx - \bar{U}[m_{y\phi}]_0^l,$$

$$C_{00} = 0 = C_{10} = C_{20},$$

$$A_{01} = \int_0^l v_1 m_y \, dx - \frac{\bar{U}}{\omega_e^2} \int_0^l v_1' N_y \, dx + \frac{\bar{U}}{\omega_e^2} [v_1 N_y]_0^l,$$

$$B_{01} = \int_0^l v_1 N_y \, dx + \bar{U} \int_0^l v_1' m_y \, dx - \bar{U} [v_1 m_y]_0^l,$$

$$A_{11} = \int_0^l v_1^2 m_y \, dx + \frac{\bar{U}^2}{\omega_e^2} \int_0^l (v_1')^2 m_y \, dx + \frac{\bar{U}}{\omega_e^2} [v_1^2 N_y]_0^l - \frac{\bar{U}^2}{\omega_e^2} [v_1 v_1' m_y]_0^l,$$

$$B_{11} = \int_0^l v_1^2 N_y \, dx + \frac{\bar{U}^2}{\omega_e^2} \int_0^l (v_1')^2 N_y \, dx - \bar{U} [v_1^2 m_y]_0^l - \frac{\bar{U}^2}{\omega_e^2} [v_1 v_1' N_y]_0^l,$$

$$A_{21} = \int_0^l v_1 m_{y\phi} \, dx - \frac{\bar{U}}{\omega_e^2} \int_0^l v_1' N_{y\phi} \, dx + \frac{\bar{U}}{\omega_e^2} [v_1 N_{y\phi}]_0^l,$$

$$B_{21} = \int_0^l v_1 N_{y\phi} \, dx + \bar{U} \int_0^l v_1' m_{y\phi} \, dx - \bar{U}[v_1 m_{y\phi}]_0^l,$$

$$C_{01} = 0 = C_{11} = C_{21},$$

$$A_{02} = \int_0^l m_{\phi y}\, dx + \frac{\bar{U}}{\omega_e^2}[N_{\phi y}]_0^l,$$

$$B_{02} = \int_0^l N_{\phi y}\, dx - \bar{U}[m_{\phi y}]_0^l,$$

$$A_{12} = \int_0^l v_1 m_{\phi y}\, dx + \frac{\bar{U}}{\omega_e^2}\int_0^l v_1' N_{\phi y}\, dx + \frac{\bar{U}}{\omega_e^2}[v_1 N_{\phi y}]_0^l - \frac{\bar{U}^2}{\omega_e^2}[v_1' m_{\phi y}]_0^l,$$

$$B_{12} = \int_0^l v_1 N_{\phi y}\, dx - \bar{U}\int_0^l v_1' m_{\phi y}\, dx - \bar{U}[v_1 m_{\phi y}]_0^l - \frac{\bar{U}^2}{\omega_e^2}[v_1' N_{\phi y}]_0^l,$$

$$C_{12} = 0 = C_{02},$$

$$A_{22} = \int_0^l I_x\, dx + \frac{\bar{U}}{\omega_e^2}[N_\phi]_0^l,$$

$$B_{22} = \int_0^l N_\phi\, dx - \bar{U}[I_x]_0^l,$$

$$C_{22} = \int_0^l \rho g\left[\frac{B^3}{12} - \overline{BC}\, S(x)\right] dx = \overline{CM}\cdot\Delta,$$

where $\overline{CM}$ is the metacentric height and Δ is the weight of water displaced.

For the rigid body modes, the amplitudes of the generalised wave forces in equation (10.35) are

$$\Xi_0 = \int_0^l (T_1 + T_2 - iT_3)\, dx,$$

$$\Xi_1 = \int_0^l v_1(T_1 + T_2 - iT_3)\, dx,$$

$$\Xi_2 = \int_0^l [z_C T_1 + (T_4 - iT_5)]\, dx.$$

We are thus in a position to write down the equations governing antisymmetric seakeeping responses and they will be found to agree with those given by Salvesen *et al.* (1970).

Suppose that the antisymmetric seakeeping motions of a ship are examined under the conventional assumption that the hull is rigid, p_0, p_1, and p_2 only being allowed to vary. If it were possible by some means to shift the line of shear centres, the vessel's predicted behaviour should not thereby be altered; for the shift could only affect distortions of the hull, i.e. the responses at $p_3, p_4, \ldots$ which are assumed to be suppressed. This is reflected in the equations since the quantity z_S no longer appears in them.[†]

---

† If, as in certain work published previously by the writers, the component $v_2(x) = z_S(x)\phi_2(x)$ is ignored, this ceases to be so.

10.5.2 *The 'boxlike ship' approximation*

It was shown in chapters 2 and 3 that for a boxlike ship in which C and S coincide (i.e. such that $\bar{z}(x) = 0$), the equations relating to the transverse bending and twisting distortions fall into two independent sets. This is because the principal coordinates relating to horizontal bending of the dry hull are unrelated to those governing twisting. That is to say, for a boxlike ship we write

$$v(x, t) = \sum_{r=0}^{\infty} p_r(t) v_r(x), \qquad \phi(x, t) = \sum_{i=0}^{\infty} q_i(t) \phi_i(x), \qquad (10.39)$$

where $p_r(t)$ and $q_i(t)$ are the rth and ith principal coordinates for the two independent sets of modes.

The equations of transverse bending are now

$$\left.\begin{aligned}
a_{00}\ddot{p}_0 &= \int_0^l Y \, dx, \\
a_{11}\ddot{p}_1 &= \int_0^l v_1 Y \, dx, \\
a_{ss}\ddot{p}_s + \sum_{r=2}^{\infty} (\alpha_{rs} + \beta_{rs})\dot{p}_r + \omega_s^2 a_{ss} p_s &= \int_0^l v_s Y \, dx \quad (s = 2, 3, \ldots),
\end{aligned}\right\} \qquad (10.40)$$

where

$$\left.\begin{aligned}
a_{ss} &= \int_0^l (\mu v_s^2 + I_z \theta_s^2) \, dx, \\
c_{ss} &= \omega_s^2 a_{ss} = \int_0^l (EI\theta_s'^2 + kAG\gamma_s^2) \, dx.
\end{aligned}\right\} \qquad (10.41)$$

The rigid modes are

$$\left.\begin{aligned}
v_0(x) &= 1, & \theta_0(x) &= 0 = \gamma_0(x), \\
v_1(x) &= 1 - \frac{x}{\bar{x}}, & \theta_1(x) &= 0 = \gamma_1(x),
\end{aligned}\right\} \qquad (10.42)$$

and ω_s is the natural frequency of pure bending.

Turning next to the twisting deflections we find that

$$\left.\begin{aligned}
\bar{a}_{00}\ddot{q}_0 &= \int_0^l K \, dx, \\
\bar{a}_{jj}\ddot{q}_j + \sum_{i=1}^{\infty} \Gamma_{ij}\dot{q}_i + \Omega_j^2 \bar{a}_{jj} q_j &= \int_0^l K\phi_j \, dx \quad (j = 1, 2, \ldots),
\end{aligned}\right\} \qquad (10.43)$$

where

$$\bar{a}_{jj} = \int_0^l I_C \phi_j^2 \, dx, \qquad \bar{c}_{jj} = \Omega_j^2 \bar{a}_{jj} = \int_0^l C(x)(\phi_j')^2 \, dx, \qquad (10.44)$$

since there is now no warping of the hull sections and

$$\phi_0(x) = 1. \qquad (10.45)$$

(The 'bar' notation is introduced merely to distinguish twisting from bending.) It will be recalled that the mode $\phi_0(x) = 1$ introduces a component $v(x) = z_C(x)\phi_0(x)$.

The generalised fluid forces, as given by equations (10.22), are now found to be

$$
\left.
\begin{aligned}
&\int_0^l v_s Y \, dx = \int_0^l v_s F_A \, dx \qquad (s = 0, 1, 2, \dots) \\
&\text{and} \\
&\int_0^l \phi_j K \, dx = \int_0^l \phi_j (K_A - z_C F_A) \, dx \qquad (j = 0, 1, 2, \dots).
\end{aligned}
\right\}
\tag{10.46}
$$

These quantities may be found in the same way as before, and then broken down into the contributions that we identified previously.

The contribution to the generalised fluid force from the hull motion only may be written

$$
-\sum_{r=0}^\infty (\ddot{p}_r A_{rs} + \dot{p}_r B_{rs} + p_r C_{rs}) - \sum_{i=0}^\infty (\ddot{q}_i D_{is} + \dot{q}_i E_{is} + q_i F_{is})
\tag{10.47}
$$

in the bending equation, for we have no assurance that the fluid actions will fail to couple the bending and twisting motions. Again the contribution to the generalised fluid force in the twisting equation, that is attributable to hull motion may be written in the form

$$
-\sum_{r=0}^\infty (\ddot{p}_r K_{rj} + \dot{p}_r L_{rj} + p_r M_{rj}) - \sum_{i=0}^\infty (\ddot{q}_i N_{ij} + \dot{q}_i R_{ij} + q_i T_{ij}).
\tag{10.48}
$$

If these contributions are isolated, the various coefficients are found to be

$$
A_{rs} = \int_0^l v_r v_s m_y \, dx + \frac{\bar{U}^2}{\omega_e^2} \int_0^l v_r' v_s' m_y \, dx - \frac{\bar{U}^2}{\omega_e^2} [v_r' v_s m_y]_0^l
$$
$$
+ \frac{\bar{U}}{\omega_e^2} \int_0^l (v_r' v_s - v_r v_s') N_y \, dx + \frac{\bar{U}}{\omega_e^2} [v_r v_s N_y]_0^l,
$$

$$
B_{rs} = \int_0^l v_r v_s N_y \, dx + \frac{\bar{U}^2}{\omega_e^2} \int_0^l v_r' v_s' N_y \, dx - \frac{\bar{U}^2}{\omega_e^2} [v_r' v_s N_y]_0^l
$$
$$
- \bar{U} \int_0^l (v_r' v_s - v_r v_s') m_y \, dx - \bar{U} [v_r v_s m_y]_0^l,
$$

$$
C_{rs} = 0,
$$

$$
D_{0s} = \int_0^l v_s m_{y\phi} \, dx - \frac{\bar{U}}{\omega_e^2} \int_0^l v_s' N_{y\phi} \, dx + \frac{\bar{U}}{\omega_e^2} [v_s N_{y\phi}]_0^l,
$$

$$
D_{is} = -\int_0^l z_C \phi_i v_s m_y \, dx - \frac{\bar{U}^2}{\omega_e^2} \int_0^l (z_C \phi_i)' v_s' m_y \, dx + \frac{\bar{U}^2}{\omega_e^2} [(z_C \phi_i)' v_s m_y]_0^l
$$
$$
+ \int_0^l \phi_i v_s m_{y\phi} \, dx + \frac{\bar{U}^2}{\omega_e^2} \int_0^l \phi_i' v_s' m_{y\phi} \, dx - \frac{\bar{U}^2}{\omega_e^2} [\phi_i' v_s m_{y\phi}]_0^l
$$

$$+ \frac{\bar{U}}{\omega_e^2} \int_0^l [z_C\phi_i v_s' - (z_C\phi_i)' v_s] N_y \, \mathrm{d}x - \frac{\bar{U}}{\omega_e^2} [z_C\phi_i v_s N_y]_0^l$$

$$+ \frac{\bar{U}}{\omega_e^2} \int_0^l (\phi_i' v_s - \phi_i v_s') N_{y\phi} \, \mathrm{d}x + \frac{\bar{U}}{\omega_e^2} [\phi_i v_s N_{y\phi}]_0^l \qquad (i \neq 0),$$

$$E_{0s} = \bar{U} \int_0^l v_s' m_{y\phi} \, \mathrm{d}x - \bar{U}[v_s m_{y\phi}]_0^l + \int_0^l v_s N_{y\phi} \, \mathrm{d}x,$$

$$E_{is} = \bar{U} \int_0^l [(z_C\phi_i)' v_s - z_C\phi_i v_s'] \, m_y \, \mathrm{d}x + \bar{U} \, [z_C\phi_i v_s m_y]_0^l$$

$$+ \bar{U} \int_0^l (\phi_i v_s' - \phi_i' v_s) m_{y\phi} \, \mathrm{d}x - \bar{U} \, [\phi_i v_s m_{y\phi}]_0^l - \int_0^l z_C\phi_i v_s N_y \, \mathrm{d}x$$

$$- \frac{\bar{U}^2}{\omega_e^2} \int_0^l (z_C\phi_i)' v_s' N_y \, \mathrm{d}x + \frac{\bar{U}^2}{\omega_e^2} [(z_C\phi_i)' v_s N_y]_0^l$$

$$+ \int_0^l \phi_i v_s N_{y\phi} \, \mathrm{d}x + \frac{\bar{U}^2}{\omega_e^2} \int_0^l \phi_i' v_s' N_{y\phi} \, \mathrm{d}x - \frac{\bar{U}^2}{\omega_e^2} [\phi_i' v_s N_{y\phi}]_0^l \qquad (i \neq 0),$$

$$F_{is} = 0,$$

$$K_{rj} = - \int_0^l z_C\phi_j v_r m_y \, \mathrm{d}x - \frac{\bar{U}^2}{\omega_e^2} \int_0^l (z_C\phi_j)' v_r' m_y \, \mathrm{d}x + \frac{\bar{U}^2}{\omega_e^2} [z_C\phi_j v_r' m_y]_0^l$$

$$+ \int_0^l \phi_j v_r m_{\phi y} \, \mathrm{d}x + \frac{\bar{U}^2}{\omega_e^2} \int_0^l \phi_j' v_r' m_{\phi y} \, \mathrm{d}x - \frac{\bar{U}^2}{\omega_e^2} [\phi_j v_r' m_{\phi y}]_0^l$$

$$+ \frac{\bar{U}}{\omega_e^2} \int_0^l [(z_C\phi_j)' v_r - z_C\phi_j v_r'] \, N_y \, \mathrm{d}x - \frac{\bar{U}}{\omega_e^2} [z_C\phi_j v_r N_y]_0^l$$

$$+ \frac{\bar{U}}{\omega_e^2} \int_0^l (\phi_j v_r' - \phi_j' v_r) N_{\phi y} \, \mathrm{d}x + \frac{\bar{U}}{\omega_e^2} [\phi_j v_r N_{\phi y}]_0^l,$$

$$L_{rj} = \bar{U} \int_0^l [z_C\phi_j v_r' - (z_C\phi_j)' v_r] m_y \, \mathrm{d}x + \bar{U} \, [z_C\phi_j v_r m_y]_0^l$$

$$+ \bar{U} \int_0^l (\phi_j' v_r - \phi_j v_r') m_{\phi y} \, \mathrm{d}x - \bar{U} \, [\phi_j v_r m_{\phi y}]_0^l - \int_0^l z_C\phi_j v_r N_y \, \mathrm{d}x$$

$$- \frac{\bar{U}^2}{\omega_e^2} \int_0^l (z_C\phi_j)' v_r' N_y \, \mathrm{d}x + \frac{\bar{U}^2}{\omega_e^2} [z_C\phi_j v_r' N_y]_0^l + \int_0^l \psi_j v_r N_{\phi y} \, \mathrm{d}x$$

$$+ \frac{\bar{U}^2}{\omega_e^2} \int_0^l \phi_j' v_r' N_{\phi y} \, \mathrm{d}x - \frac{\bar{U}^2}{\omega_e^2} [\phi_j v_r' N_{\phi y}]_0^l,$$

$$M_{rj} = 0$$

$$N_{0j} = - \frac{\bar{U}^2}{\omega_e^2} \int_0^l z_C' \phi_j' m_{\phi y} \, \mathrm{d}x - \int_0^l z_C\phi_j m_{y\phi} \, \mathrm{d}x - \frac{\bar{U}}{\omega_e^2} \int_0^l \phi_j' N_\phi \, \mathrm{d}x$$

$$+ \frac{\bar{U}}{\omega_e^2} [\phi_j N_\phi]_0^l + \frac{\bar{U}}{\omega_e^2} \int_0^l (z_C\phi_j)' N_{y\phi} \, \mathrm{d}x - \frac{\bar{U}}{\omega_e^2} [z_{C'} \phi_j N_{y\phi}]_0^l,$$

$$N_{ij} = \int_0^l z_C^2 \phi_i \phi_j m_y \, dx + \frac{\bar{U}^2}{\omega_e^2} \int_0^l (z_C \phi_i)'(z_C \phi_j)' m_y \, dx$$

$$- \frac{\bar{U}^2}{\omega_e^2} [(z_C \phi_i)' z_C \phi_j m_y]_0^l - \int_0^l z_C \phi_i \phi_j m_{\phi y} \, dx$$

$$- \frac{\bar{U}^2}{\omega_e^2} \int_0^l (z_C \phi_i)' \phi_j' m_{\phi y} \, dx$$

$$+ \frac{\bar{U}^2}{\omega_e^2} [(z_C \phi_i)' \phi_j m_{\phi y}]_0^l + \int_0^l \phi_i \phi_j I_x \, dx + \frac{\bar{U}^2}{\omega_e^2} \int_0^l \phi_i' \phi_j' I_x \, dx$$

$$- \frac{\bar{U}^2}{\omega_e^2} [\phi_i' \phi_j I_x]_0^l - \int_0^l z_C \phi_i \phi_j m_{y\phi} \, dx$$

$$- \frac{\bar{U}^2}{\omega_e^2} \int_0^l \phi_i' (z_C \phi_j)' m_{y\phi} \, dx + \frac{\bar{U}^2}{\omega_e^2} [\phi_i' z_C \phi_j m_{y\phi}]_0^l$$

$$+ \frac{\bar{U}}{\omega_e^2} \int_0^l [(z_C \phi_i)' z_C \phi_j - z_C \phi_i (z_C \phi_j)'] N_y \, dx + \frac{\bar{U}}{\omega_e^2} [z_C^2 \phi_i \phi_j N_y]_0^l$$

$$+ \frac{\bar{U}}{\omega_e^2} \int_0^l [z_C \phi_i \phi_j' - (z_C \phi_i)' \phi_j] N_{\phi y} \, dx - \frac{\bar{U}}{\omega_e^2} [z_C \phi_i \phi_j N_{\phi y}]_0^l$$

$$+ \frac{\bar{U}}{\omega_e^2} \int_0^l (\phi_i' \phi_j - \phi_i \phi_j') N_\phi \, dx + \frac{\bar{U}}{\omega_e^2} [\phi_i \phi_j N_\phi]_0^l$$

$$+ \frac{\bar{U}}{\omega_e^2} \int_0^l [\phi_i (z_C \phi_j)' - \phi_i' z_C \phi_j] N_{y\phi} \, dx - \frac{\bar{U}}{\omega_e^2} [z_C \phi_i \phi_j N_{y\phi}]_0^l \quad (i \neq 0),$$

$$R_{0j} = \bar{U} \int_0^l \phi_j' I_x \, dx - \bar{U} [\phi_j I_x]_0^l - \bar{U} \int_0^l (z_C \phi_j)' m_{y\phi} \, dx$$

$$+ \bar{U} [z_C \phi_j m_{y\phi}]_0^l + \int_0^l \phi_j N_\phi \, dx - \int_0^l z_C \phi_j N_{y\phi} \, dx,$$

$$R_{ij} = \bar{U} \int_0^l [z_C \phi_i (z_C \phi_j)' - (z_C \phi_i)' z_C \phi_j] m_y \, dx - \bar{U} [z_C^2 \phi_i \phi_j m_y]_0^l$$

$$+ \bar{U} \int_0^l [(z_C \phi_i)' \phi_j - z_C \phi_i \phi_j'] m_{\phi y} \, dx + \bar{U} [z_C \phi_i \phi_j m_{\phi y}]_0^l$$

$$+ \bar{U} \int_0^l (\phi_i \phi_j' - \phi_i' \phi_j) I_x \, dx - \bar{U} [\phi_i \phi_j I_x]_0^l$$

$$+ \bar{U} \int_0^l [\phi_i' z_C \phi_j - \phi_i (z_C \phi_j)'] m_{y\phi} \, dx + \bar{U} [z_C \phi_i \phi_j m_{y\phi}]_0^l$$

$$+ \int_0^l z_C^2 \phi_i \phi_j N_y \, dx + \frac{\bar{U}^2}{\omega_e^2} \int_0^l (z_C \phi_i)'(z_C \phi_j)' N_y \, dx$$

$$- \frac{\bar{U}^2}{\omega_e^2} [(z_C \phi_i)' z_C \phi_j N_y]_0^l - \int_0^l z_C \phi_i \phi_j N_{\phi y} \, dx$$

$$- \frac{\bar{U}^2}{\omega_e^2} \int_0^l (z_C \phi_i)' \phi_j' N_{\phi y} \, dx + \frac{\bar{U}^2}{\omega_e^2} [(z_C \phi_i)' \phi_j N_{\phi y}]_0^l$$

$$+ \int_0^l \phi_i \phi_j N_\phi \, dx + \frac{\bar{U}^2}{\omega_e^2} \int_0^l \phi_i' \phi_j' N_\phi \, dx - \frac{\bar{U}^2}{\omega_e^2} [\phi_i' \phi_j N_\phi]_0^l$$

$$-\int_0^l z_C\phi_i\phi_j N_{y\phi}\,\mathrm{d}x - \frac{\bar{U}^2}{\omega_e^2}\int_0^l \phi_i'(z_C\phi_j)'N_{y\phi}\,\mathrm{d}x$$

$$+\frac{\bar{U}^2}{\omega_e^2}[\phi_i' z_C\phi_j N_{y\phi}]_0^l \qquad (i\neq 0).$$

The only non-zero restoring terms appear in the twisting equations, being such that

$$T_{ij}=\int_0^l \rho g\phi_i\phi_j\left[\frac{B^3}{12}-\overline{BC}\,S(x)\right]\mathrm{d}x.$$

Finally we have the excitation terms that are determined by the impinging waves. For deep water the bending equation has the generalised force amplitude

$$\Xi_s=\int_0^l v_s(T_1+T_2-\mathrm{i}T_3)\,\mathrm{d}x \qquad (s=0,1,2,\ldots),$$

while that of the twisting equation is

$$\Xi_j=\int_0^l \phi_j[T_4-\mathrm{i}T_5-z_C(T_2-\mathrm{i}T_3)]\,\mathrm{d}x \qquad (j=0,1,2,\ldots).$$

In this case the principal coordinates $p_r(t)$ are associated with distortions of frequency ω_e at the shear centre S (which coincides with the centre of mass C) whilst the principal coordinates $q_i(t)$ correspond to twisting with frequency ω_e about the same axis.

10.6 Hydrodynamic coefficients

The hydrodynamic coefficients – i.e. the added masses and fluid damping coefficients – of a ship section having an arbitrary shape may be determined theoretically by a generalisation of Ursell's (1949a,b) method; the approach has been used by Vugts (1968) and de Jong (1973) who studied the swaying and rolling of cylinders. In that theory it is assumed that the fluid is inviscid and the motions are small. The fact that viscosity and large amplitude may have dominant influences on the roll coefficients is ignored.

The hydrodynamic coefficients will now be discussed for the chine, rectangular, triangular, fine and bulbous sections illustrated in figs. 7.16(a), 7.17(a), ..., 7.20(a) respectively. They have again been calculated using a Lewis-form fit, the conformal transformation technique and, in some cases, Frank's close-fit method. The characteristics and the number of parameters used in the conformal mapping are as described in section 7.1.6. The non-dimensional added mass and damping coefficients in sway, the added inertia and damping coefficients in roll, together with their cross-coupling coefficients, have been calculated for each section using a wide range of non-dimensional frequency $\omega(B/2g)^{\frac{1}{2}}$. In each case the overall beam is B, the draught is T and the section area is S.

Fig. 10.4. The variation of (a) dimensionless added mass for sway, (b) dimensionless damping for sway, (c) dimensionless added moment of inertia for roll, (d) dimensionless damping for roll, (e) dimensionless added inertia of cross-coupling ($m_{\phi y} = m_{y\phi}$) and (f) dimensionless cross-coupling damping ($N_{\phi y} = N_{y\phi}$) for the chine, rectangular and triangular sections. The various quantities are plotted against non-dimensional frequency $\omega(B/2g)^{\frac{1}{2}}$. The full-line curves refer to the conformal mapping calculations and the broken-line curves relate to the Lewis-form results.

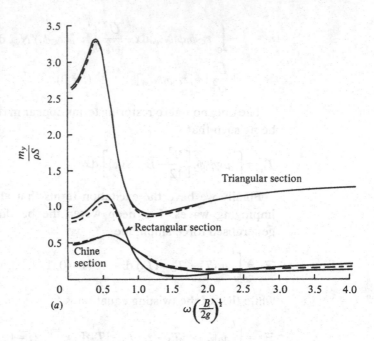

(a)

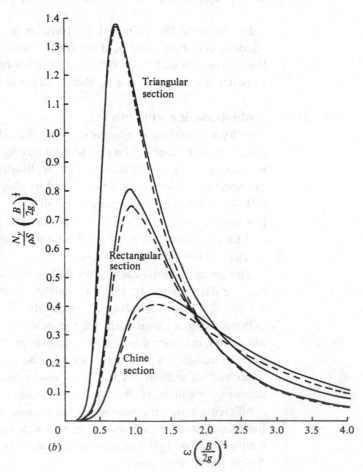

(b)

Fig. 10.4. (Continued)

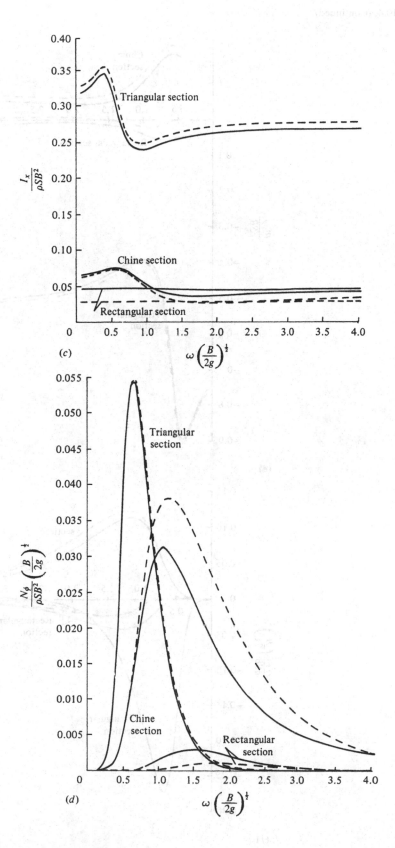

(c)

(d)

Fig. 10.4. (Continued)

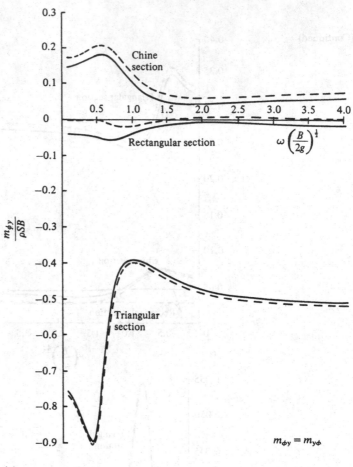

(e)

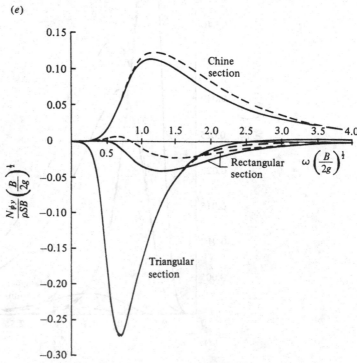

(f)

Fig. 10.5. The variation of (a) dimensionless added mass for sway, (b) dimensionless damping for sway, (c) dimensionless added moment of inertia for roll, (d) dimensionless damping for roll, (e) dimensionless inertia for cross-coupling ($m_{\phi y} = m_{y\phi}$) and (f) dimensionless cross-coupling damping ($N_{\phi y} = N_{y\phi}$) for the fine and bulbous sections. The various quantities are plotted against non-dimensional frequency $\omega(B/2g)^{\frac{1}{2}}$. The full-line curve shows results of the conformal mapping calculations, the broken line refers to the Lewis-form results, the chained lines with crosses relate to the Frank close-fit method applied to the bulbous section and the chained curves illustrate the Frank close-fit method for the fine section.

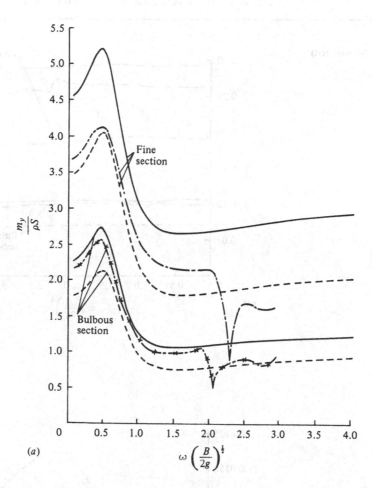

(a)

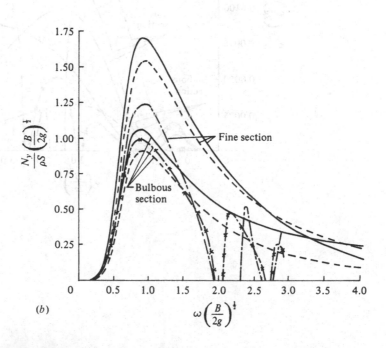

(b)

Fig. 10.5. (Continued)

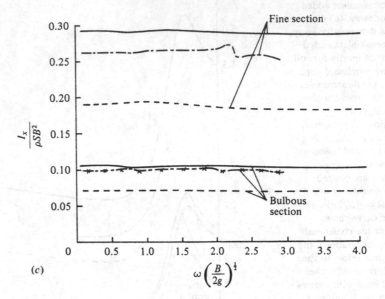

(c)

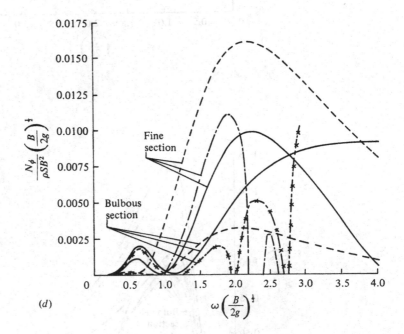

(d)

Fig. 10.5. (Continued)

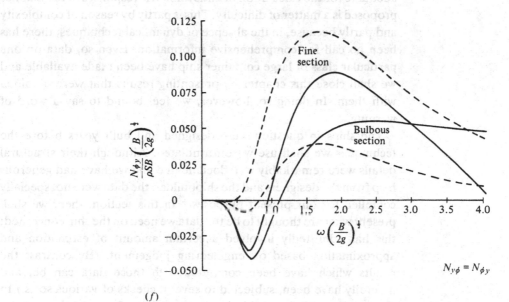

$$\omega \left(\frac{B}{2g} \right)^{\frac{1}{2}}$$

(e)

$$\frac{m_{\phi y}}{\rho SB}$$

Bulbous section

Fine section

$$m_{y\phi} = m_{\phi y}$$

0.125
0.100
0.075
0.050
0.025
0
-0.025
-0.050

Fine section

Bulbous section

$$\frac{N_{\phi y}}{\rho SB} \left(\frac{B}{2g} \right)^{\frac{1}{4}}$$

$$\omega \left(\frac{B}{2g} \right)^{\frac{1}{2}}$$

$$N_{y\phi} = N_{\phi y}$$

(f)

(i) *Chine, rectangular and triangular sections.* Figs. 10.4(*a*)–(*f*) show
the calculated hydrodynamic coefficients of the chine, rectangular and
triangular sections. The curves are shown for the Lewis-form approach
and the conformal transformation technique. The results for the
triangular section show the closest agreement, the chine and rectangular results displaying slightly greater discrepancies.

(ii) *Fine and bulbous sections.* Figs. 10.5 (*a*)–(*f*) show the calculated
hydrodynamic coefficients for the fine and bulbous sections. These
have been determined using Lewis forms, conformal mapping and by
the Frank close-fit method. The results in which Frank's method has
been used were derived by Faltinsen (1969) and it is again noticeable
that they suffer from discontinuities at certain irregularly spaced
frequencies which occur more often with increasing frequency. Unless
some smoothing technique is employed these irregularities will
produce difficulties in the calculation of responses. On the whole, it will
be seen that results found with the conformal mapping and the Frank
method show better agreement between themselves than with those
obtained from the Lewis-form calculations.

10.7   Results for a container ship

It has already been pointed out that the acquisition of data that are
adequate for the calculation of antisymmetric responses in the manner
proposed is a matter of difficulty. This is partly by reason of complexity
and partly because, in the absence of dynamical techniques, there has
been no call for comprehensive information. Even so, data on one
particular class of large container ship have been made available and
we shall close this chapter by presenting results that were obtained
with them. In doing so, however, we feel bound to say a word of
warning.

The ships in question were designed and built years before the
techniques we shall use were formulated. Although their structural
details were remarkably well documented and we have had generous
help from the designers and the shipbuilder, the data were not specially
compiled for our present purposes.† In this section, then, we shall
present what are thought to be the data we need on the ship concerned;
this has admittedly involved a certain amount of estimation and
approximation based on engineering judgement. (By contrast the
results which have been computed with those data can be, and
naturally have been, subjected to several checks of various sorts.) In
short, the remainder of this chapter contains a first tentative presentation of antisymmetric responses for a ship.

---

† Although much the same could be said about the destroyer and tanker, our needs
were easier to meet for symmetric responses than they are for antisymmetric.

10.7.1   *The ship*

The main dimensions of the ship are given in Table 10.1. The vessel itself is sketched in fig. 10.6(*a*)–(*c*), along with relevant data for the fully loaded condition.†

The curves of fig. 10.6(*a*) require little comment. They show the variations of

mass per unit length $\mu(x)$,
beam $B(x)$ at the waterline and draught $T(x)$,
area of section immersed $S(x)$,
second moment of area about a vertical line in the plane of symmetry $I(x)$.

Fig. 10.6(*a*). The variations of (i) mass per unit length, (ii) breadth and draught, (iii) immersed sectional area and (iv) second moment of area about a vertical axis of the container ship in the loaded condition.

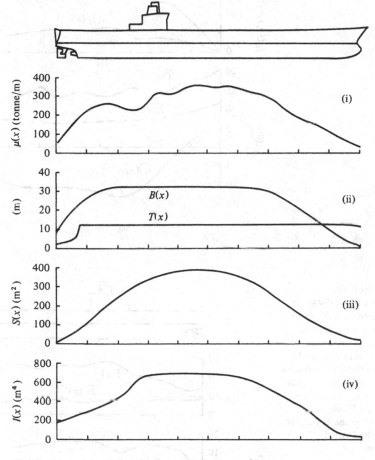

Table 10.1. *Details of a large container ship*

Length overall	287.75 m
Approx. length at waterline (taken as length *l*)	281.00 m
Breadth moulded	32.26 m
Draught in loaded condition	12.2 m
Weight loaded	67 150 tonne f

† No allowance is made for stiffening by the containers.

Fig. 10.6(*b*). The variations of (i) shear area for distortion athwartships, (ii) moment of inertia per unit length about longitudinal axes and (iii) position of the centre of mass, centre of buoyancy and shear centre with respect to the keel for the container ship in the loaded condition.

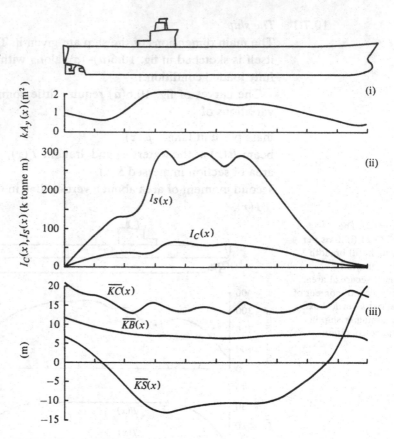

Fig. 10.6(*c*). The variations of (i) distances between the shear centre and centre of mass and between the centre of buoyancy and centre of mass, (ii) torsional stiffness divided by shear modulus and (iii) warping stiffness divided by Young's modulus for the container ship in the loaded condition.

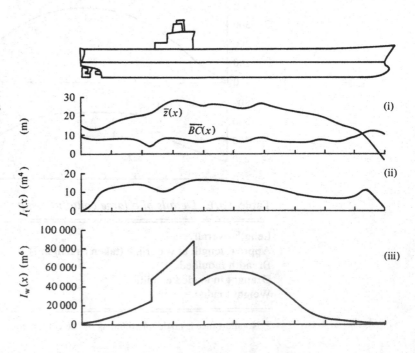

In fig. 10.6(b), the uppermost curve is that of shear area $kA_y(x)$. The next graph shows the variation of $I_C(x)$, and this has been taken as 'given data' from which the variation $I_S(x)$ has been calculated. In the last set of three curves, the positions are given of the centre of buoyancy B, centre of mass C and shear centre S with respect to the keel K.

The uppermost curves in fig. 10.6(c) are self-explanatory. The distribution of torsional stiffness $C(x)$ may be expressed in the form

$$C(x) = GI_t(x),$$

where the shear modulus $G = 8.28 \times 10^7 \text{ kN/m}^2$. The variation of $I_t(x)$ is shown in the second graph of fig. 10.6(c). Likewise the warping stiffness $C_1(x)$ may be written

$$C_1(x) = EI_w(x),$$

where Young's modulus $E = 2.07 \times 10^8 \text{ kN/m}$. The last graph in fig. 10.6(c) shows the variation of $I_w(x)$; it will be seen that no attempt has been made to 'fair' this curve. (As we shall see there is some doubt about the warping stiffness and so there is little point in attending to minor details at this stage.)

10.7.2 Modal properties of the structure

The rigid body modes are

sway $(r = 0)$,

yaw $(r = 1)$,

roll $(r = 2)$.

Since the container ship is not even approximately 'boxlike', its distortion modes combine lateral bending and twisting. These may be computed along with the natural frequencies by means of an extension of the Prohl–Myklestad technique that is discussed by Bishop, Price & Temarel (1979). Twenty slices were used in idealising the hull. The calculations were made with allowance for shear deflection and for warping stiffness but with no allowance for rotatory inertia.

Table 10.2. *Natural frequencies of antisymmetric oscillation of the container ship hull (dry)*

Modal index, r		3	4	5	6	7[b]	8[b]
Natural frequency[a] (rad/s)	$C_1(x) \neq 0$	2.23	3.89	7.54	10.65	14.18	15.60
	$C_1(x) = 0$	1.92	2.07	3.11	3.82	4.81	5.49

[a] For the rigid body modes, $r = 0, 1, 2$, the natural frequency $\omega_r = 0$.
[b] Modes 7 and 8 were not used in the calculations referred to in section 10.7.3.

If the calculations of principal modes and natural frequencies are repeated but with no allowance made for warping stiffness (i.e. taking $C_1(x) = 0$), the results are somewhat different. The natural frequencies are given in Table 10.2. It will be seen that as the order of the principal mode is increased so the effects of allowing for warping stiffness become very pronounced indeed.

Two tentative conclusions may be drawn from the results given in Table 10.2. The first is that we certainly cannot discount the possibility of serious resonant encounter problems in an actual seaway. The lowest non-zero natural frequency of antisymmetric distortion of the loaded container ship is even lower (in both cases) than that of symmetric distortion of the loaded tanker (see Table 4.4). The second conclusion is that the allowance for warping stiffness appears to be so significant that it is likely to determine how many modes should be admitted into the computations when working with a given wave encounter spectrum.

This last point is underlined particularly heavily by inspection of the mode shapes. Indeed so great is the difference made by allowing for warping stiffness, one is led to question the theory or the data, or both. Figs. 10.7(a)–(d) show the mode shapes $v_r(x)$, $\phi_r(x)$, modal bending and twisting moments $M_r(x)$, $T_r(x)$ and modal shearing force $V_r(x)$ for $r = 3, 4, 5$ and 6 that are found when $C_1(x)$ is taken into account. Figs. 10.8(a)–(d) show the corresponding quantities found when warping stiffness is ignored. At least it can be said with some confidence that stiffness distributions for antisymmetric distortions should be accorded much more attention, both experimental and theoretical, than they command at present.

When confronted by striking discrepancies like those found on comparing figs. 10.7 and 10.8, one naturally questions the correctness of the computer program used. As discussed by Bishop *et al.* (1979), however, the program in question was validated against an 'exact' solution which was specially derived as an exercise in theoretical mechanics (rather as we employed the 'uniform ship' in section 4.2.2). Moreover the orthogonality of the modes was checked as shown in Table 10.3 for $C_1(x) \neq 0$ and in Table 10.4 for the case in which $C_1(x)$ is ignored. It will be seen that the generalised mass and generalised stiffness matrices are essentially diagonal in both cases.†

---

† In computing Tables 10.3 and 10.4, the quantity $v_2(x)$ was taken as zero instead of $z_S(x)$ because a finite value of $v_2(x)$ is not reflected in any contribution to the energy of the hull.

Fig. 10.7. Principal modes of the loaded container ship. The full lines in curves (i) show the variations of deflection and the broken lines in curves (i) show the variations of twist. The curves (ii) show the modal functions representing bending moment, shearing force and twisting moment. The modal indices are $r = 3$ in curves (a), $r = 4$ in (b), $r = 5$ in (c) and $r = 6$ in (d).

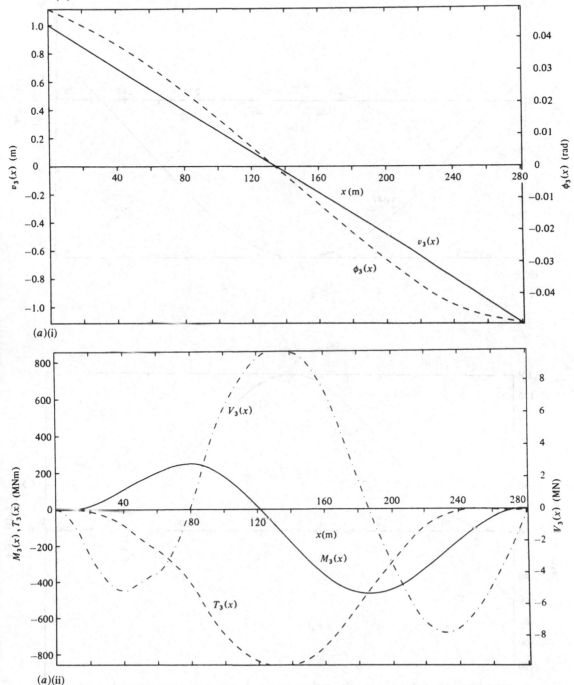

(a)(i)

(a)(ii)

Fig. 10.7. (Continued)

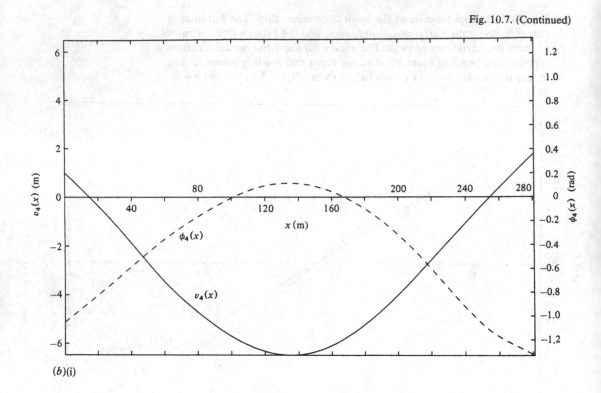

(b)(i)

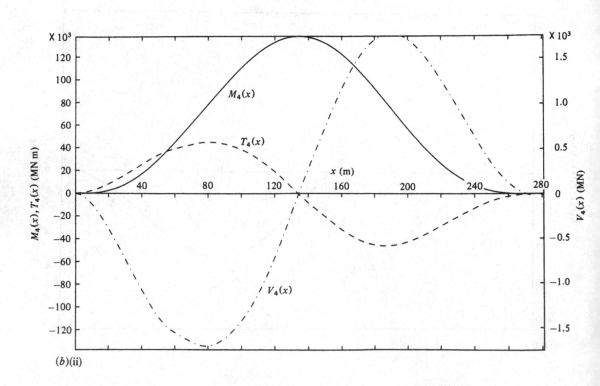

(b)(ii)

Fig. 10.7. (Continued)

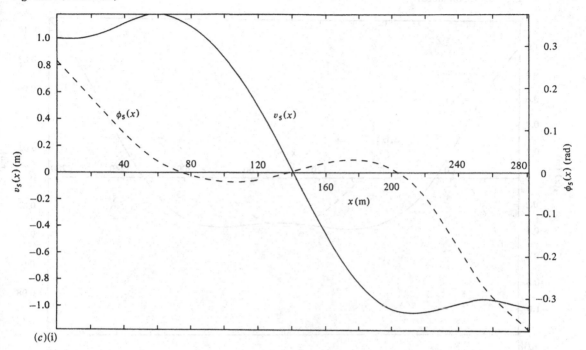

(c)(i)

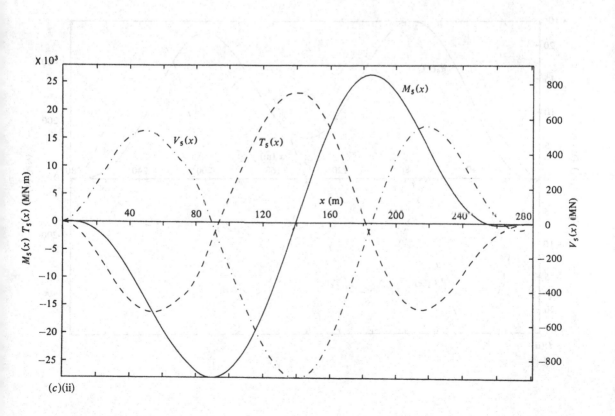

(c)(ii)

Fig. 10.7. (Continued)

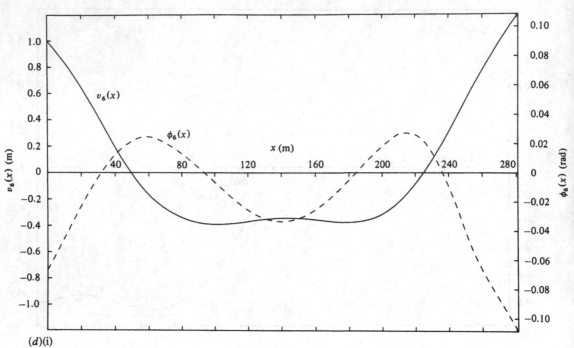

$(d)(i)$

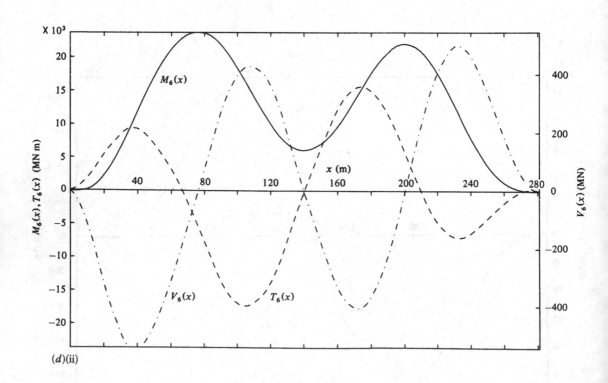

$(d)(ii)$

Fig. 10.8. Curves representing the same quantities as those in fig. 10.7 but computed without allowance for warping stiffness.

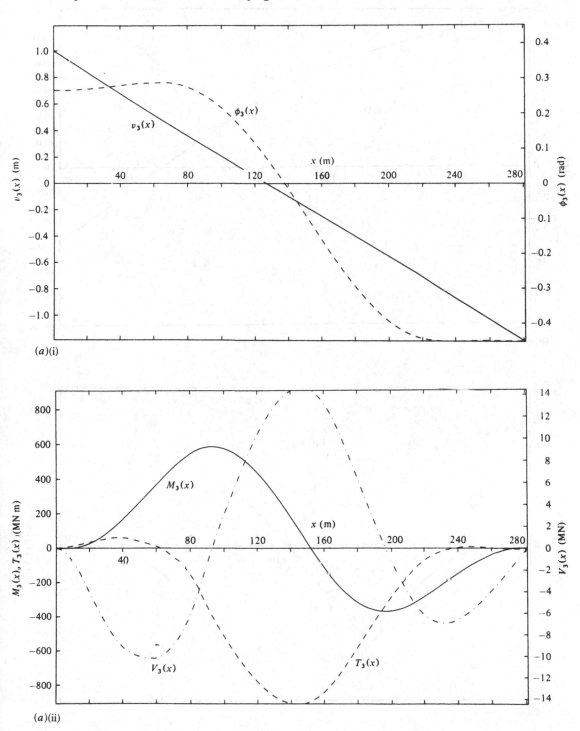

(a)(i)

(a)(ii)

Fig. 10.8. (Continued)

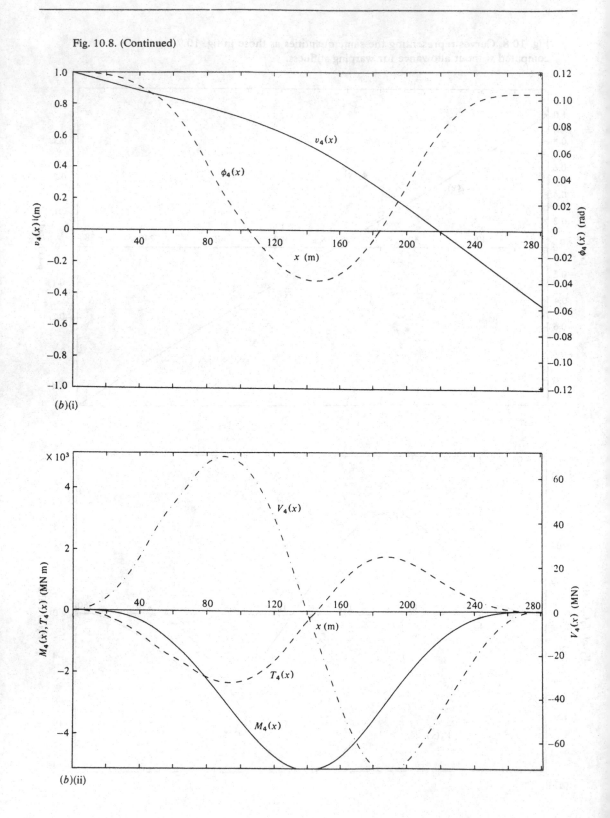

$(b)(i)$

$(b)(ii)$

Fig. 10.8. (Continued)

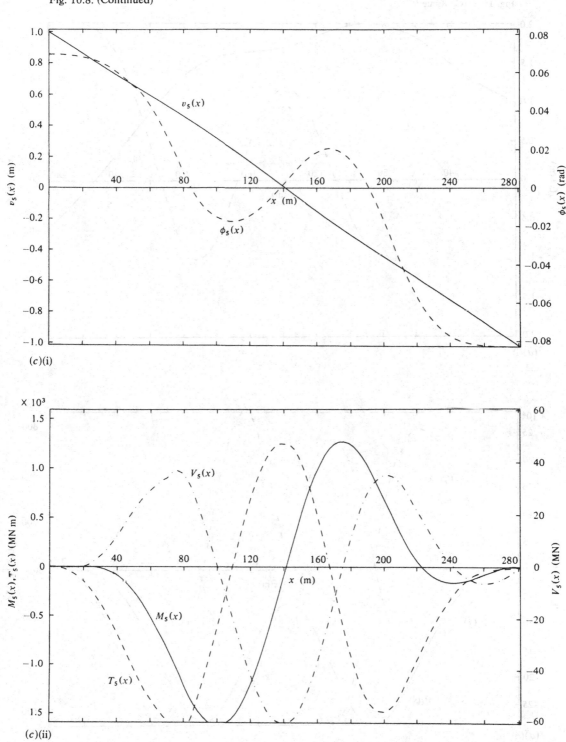

(c)(i)

(c)(ii)

Fig. 10.8. (Continued)

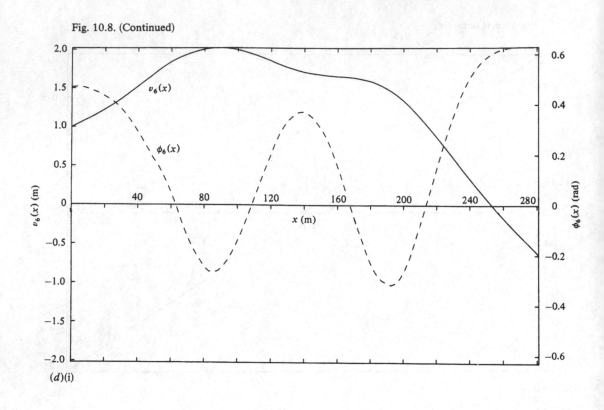

(d)(i)

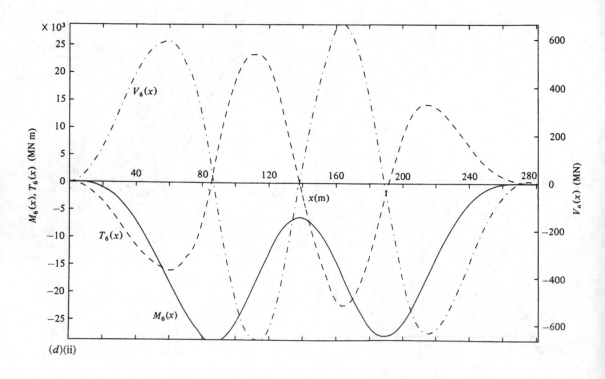

(d)(ii)

It is of interest to digress briefly in order to take up a matter of some significance in Table 10.3(b), namely its symmetry. The value of c_{rs} is given by equation (3.53); that is

$$c_{rs} = \int_0^l (EI\theta_r'\theta_s' + kA_yG\gamma_r\gamma_s + T_r\phi_s') \, dx.$$

This is not obviously symmetric in r and s. Now

$$T_r = C\phi_r' - (C_1\phi_r'')'$$

and so

$$c_{rs} = \int_0^l (EI\theta_r'\theta_s' + kA_yG\gamma_r\gamma_s + C\phi_r'\phi_s') \, dx - \int_0^l (C_1\phi_r'')'\phi_s' \, dx.$$

Integration by parts shows that

$$\int_0^l (C_1\phi_r'')'\phi_s' \, dx = [C_1\phi_r''\phi_s']_0^l - \int_0^l C_1\phi_r''\phi_s'' \, dx.$$

Table 10.3(a). *Values of generalised mass a_{rs} for the container ship (in tonne m^2) when warping stiffness $C_1(x)$ is allowed for $[v_2(x) = 0]$*

Mode	0	1	2	3	4	5	6
0	67 204	−91	−1 507 819	−8	407	284	60
1	−91	16 430	−11 064	−4	−1477	−160	−113
2	−1 507 819	−11 064	45 000 000	−38	226	−5482	−1019
3	−8	−4	−38	8407	−56	−102	−18
4	407	−1477	226	−56	5 455 816	5283	1809
5	284	−160	−5482	−102	5283	144 722	398
6	60	−113	−1019	−18	1809	398	28 008

Table 10.3(b). *Values of generalised stiffness c_{rs} for the container ship (in MNm) when warping stiffness $C_1(x)$ is allowed for $[v_2(x) = 0]$*

Mode	0	1	2	3	4	5	6
0	0	0	0	0	0	0	0
1	0	0	0	0	0	0	0
2	0	0	0	0	0	0	0
3	0	0	0	42 004	−1198	1504	−382
4	0	0	0	−1198	82 509 630	−6620	43 811
5	0	0	0	1504	−6620	8 230 505	22
6	0	0	0	−382	43 811	22	3 159 378

The integrated term vanishes because, as equation (2.34) shows, $\phi_r''(0) = 0 = \phi_r''(l)$ and hence

$$c_{rs} = \int_0^l (EI\theta_r'\theta_s' + kA_y G\gamma_r\gamma_s + C\phi_r'\phi_s' + C_1\phi_r''\phi_s'') \, dx.$$

This is a symmetric function in r and s.

To return to the mode shapes given in figs. 10.7 and 10.8 we may note that some are dominated by bending and others are dominated by twisting. Even though hard and fast rules would be hard to frame for deciding which is which, we can readily make the distinctions drawn in Table 10.5. (We have referred to a mode as being 'twisting dominated' if, for a 1 m deflection $v_r(0)$ at the stern, the hull rotates more than 30° at any section. If all rotation is less than 10°, on the other hand, we take the mode as 'bending dominated'.)

Table 10.4(a). *Values of generalised mass a_{rs} for the container ship (in tonne m^2) when warping stiffness $C_1(x)$ is ignored $[v_2(x) = 0]$*

Mode	0	1	2	3	4	5	6
0	67 204	−91	−1 507 819	−26	−47	49	−232
1	−91	16 430	−11 064	−10	145	−18	759
2	−1 507 819	−11 064	45 410 651	385	−251	−1147	−29
3	−26	−10	385	13 189	26	−27	192
4	−47	145	−251	26	113 407	−125	341
5	49	−18	−1147	−27	−125	26 673	−524
6	−232	759	−29	192	341	−524	2 913 280

Table 10.4(b). *Values of generalised stiffness c_{rs} for the container ship (in MNm) when warping stiffness $C_1(x)$ is ignored $[v_2(x) = 0]$*

Mode	0	1	2	3	4	5	6
0	0	0	0	0	0	0	0
1	0	0	0	0	0	0	0
2	0	0	0	0	0	0	0
3	0	0	0	48 688	12	12	184
4	0	0	0	12	484 954	−104	−1892
5	0	0	0	12	−104	258 994	−1716
6	0	0	0	184	−1892	−1716	42 378 354

Table 10.5. *Bending-dominated (B) and twisting-dominated (T) principal modes of the container ship*

Modal index r	3	4	5	6	7	8
$C_1(x) \neq 0$ (fig. 10.7)	B	T	?	B	B	?
$C_1(x) = 0$ (fig. 10.8)	?	B	B	T	?	T

In the absence of any solid information on structural damping, we have no alternative but to guess it. If the modal damping factors are takes as $\nu_3 = 0.010$, $\nu_4 = 0.012$, $\nu_5 = 0.015$ and $\nu_6 = 0.019$, equation (4.6) suggests that the generalised damping coefficients b_{rr} should be roughly those given in Table 10.6.

10.7.3 Response amplitude operators

For the sake of definiteness results will be given for the container ship when it proceeds at 26 knots (13.38 m/s) in waves of 1 m amplitude with a heading of 135° (i.e. in a 45° bow sea).† The more complete version of strip theory (Theory B) is employed. The independent variable will be taken as l/λ, the ratio of ship length to wave length. In all cases, $C_1(x) \neq 0$ and allowance is made for shear flexibility (though not for rotatory inertia).

Fig. 10.9(a) shows the variations of $|p_0|$ and $|p_1|$ with l/λ. The amplitude of sway in waves of infinite length is about $1/\sqrt{2}$m, i.e. 0.71 m approximately, while that of yaw is zero. Both sway and yaw display a very sharply defined coupling with resonant roll at $l/\lambda \approx 0.11$ and yaw passes through a maximum value when $l/\lambda \approx 0.88$. Fig. 10.9(b) shows how $|p_2|$ varies over the same range of l/λ. The sharply defined resonant roll condition when $l/\lambda \approx 0.1$ is immediately apparent.

The variations of $|p_3|$, $|p_4|$, $|p_5|$ and $|p_6|$ are shown in figs. 10.9(c) to (f) respectively. The amplitudes of $|p_3|$, $|p_5|$ and $|p_6|$ display sharply defined resonances at $l/\lambda \approx 0.1$, but all show more significant and less well defined maxima at higher values of l/λ. These latter maxima are like those we found in chapter 8, associated with loops in polar plots.

It is of interest to examine the response amplitude operators of fig. 10.9 over a somewhat extended range of l/λ. Figs. 10.10(a)–(f) show how the quantities $|p_r|$ vary over the range $0 \le l/\lambda \le 36$. It will be seen

Table 10.6. *Values of generalised damping coefficient b_{rr} of the container ship (in tonne m^2/s)*

Modal index r	3	4	5	6
$C_1(x) \neq 0$	375	509 355	32 736	11 335
$C_1(x) = 0$	506	5634	2484	422 892

† The rudder is assumed to remain undeflected. In practice, motions imparted to the rudder in attempts to keep the ship on a prescribed heading produce substantial bending, shearing and twisting of the hull.

Fig. 10.9. The amplitudes $|p_r|$ of the principal coordinates of the loaded container ship travelling in a long crested sea of amplitude 1 m with heading angle $\chi = 135°$ and speed 13.38 m/s (i.e. 26 knots) according to Theory B. In curve (a) $r = 0, 1$, while curves (b)–(f) correspond to $r = 2, 3, 4, 5, 6$ respectively.

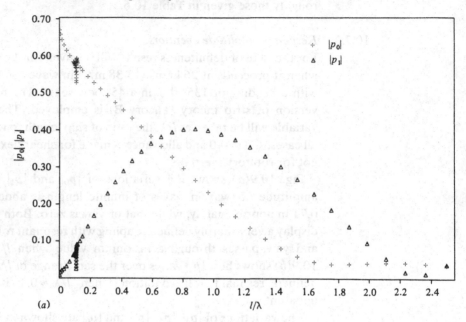

(a)

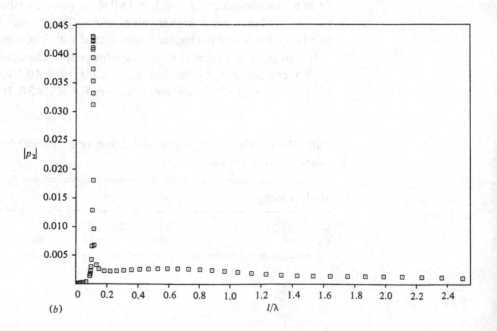

(b)

Fig. 10.9. (Continued)

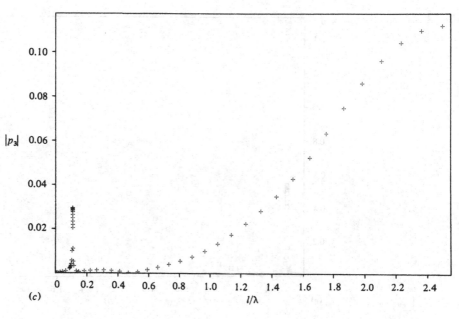

(c)

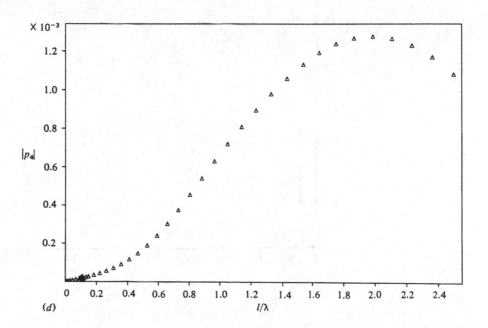

(d)

Fig. 10.9. (Continued)

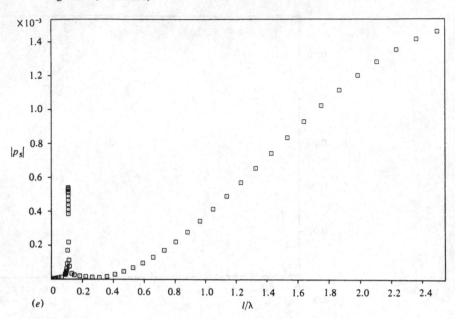

(e)

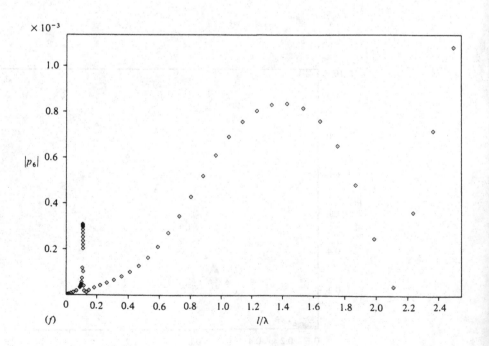

(f)

Fig. 10.10. The amplitudes $|p_r|$ of the principal coordinates of the loaded container ship as in fig. 10.9 but shown for an extended range of l/λ values.

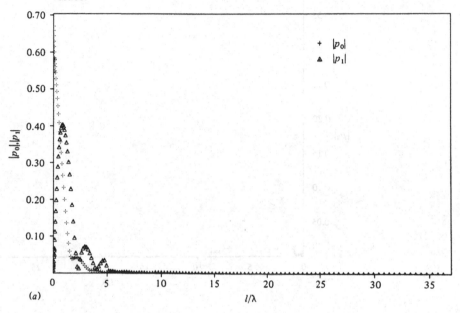

(a)

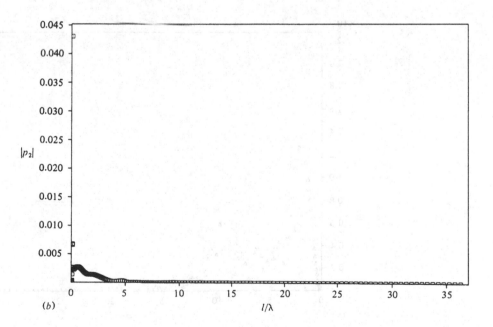

(b)

Fig. 10.10. (Continued)

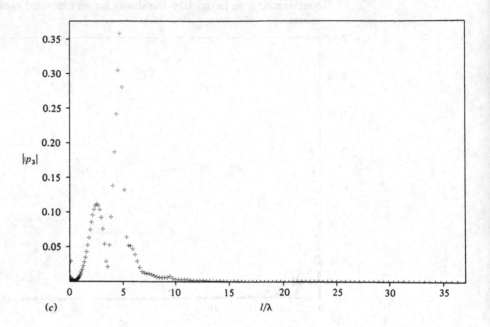

(c)

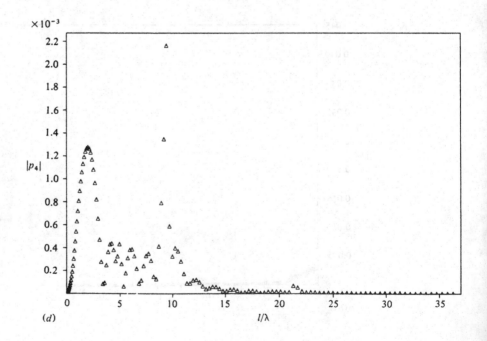

(d)

Fig. 10.10. (Continued)

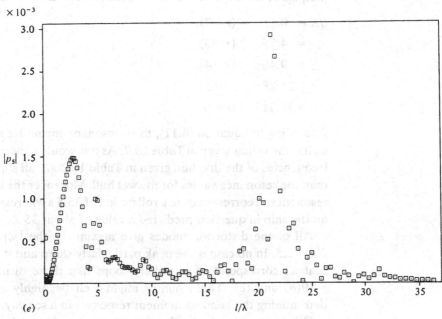

(e)

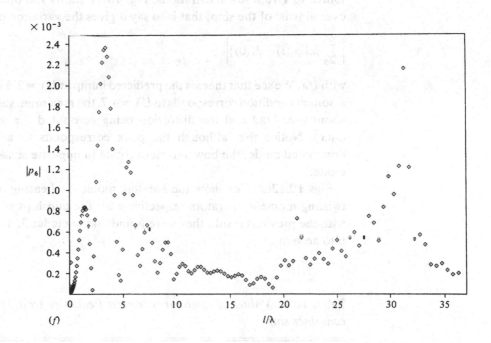

(f)

that, apart from the effects of coupling, resonances occur at

$$
\begin{aligned}
l/\lambda = 0.11 \quad & (r=2), \\
= 4.75 \quad & (r=3), \\
= 9.46 \quad & (r=4), \\
= 21.28 \quad & (r=5), \\
= 31.31 \quad & (r=6).
\end{aligned}
$$

According to equation (8.11), then, resonant encounter occurs when ω_e has the values given in Table 10.7. As one would expect, the natural frequencies of the dry hull given in Table 10.2 are all slightly greater than the resonance values for the wet hull. Moreover the frequency of resonant roll corresponds to a roll period of 35.5 s whereas model tests on the ship in question predicted a value of about 35 s.

All of the distortion modes give maxima in the $|p_r|$ at, or near, $l/\lambda = 2.5$. In no case is the peak particularly sharp and it seems likely that all correspond to subsidiary loops like those displayed in fig. 8.39(b) and (c). These humps might well be highly significant in determining the bending moment responses in a seaway.

Other response amplitude operators can readily be found once the complex values of the p_r have been found as functions of l/λ. It will suffice to give a few illustrations. Fig. 10.11 shows the operator for overall twist of the ship; that is to say it gives the variation of

$$
\left| \sum_{r=3}^{6} p_r[\phi_r(l) - \phi_r(0)] \right|
$$

with l/λ. We see that there is the predicted hump at $l/\lambda \approx 2.5$ but that a resonant condition corresponds to $l/\lambda \approx 4.7$, the maximum value being about 0.037 rad and the distortion being accounted for mainly by $\phi_3(x)$. Notice that although the peak corresponds to a *bending-*dominated mode, the bow and stern rotate in opposite senses in that mode.

Figs 10.12(a)–(c) show the bending moment, shearing force and twisting moment operators respectively for the midships section. As with the previous result, they were found taking modes 3, 4, 5, 6 only into account.

Table 10.7. *Values of resonant encounter frequency (rad/s) for the container ship*

Index of dominant mode, r	2	3	4	5	6
Resonant encounter frequency ω_e	0.177	2.02	3.44	6.66	9.24

Consider first the bending moment curve. Fig. 10.12(a) shows a substantial hump at $l/\lambda \approx 1.9$ and a sharp resonance at $l/\lambda \approx 9.5$. This is entirely in line with what one would expect from figs. 10.7(b)(ii) and 10.10(d). That is to say the $r = 4$ mode accounts for the substantial bending moment amidships because $M_4(l/2)$ is large, whereas it will be found that $M_3(l/2)$, $M_5(l/2)$ and $M_6(l/2)$ are all comparatively small.

The shearing force curve, fig. 10.12(b), has two significant peaks, both sharply defined. They correspond to $l/\lambda \approx 4.7$ and 21.3, i.e. to resonances of the $r = 3, 5$ modes. On consulting figs. 10.7 we see that this is what we should expect since $V_3(l/2)$ and $V_5(l/2)$ are both large while $V_4(l/2)$ and $V_6(l/2)$ are both small.

Fig. 12(c) shows that the twisting moment amidships is sharply resonant in the $r = 3$ mode. Fig. 10.7(a)(ii) gives the explanation for this while fig. 10.7(c)(ii) also suggests that a second sharp peak should appear, corresponding to the $r = 5$ mode, at $l/\lambda \approx 21.3$. But the resonance value of $|p_5|$ is much less than that of $|p_3|$, as may be seen from figs. 10.10(c) and (e), and so the latter peak is much lower than the first.

Fig. 10.11. The amplitude of overall hull twist for the loaded container ship under the conditions relevant to fig. 10.9.

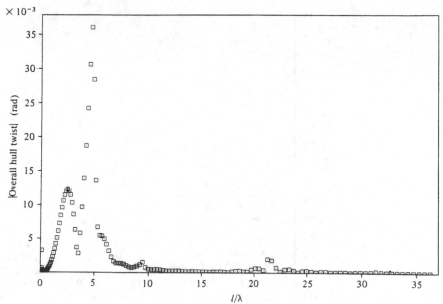

Fig. 10.12. The amplitudes of (*a*) bending moment, (*b*) shearing force and (*c*) twisting moment amidships for the loaded container ship travelling at 13.38 m/s in a 45° bow sea ($\chi = 135°$) of amplitude 1 m.

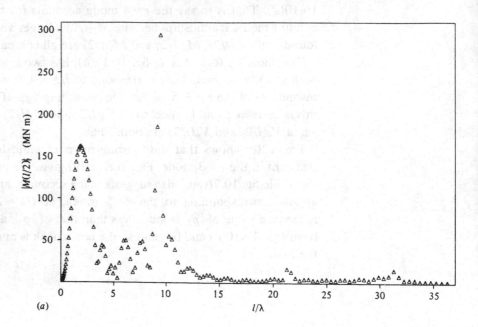

(*a*)

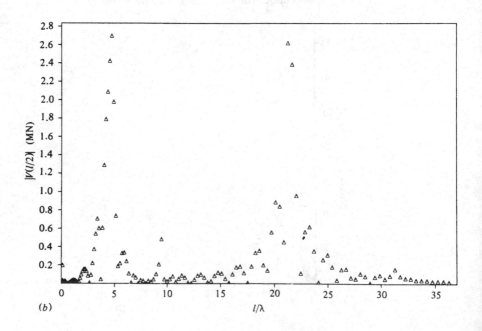

(*b*)

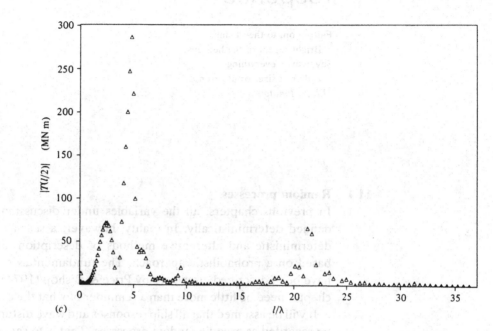

Fig. 10.12. (Continued)

11 Statistical analysis of ship response

Fair moon, to thee I sing,
  Bright regent of the heavens,
Say, why is everything
  Either at sixes or at sevens?
H.M.S. Pinafore

11.1 Random processes

In previous chapters, all the variables under discussion have been defined deterministically. In reality, however, a seaway is far from deterministic and alternative methods of description are required, based on a probabilistic approach. The fundamentals of the theory have been described elsewhere by Price & Bishop (1974) so that this chapter need be little more than a reminder of what the essentials are.

It will be assumed that all ship responses and wave disturbances may be regarded as ergodic random processes. That is to say:

(*a*) the random processes are stationary, so that their statistical properties (mean values, mean square values, etc.) remain unchanged with time;

(*b*) the statistical properties evaluated across the ensemble at some instant are the same as those derived by taking temporal averages along a single 'typical' realisation.

In practice, the time over which these conditions are met is limited and statistical properties based on the assumptions of ergodicity provide a description of the random process only for the short term. Such data determined from a single record of short duration (several minutes) are usually referred to as 'short-term statistics', whereas data derived from a collection of records taken over a period of one year (say) are referred to as 'long-term statistics'.

The interested reader who requires more information about the nature and properties of random processes should refer to Bendat (1958), Lee (1960), Middleton (1960), Crandall & Mark (1963), Robson (1963), Papoulis (1965), Price & Bishop (1974) or Newland (1975). The literature on random process theory and its applications is now quite extensive.

11.1.1 *Input–output relationships for random processes*

In section 9.2 it was recalled that the input–output relationship for a deterministic linear system may be expressed in the form of a convolution integral. It is

$$q(t) = \int_{-\infty}^{\infty} h(\tau) Q(t - \tau) \, d\tau, \tag{11.1}$$

where the input function $Q(t - \tau) = 0$ for $\tau > t$ and the unit-impulse response satisfies the condition $h(\tau) = 0$ for $\tau < 0$. The function $h(\tau)$ satisfies the relationship

$$\left. \begin{aligned} h(t) &= \int_{-\infty}^{\infty} H(\omega) \, e^{-i\omega t} \, d\omega, \\ H(\omega) &= \frac{1}{2\pi} \int_{-\infty}^{\infty} h(t) \, e^{i\omega t} \, dt. \end{aligned} \right\} \tag{11.2}$$

That is to say, the unit-impulse response $h(t)$ and receptance $H(\omega)$ form a Fourier transform pair. These previous results remain valid for any one realisation even when the inputs and outputs are random processes, provided that the system is linear.

If the time average is taken of both sides of the convolution integral equation in which the input $Q(t)$ and output $q(t)$ are random processes it is found that

$$\langle q(t) \rangle = \left\langle \int_{-\infty}^{\infty} h(\tau) Q(t - \tau) \, d\tau \right\rangle = \int_{-\infty}^{\infty} h(\tau) \langle Q(t - \tau) \rangle \, d\tau.$$

The sign $\langle \rangle$ indicates that a temporal average has been taken and it implies only an operation on the variable t. For an ergodic random process, however, this expected mean value is constant and so

$$\langle q(t) \rangle = \langle Q(t) \rangle \int_{-\infty}^{\infty} h(\tau) \, d\tau = \langle Q(t) \rangle H(0).$$

This result shows that the average of the output is proportional to the average of the input, both of which we shall assume zero henceforth. Supposing that the input and output refer to a 'typical' realisation of an ergodic random process, then, we have established not only an important result of statistical theory but shown that the receptance $H(\omega)$ plays an important role.

Extending the concept of temporal averaging to the product of the outputs at times t and $t + \tau$, we may derive the 'autocorrelation function' of the output. It is defined as

$$R_{qq}(\tau) = \langle q(t) q(t + \tau) \rangle = \langle q(t - \tau) q(t) \rangle = R_{qq}(-\tau). \tag{11.3}$$

Evidently the autocorrelation function is a real and even function of τ. At $\tau = 0$, it reduces to the mean square value of the output given by

$$R_{qq}(0) = \langle q^2(t) \rangle.$$

If equation (11.1) is substituted into (11.3), the autocorrelation function is found in the form

$$R_{qq}(\tau) = \left\langle \int_{-\infty}^{\infty} h(\tau_1) Q(t - \tau_1) \, d\tau_1 \int_{-\infty}^{\infty} h(\tau_2) Q(t + \tau - \tau_2) \, d\tau_2 \right\rangle,$$

where the dummy variables have been written as τ_1 and τ_2 so as to preserve their identity. When the integrals are suitably well behaved, reversing the order of integration and temporal averaging gives

$$R_{qq}(\tau) = \int_{-\infty}^{\infty} \int_{-\infty}^{\infty} h(\tau_1)h(\tau_2)\langle Q(t-\tau_1)Q(t+\tau-\tau_2)\rangle \, d\tau_1 \, d\tau_2$$

$$= \int_{-\infty}^{\infty} \int_{-\infty}^{\infty} h(\tau_1)h(\tau_2)R_{QQ}(\tau+\tau_1-\tau_2) \, d\tau_1 \, d\tau_2, \qquad (11.4)$$

where the autocorrelation function of the input ergodic random process is defined as

$$R_{QQ}(\tau+\tau_1-\tau_2) = \langle Q(t-\tau_1)Q(t+\tau-\tau_2)\rangle = \langle Q(t)Q(t+\tau+\tau_1-\tau_2)\rangle.$$

Just as the impulsive response function and the receptance are related as a Fourier transform pair as shown in equation (11.2), so the autocorrelation function has a counterpart in the frequency domain. This is referred to as the 'mean square spectral density' function (or 'spectral density function' for short). For the output $q(t)$, it satisfies the relationship

$$R_{qq}(\tau) = \int_{-\infty}^{\infty} S_{qq}(\omega) \, e^{i\omega\tau} \, d\omega, \qquad (11.5)$$

with a Fourier transform

$$S_{qq}(\omega) = \frac{1}{2\pi} \int_{-\infty}^{\infty} R_{qq}(\tau) \, e^{-i\omega t} \, d\tau. \qquad (11.6)$$

By substituting equation (11.4) into (11.6) we find that

$$S_{qq}(\omega) = \frac{1}{2\pi} \int_{-\infty}^{\infty} \left[\int_{-\infty}^{\infty} \int_{-\infty}^{\infty} h(\tau_1)h(\tau_2)R_{QQ}(\tau+\tau_1-\tau_2) \, d\tau_1 \, d\tau_2 \right] e^{-i\omega\tau} \, d\tau$$

$$= \int_{-\infty}^{\infty} h(\tau_1) \, e^{i\omega\tau_1} \, d\tau_1 \int_{-\infty}^{\infty} h(\tau_2) \, e^{-i\omega\tau_2} \, d\tau_2$$

$$\times \left[\frac{1}{2\pi} \int_{-\infty}^{\infty} R_{QQ}(\tau+\tau_1-\tau_2) \, e^{-i\omega(\tau+\tau_1-\tau_2)} \, d(\tau+\tau_1-\tau_2) \right]$$

$$= H^*(\omega)H(\omega)S_{QQ}(\omega)$$

$$= |H(\omega)|^2 S_{QQ}(\omega), \qquad (11.7)$$

where $H^*(\omega)$ is the complex conjugate of the receptance $H(\omega)$. The input spectral density function is such that

$$S_{QQ}(\omega) = \frac{1}{2\pi} \int_{-\infty}^{\infty} R_{QQ}(\tau) \, e^{-i\omega\tau} \, d\tau.$$

The important result in equation (11.7) states that if the input spectral density function is known and the modulus of the receptance has been found for a deterministic sinusoidal input then the spectral density function of the output can be evaluated.

Since the autocorrelation function is an even function of τ, it follows from equation (11.6) that

$$S_{qq}(\omega) = \frac{1}{\pi} \int_0^\infty R_{qq}(\tau) \cos \omega\tau \, \mathrm{d}\tau.$$

That is to say the spectral density function of the output is a real even function of frequency, ω. This property reduces equation (11.5) to

$$R_{qq}(\tau) = \int_{-\infty}^\infty S_{qq}(\omega) \cos \omega\tau \, \mathrm{d}\omega = \int_0^\infty 2S_{qq}(\omega) \cos \omega\tau \, \mathrm{d}\omega$$

$$= \int_0^\infty \Phi_{qq}(\omega) \cos \omega\tau \, \mathrm{d}\omega, \tag{11.8}$$

where the physically realisable one-sided spectral density function is

$$\Phi_{qq}(\omega) = \begin{cases} 2S_{qq}(\omega) & (\omega \geq 0), \\ 0 & (\omega < 0). \end{cases}$$

It is this function which may be measured experimentally in the frequency range $0 \leq \omega \leq \infty$; it is only for mathematical convenience that the function $S_{qq}(\omega)$ is used in the range $-\infty < \omega < \infty$.

In terms of physically realisable spectral densities the input–output relationship of equation (11.7) becomes

$$\Phi_{qq}(\omega) = |H(\omega)|^2 \Phi_{QQ}(\omega) \qquad (0 \leq \omega \leq \infty), \tag{11.9}$$

since $|H(\omega)|^2$ is also an even function in ω. The mean square value of the output response is

$$\langle q^2(t) \rangle = R_{qq}(0) = \int_0^\infty \Phi_{qq}(\omega) \, \mathrm{d}\omega = \int_0^\infty |H(\omega)|^2 \Phi_{QQ}(\omega) \, \mathrm{d}\omega. \tag{11.10}$$

Fig. 11.1 illustrates the formation of $\Phi_{qq}(\omega)$.

The previous arguments can be extended to derive other statistical relationships between the input, $Q(t)$, and the output, $q(t)$. The cross-correlation function is defined as

$$R_{Qq}(\tau) = \langle Q(t)q(t+\tau) \rangle = \langle Q(t-\tau)q(t) \rangle = R_{qQ}(-\tau).$$

This is a real function of τ only. By substituting from equation (11.1) and reversing the order of integration and temporal averaging we find that

$$R_{Qq}(\tau) = \int_{-\infty}^\infty h(\tau_1) \langle Q(t)Q(t+\tau-\tau_1) \rangle \, \mathrm{d}\tau_1 = \int_{-\infty}^\infty h(\tau_1) R_{QQ}(\tau-\tau_1) \, \mathrm{d}\tau_1.$$

Again, it can be shown that there exists a cross-spectral density function defined as

$$S_{Qq}(\omega) = \frac{1}{2\pi} \int_{-\infty}^\infty R_{Qq}(\tau) \, \mathrm{e}^{-\mathrm{i}\omega\tau} \, \mathrm{d}\tau$$

$$= \int_{-\infty}^\infty h(\tau_1) \, \mathrm{e}^{-\mathrm{i}\omega\tau_1} \, \mathrm{d}\tau_1 \left[\frac{1}{2\pi} \int_{-\infty}^\infty R_{QQ}(\tau-\tau_1) \, \mathrm{e}^{-\mathrm{i}\omega(\tau-\tau_1)} \, \mathrm{d}(\tau-\tau_1) \right]$$

$$= H(\omega) S_{QQ}(\omega). \tag{11.11}$$

That is to say the cross-spectral density function is the product of the complex receptance and the spectral density function defined in the frequency range $-\infty < \omega < \infty$.

11.2   Representation of an irregular seaway

The wave elevation at a fixed point in a simple deterministic sinusoidal wave may be specified as

$$\zeta(t) = a \cos(\omega t + \alpha),$$

Differences of location produce differences in the value of the phase angle α. As discussed by Price & Bishop (1974), this wave elevation can be thought of as just one realisation of the ergodic random process

$$X(t) = a_x \cos(\omega_x t + A_x),$$

where a_x, ω_x are determinate quantities satisfying the wave equation relationships derived in chapter 6 and A_x is a phase angle the probability of whose value is equally distributed over the range $(0, 2\pi)$.

Fig. 11.1. Arrival at the output mean square spectral density function by multiplying the input spectrum by the square of the RAO. The area beneath the curve $\Phi_{qq}(\omega)$ is $\langle q^2(t) \rangle$.

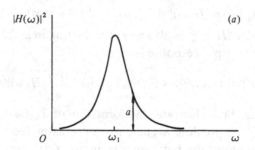

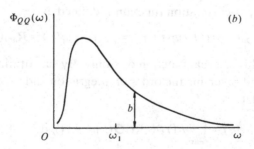

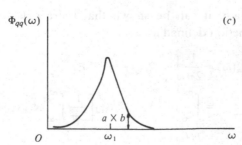

Such a random process has a mean value

$$\langle X(t)\rangle = 0$$

and a mean square value

$$\langle X^2(t)\rangle = \tfrac{1}{2}a_x^2 = R_{XX}(0) = \int_0^\infty \Phi_{XX}(\omega)\,d\omega.$$

In section 6.4.6 it was shown that the total average wave energy per unit horizontal surface area is also proportional to $a_x^2/2$. It therefore follows that the area beneath the spectral density curve is proportional to the total average wave energy per unit horizontal surface area. Thus the spectral density function gives a measure of the energy content in the random process and it is for this reason that it is sometimes referred to as the 'energy spectrum' or 'power spectrum'.

Longuet-Higgins (1952) suggested that an irregular sea surface can be represented as the sum of a large number of regular waves, each component satisfying the relationships discussed in chapter 6 and having a deterministic amplitude, frequency, wave number and direction but with an independent randomly distributed phase angle. Thus the ergodic random process describing the surface wave elevation is given by

$$\zeta(t) = \sum_{i=1}^n X_i(t), \tag{11.12}$$

where

$$X_i(t) = a_i \cos(\omega_i t + A_i),$$

A_i being an independent random variable which is evenly distributed over the range $(0, 2\pi)$.

The mean value of the surface elevation is

$$\langle \zeta(t)\rangle = \sum_{i=1}^n \langle X_i(t)\rangle = 0,$$

the mean square value is

$$\langle \zeta^2(t)\rangle = \sum_{i=1}^n \langle X_i^2(t)\rangle = \sum_{i=1}^n \tfrac{1}{2}a_i^2$$

and the autocorrelation function is

$$R_{\zeta\zeta}(\tau) = \sum_{i=1}^n R_{ii}(\tau) = \int_0^\infty \sum_{i=1}^n \Phi_{ii}(\omega) \cos \omega\tau\,d\omega.$$

Thus the mean square value may be expressed variously as

$$\langle \zeta^2(t)\rangle = R_{\zeta\zeta}(0) = \sum_{i=1}^n R_{ii}(0) = \int_0^\infty \sum_{i=1}^n \Phi_{ii}(\omega)\,d\omega = \int_0^\infty \Phi_{\zeta\zeta}(\omega)\,d\omega$$

$$= \sum_{i=1}^n \tfrac{1}{2}a_i^2. \tag{11.13}$$

We see that the area beneath a curve of $\Phi_{\zeta\zeta}(\omega)$ gives a measure of the total average wave energy per unit horizontal surface area of all the components in the unidirectional seaway. The area represented by the quantity $\Phi_{\zeta\zeta}(\omega)\,\mathrm{d}\omega$ is that contribution to the average wave energy per unit horizontal surface area at frequency ω made by the bandwidth $\mathrm{d}\omega$. In future, this wave energy spectrum will simply be denoted by $\Phi(\omega)$ and it will be taken as understood that it refers to wave elevation.

11.2.1   *Transformation of spectra*

It was seen in section 6.5.1 that the sinusoidal surface profile, defined with respect to the moving ship, was a function of wave encounter frequency ω_e. Similarly, it is sometimes desirable to specify the spectral density function as dependent on variables other than frequency of encounter – variables such as $\ln \omega_e$, $\ln \lambda$ – rather than the absolute wave frequency ω. Now in all cases, the energy of the waves is the same whether it is expressed in terms of ω or of an arbitrary function β, say; i.e.

$$\langle \zeta^2(t) \rangle = \int_0^\infty \Phi(\omega)\,\mathrm{d}\omega = \int_0^\infty \Phi(\beta)\,\mathrm{d}\beta,$$

where there exists a relationship between ω and β of the form

$$\omega = g(\beta) \quad \text{and} \quad \beta = h(\omega).$$

Remembering the physical interpretation of $\Phi(\omega)\,\mathrm{d}\omega$, we see that

$$\Phi(\omega)\,\mathrm{d}\omega = \Phi(\beta)\,\mathrm{d}\beta.$$

It follows that

$$\Phi(\beta) = \left[\frac{\Phi(\omega)}{|\mathrm{d}\beta/\mathrm{d}\omega|} \right]_{\omega=g(\beta)} = \left[\frac{\Phi(\omega)}{|h'(\omega)|} \right]_{\omega=g(\beta)}, \tag{11.14}$$

where the modulus sign is introduced because the spectral density function must always be positive by definition.

By way of example, consider deep water waves, for which the frequency of wave encounter ω_e is

$$\omega_e = \omega - \bar{U}\frac{\omega^2}{g}\cos\chi.$$

Here, $\omega_e \equiv \beta$, so the expression for ω_e is the function $h(\omega)$. Now

$$\frac{\mathrm{d}\omega_e}{\mathrm{d}\omega} = 1 - 2\bar{U}\frac{\omega}{g}\cos\chi$$

and so

$$\Phi(\omega_e) = \frac{\Phi(\omega)}{\left|1 - 2\bar{U}\dfrac{\omega}{g}\cos\chi\right|}$$

for the real, positive value of ω given by

$$\omega = \frac{1 \pm [1 - 4(\bar{U}/g)\omega_e \cos\chi]^{\frac{1}{2}}}{2(\bar{U}/g)\cos\chi}.$$

The variation of the wave encounter spectrum for head seas with ship speed is illustrated in fig. 11.2 where it is seen that the effect of the transformation is to produce a flattened curve extending over a wider range of frequencies as the ship speed increases.

Alternatively, for deep water waves, the wave number

$$k = \frac{2\pi}{\lambda} = \frac{\omega^2}{g}.$$

and so

$$\lambda = \frac{2\pi g}{\omega^2}.$$

It follows that

$$\ln \lambda = \ln 2\pi g - 2 \ln \omega$$

whence

$$d(\ln \lambda) = -\frac{2}{\omega} d\omega$$

Fig. 11.2. The variation of the wave encounter spectrum with ship speed.

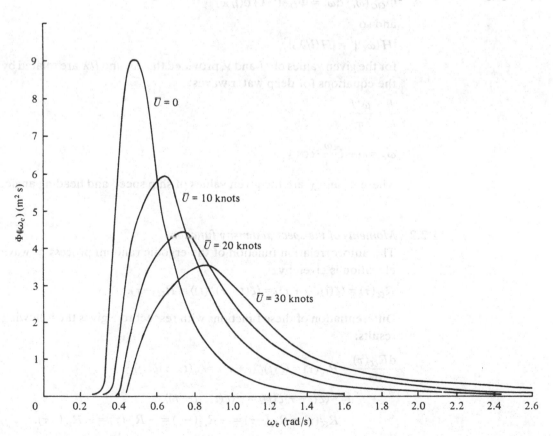

and

$$\Phi(\ln \lambda) = \frac{\Phi(\omega)}{|d(\ln \lambda)/d\omega|} = \frac{\omega}{2}\Phi(\omega)$$

evaluated at $\omega = (2\pi g/\lambda)^{\frac{1}{2}}$.

When it is applied to a ship under way the input–output relationship of equation (11.9) has to be interpreted in a special way. The driving frequency ω then becomes the frequency of encounter, so that

$$\Phi_{qq}(\omega_e) = |H(\omega_e)|^2 \Phi_{QQ}(\omega_e). \tag{11.15}$$

Now this relationship holds good regardless of what the measured variable is; indeed the variable may be replaced by more than one. Thus $(l/\lambda, \bar{U}, \chi)$ may be used instead of ω_e.

Suppose that, for some particular values of $\bar{U}$ and χ, we write

$$\Phi_{qq}(l/\lambda) = |H(l/\lambda)|^2 \Phi_{QQ}(l/\lambda). \tag{11.16}$$

The energy content of the input and output remain the same so that

$$\Phi_{qq}(\omega_e)\, d\omega_e = \Phi_{qq}(l/\lambda)\, d(l/\lambda).$$

It follows that

$$|H(\omega_e)|^2 \Phi_{QQ}(\omega_e)\, d\omega_e = |H(l/\lambda)|^2 \Phi_{QQ}(l/\lambda)\, d(l/\lambda).$$

But for the input

$$\Phi_{QQ}(\omega_e)\, d\omega_e = \Phi_{QQ}(l/\lambda)\, d(l/\lambda)$$

and so

$$|H(\omega_e)|^2 = |H(l/\lambda)|^2$$

for the given values of $\bar{U}$ and χ, provided that ω_e and l/λ are related by the equations for deep water waves:

$$\frac{l}{\lambda} = \frac{\omega^2 l}{2\pi g},$$

$$\omega_e = \omega - \bar{U}\frac{\omega^2}{g}\cos\chi,$$

where $\bar{U}$ and χ are the given values of ship speed and heading angle.

11.2.2 Moments of the spectral density function

The autocorrelation function of the ergodic random process of wave elevation is given by

$$R_{\zeta\zeta}(\tau) = \langle \zeta(t)\zeta(t+\tau)\rangle = \langle \zeta(t-\tau)\zeta(t)\rangle = R_{\zeta\zeta}(-\tau).$$

Differentiation of these functions with respect to τ gives the following results:

$$\frac{dR_{\zeta\zeta}(\tau)}{d\tau} = \dot{R}_{\zeta\zeta}(\tau) = \langle \zeta(t)\dot{\zeta}(t+\tau)\rangle = \langle \zeta(t-\tau)\dot{\zeta}(t)\rangle$$

$$= -\langle \dot{\zeta}(t-\tau)\zeta(t)\rangle = -\langle \dot{\zeta}(t)\zeta(t+\tau)\rangle$$

$$= R_{\zeta\dot{\zeta}}(\tau) = R_{\zeta\dot{\zeta}}(-\tau) = -R_{\dot{\zeta}\zeta}(-\tau) = -R_{\dot{\zeta}\zeta}(\tau) = -\dot{R}_{\zeta\zeta}(-\tau).$$

The derivative $\dot{R}_{\zeta\zeta}(\tau)$ is thus an odd function of τ and this implies that, at $\tau = 0$,

$$\dot{R}_{\zeta\zeta}(0) = 0 = \langle \zeta(t)\dot{\zeta}(t) \rangle.$$

A second differentiation of the autocorrelation function shows that

$$\ddot{R}_{\zeta\zeta}(\tau) = \langle \zeta(t)\ddot{\zeta}(t+\tau) \rangle = -\langle \dot{\zeta}(t)\dot{\zeta}(t+\tau) \rangle = -R_{\dot{\zeta}\dot{\zeta}}(\tau)$$

which, at $\tau = 0$, gives

$$\ddot{R}_{\zeta\zeta}(0) = -\langle \dot{\zeta}^2(t) \rangle.$$

But from equation (11.8) it is seen that the autocorrelation function of the wave elevation may be expressed in terms of the wave spectrum:

$$R_{\zeta\zeta}(\tau) = \int_0^\infty \Phi(\omega) \cos \omega\tau \, d\omega$$

and when this expression is differentiated twice it gives

$$\ddot{R}_{\zeta\zeta}(\tau) = -R_{\dot{\zeta}\dot{\zeta}}(\tau) = -\int_0^\infty \omega^2 \Phi(\omega) \cos \omega\tau \, d\omega.$$

In the same way, further differentiation shows that

$$\ddddot{R}_{\zeta\zeta}(\tau) = \langle \ddot{\zeta}(t)\ddot{\zeta}(t+\tau) \rangle = R_{\ddot{\zeta}\ddot{\zeta}}(\tau) = \int_0^\infty \omega^4 \Phi(\omega) \cos \omega\tau \, d\omega.$$

The moments of the spectral density function are defined as

$$\left. \begin{array}{l} m_0 = \displaystyle\int_0^\infty \Phi(\omega) \, d\omega = R_{\zeta\zeta}(0) = \langle \zeta^2(t) \rangle, \\[2mm] m_2 = \displaystyle\int_0^\infty \omega^2 \Phi(\omega) \, d\omega = -\ddot{R}_{\zeta\zeta}(0) = \langle \dot{\zeta}^2(t) \rangle, \\[2mm] m_4 = \displaystyle\int_0^\infty \omega^4 \phi(\omega) \, d\omega = \ddddot{R}_{\zeta\zeta}(0) = \langle \ddot{\zeta}^2(t) \rangle. \end{array} \right\} \quad (11.17)$$

They are the mean square values of the elevation, the velocity and the acceleration of elevation respectively. Generally, the nth moment of the spectral density function is defined as

$$m_n = \int_0^\infty \omega^n \Phi(\omega) \, d\omega.$$

11.2.3 *Other properties of the spectral density function*
Many other spectral density relationships may be derived from the wave spectrum. For example, consider the variation of pressure at a given point in deep water waves. It is given in section 6.4.5 as

$$\Delta p(X, Z, t) = \rho g \, e^{kZ} \zeta(X, t).$$

The autocorrelation function of the pressure variation is

$$R_{pp}(\tau) = \langle \Delta p(X, Z, t) p(X, Z, t+\tau) \rangle = \rho^2 g^2 \, e^{2kZ} \langle \zeta(X, t)\zeta(X, t+\tau) \rangle$$
$$= \rho^2 g^2 \, e^{2kZ} R_{\zeta\zeta}(\tau).$$

The spectral density function for pressure variation is therefore given by

$$\Phi_{pp}(\omega) = \rho^2 g^2 e^{2kZ} \Phi(\omega).$$

Another relationship, derived by Price & Bishop (1974), for the wave slope spectrum of deep water waves is

$$\Phi_{\zeta'\zeta'}(\omega) = \frac{\omega^4}{g^2} \Phi(\omega),$$

where ζ' denotes the random process of wave slope.

In section 6.4.4 the horizontal and vertical component velocities of the wave fluid particles are shown to be

$$u = \frac{akg}{\omega} \frac{\cosh[k(Z+d)]}{\cosh kd} \cos(kX - \omega t),$$

$$w = \frac{akg}{\omega} \frac{\sinh[k(Z+d)]}{\cosh kd} \sin(kX - \omega t),$$

where $\omega^2 = gk \tanh kd$. By employing the simplifying procedure we used for the wave elevation, it can be shown that the spectral density functions for these component velocities are

$$\Phi_{uu}(\omega) = \frac{\omega^2 \cosh^2[k(Z+d)]}{\sinh^2 kd} \Phi(\omega)$$

and

$$\Phi_{ww}(\omega) = \frac{\omega^2 \sinh^2[k(Z+d)]}{\sinh^2 kd} \Phi(\omega)$$

respectively. The mean square value of the horizontal component velocity, for example, takes the form

$$\langle u^2(t) \rangle = \int_0^\infty \frac{\omega^2 \cosh^2[k(Z+d)]}{\sinh^2 kd} \Phi(\omega) \, d\omega$$

since the wave number k is a function of frequency ω. Such an analysis may be extended to the horizontal and vertical component accelerations of the wave fluid particles, producing the acceleration spectral relationships

$$\Phi_{\dot{u}\dot{u}}(\omega) = \omega^2 \Phi_{uu}(\omega) \quad \text{and} \quad \Phi_{\dot{w}\dot{w}}(\omega) = \omega^2 \Phi_{ww}(\omega).$$

As these examples show, a knowledge of the wave spectrum permits us to determine the spectral density functions for other variables. These latter functions are frequently referred to in ship motion analysis and this is one reason why the wave spectral density function is of such importance.

11.3 The wave spectrum

It is generally believed that sea waves are generated by the wind. Miles (1957, 1960) and Phillips (1966) derived a relationship between the pressure in a turbulent wind and the frequency of the generated wave. The shorter waves grow until they break and so start to limit their energy by dissipation. After the wind has blown for a long time the sea becomes 'fully developed' and the wave spectrum $\Phi(\omega)$ acquires a maximum at a frequency ω_m which is thought to be roughly inversely proportional to the wind speed.

Oceanographers have proposed a number of empirical formulae for wave spectra, to describe the seaway as a function of wind speed. Those of Neumann (1954), Darbyshire (1955, 1959), the British Towing Tank Panel and others have now been superseded by the Pierson & Moskowitz (1963) spectrum. This was obtained from the analysis of extensive wave data relating to 'fully developed' sea conditions in the North Atlantic; its general form, originally discussed by Bretschneider (1961), is

$$\Phi(\omega) = \frac{A}{\omega^5} \exp(-B/\omega^4), \tag{11.18}$$

where ω is the frequency of the waves in rad/s. The constants are given by

$$A = 8.1 \times 10^{-3} g^2, \qquad B = 0.74(g/V)^4,$$

V being the wind speed in m/s at a height 19.5 m above the calm water level. The curve represented by equation (11.18) has a characteristic shape. It rises from zero at the origin, reaches a maximum and then gradually subsides to zero as $\omega \to \infty$.

Most of the one-dimensional wave spectra adopted in the literature may be expressed in the form

$$\Phi(\omega) = \frac{A}{\omega^\alpha} \exp(-B/\omega^\beta),$$

where α and β are empirical coefficients and A and B are constants, the wind speed parameter being incorporated into the latter constant as discussed previously in equation (11.18). This last equation may be written in the non-dimensional form

$$\frac{\Phi(\omega)}{\Phi(\omega_m)} = \left(\frac{\omega}{\omega_m}\right)^{-\alpha} \exp\left\{\frac{\alpha}{\beta}\left[1 - \left(\frac{\omega}{\omega_m}\right)^{-\beta}\right]\right\},$$

where $\omega_m = (B\beta/\alpha)^{1/\beta}$ is the frequency at which the spectrum has its maximum value and thus satisfies the expression $d\Phi(\omega)/d\omega = 0$.

When $\alpha = 5$ and $\beta = 4$ the Pierson–Moskowitz (1963) spectrum is obtained; on the other hand $\alpha = 6$ and $\beta = 2$ give the Neumann (1954) wave spectrum. These spectra are illustrated in fig. 11.3 and compared with other spectra, namely that of Darbyshire (1955), mean JONSWAP (Joint North Sea Wave Project, see section 11.3.5), mean *Famita* (57° 30′ N, 3° 00′ E) as discussed by Saetre (1974) and the storm spectrum as reported by Bretschneider *et al.* (1962) using results from the Ocean Weather Ship *Weather Reporter*. Storm spectra from the North Sea and the North Atlantic have the same form and their shapes are similar to the mean JONSWAP spectrum; but it will be seen from fig. 11.3 that the shape of this latter curve differs substantially from the Pierson–Moskowitz spectrum which is usually adopted in theoretical investigations.

If the transformation technique of section 11.3.1 is employed, the spectrum may be expressed in terms of wave number k, rather than frequency ω, assuming that we are concerned with deep water waves. In fact the corresponding Pierson–Moskowitz wave number spectrum is found to be

$$\Phi(k) = \frac{A^*}{k^3} e^{-B^*/k^2},$$

where the wave number $k\ (=2\pi/\lambda)$ is in m^{-1} and the constants are $A^* = A/2g^2$ and $B^* = B/g^2$.

Equation (11.18) has been adopted as the basis of both the International Towing Tank Conference (ITTC) and the International Ship Structure Congress (ISSC) 'standard' wave spectra.

11.3.1 Some properties of the standard wave spectra
The maximum value of the Pierson–Moskowitz spectrum defined in equation (11.18) occurs when

$$\omega = \omega_{\mathrm{m}} = \left(\frac{4B}{5}\right)^{\frac{1}{4}} = 0.88\, g/V,$$

the maximum being
$$\Phi(\omega_{\mathrm{m}}) = A(0.8B)^{-5/4}\, e^{-5/4}.$$

That is to say, the maximum is proportional to V^5 and it occurs at a frequency whose value, ω_{m}, is proportional to V^{-1}.

Fig. 11.3. Comparison of the principal non-dimensional wave spectra adopted. (ω_{m} is the frequency at which each spectrum has a maximum value.)

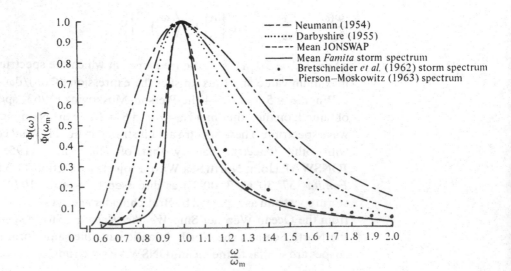

It can be shown that the moments of the spectrum (11.18) are

$$m_0 = \frac{A}{4B}, \qquad m_2 = \frac{A}{4}\left(\frac{\pi}{B}\right)^{\frac{1}{2}} \quad \text{and} \quad m_4 = \frac{A}{4}\Gamma(0), \tag{11.19}$$

whence

$$A = \frac{4}{\pi}\frac{m_2^2}{m_0} \quad \text{and} \quad B = \frac{1}{\pi}\frac{m_2^2}{m_0^2}. \tag{11.20}$$

Here the gamma function is defined as

$$\Gamma(\alpha) = \int_0^\infty y^{\alpha-1}\, e^{-y}\, dy$$

with the properties

$$\Gamma(1+\alpha) = \alpha\Gamma(\alpha), \qquad \Gamma(0+) = \infty, \qquad \Gamma(+\infty) = \infty,$$

where $0+$ indicates that α is permitted to approach zero from above. If the high-frequency end of the spectrum is discarded on the ground that a finite spectral density at very high wave frequency is physically unacceptable, some form of truncation must be devised. If the spectrum is truncated at some multiple n of the peak frequency ω_m, i.e. at $n\omega_m$, then the moment

$$m_4 = \int_0^{n\omega_m} \frac{A}{\omega}\, e^{-B/\omega^4}\, d\omega = \frac{A}{4}\int_{B(n\omega_m)^{-4}}^\infty \frac{e^{-x}}{x}\, dx = \frac{A}{4}E_1\left[\frac{B}{(n\omega_m)^4}\right], \tag{11.21}$$

where $E_1(\ )$ is an exponential integral form tabulated by Abramowitz & Stegun (1964) with a series expansion

$$E_1(z) = -\gamma - \ln z - \sum_{s=1}^\infty \frac{(-1)^s z^s}{ss!},$$

where the Euler constant $\gamma = 0.577216$. In this particular context $z = B/(n\omega_m)^4 = 5/4n^4$ and so $z \ll 1$ if $n > 2$. Accordingly the dominant contribution to E_1 comes from the first two terms, a reasonable approximation being

$$E_1(z) \simeq -\gamma - \ln z = -\gamma - \ln 1.25 + 4\ln n \qquad (n > 2)$$
$$= -0.8004 + 4\ln n.$$

Table 11.1 shows the approximate variation of $4m_4/A$ with truncation point.

If the same truncation point is adopted for the calculations of the moments m_0 and m_2 (even though these are integrable over the entire frequency range $0 \le \omega \le \infty$) we find that

$$m_0 = \int_0^{n\omega_m} \Phi(\omega)\, d\omega = \frac{A}{4B}\exp\left(-\frac{5}{4n^4}\right) \tag{11.22}$$

and if $n \to \infty$,

$$m_0 = A/4B.$$

This is the result found for the original spectrum without truncation.

The second moment becomes

$$m_2 = \int_0^{n\omega_m} \omega^2 \Phi(\omega)\, d\omega = \frac{A}{4}\left(\frac{\pi}{B}\right)^{\frac{1}{2}}\left\{1 - 2\,\text{erf}\left[\left(\frac{5}{2n^4}\right)^{\frac{1}{2}}\right]\right\}, \qquad (11.23)$$

where the error function

$$\text{erf}\,(x) = \frac{1}{(2\pi)^{\frac{1}{2}}}\int_0^x \exp\,(-y^2/2)\, dy,$$

with the properties

$$\text{erf}\,(-x) = -\text{erf}\,(x) \quad \text{and} \quad \text{erf}\,(\infty) = 0.5.$$

Again if $n \to \infty$

$$m_2 = \frac{A}{4}\left(\frac{\pi}{B}\right)^{\frac{1}{2}}$$

and this, too, is the result for the original spectrum without truncation.

With these results, the 'bandwidth parameter' ε satisfies the equation

$$\varepsilon^2 = 1 - \frac{m_2^2}{m_0 m_4} = 1 - \frac{\pi\{1 - 2\,\text{erf}\,[(2z)^{\frac{1}{2}}]\}^2}{\exp\,(-z)E_1(z)}, \qquad (11.24)$$

where $z = 5/4n^4$. (The significance of the bandwidth parameter is that the closer ε is to zero the more is the spectrum akin to a single localised peak, the random process then being referred to as 'narrow band'. At the other extreme, as $\varepsilon \to 1$, the spectrum resembles a flattened hump and the random process is referred to as 'broad band'.) The relationship between ε and n is shown in fig. 11.4.

If m_0 and m_2 are evaluated over the entire frequency range $0 \leqslant \omega \leqslant \infty$ and only m_4 is found from the truncated spectrum, the bandwidth relationship reduces to

$$\varepsilon^2 = 1 - \frac{\pi}{E_1(z)}$$

which, to be greater than zero, requires that $E_1(z) > \pi$, or $n \geqslant 2.68$.

Table 11.1. *Variation of m_4 for a Pierson–Moskowitz spectrum with n when the spectrum is truncated at $\omega = n\omega_m$*

n	$4m_4/A$
2	1.9722
3	3.5940
4	4.7448
5	5.6374
6	6.3666

This relationship for a truncated Pierson–Moskowitz spectrum, too, is shown in fig. 11.4. Note that these relationships between ε and n are both independent of the actual values of the constants A and B.

From actual wave records of two different sea conditions, Cartwright & Longuet-Higgins (1956) determined bandwidth parameter values of 0.57 and 0.67, these being equivalent to truncation points of approximate values $n = 3$ and $n = 4$ respectively. Since ω_m decreases with increased severity of the sea condition, the value chosen for n will accordingly be increased in order that a reasonable frequency range may be obtained over which the integration can be performed.

From wave data recorded at Weather Station *India*, Hoffman (1975) formed a family of wave spectra and placed them in ten groups, each containing eight wave spectra. For this family, Table 11.2 illustrates the variation of the measured bandwidth parameter ε for each group with the corresponding average characteristic mean period T_1 and significant wave height $h_{\frac{1}{3}}$ which is defined as the average of the one-third highest waves recorded. The values of n which would give the measured values of ε, i.e. giving appropriate truncation points have been calculated using the above, more complicated expression for ε^2. These values indicate how n should be selected. The increase of spectral bandwidth parameter ε with significant wave height indicates the presence of swells in the higher numbered groups.

Fig. 11.4. Relationship between bandwidth parameter ε and n.

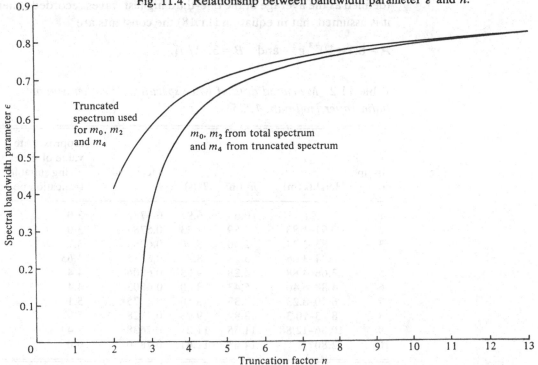

Price & Bishop (1974) have shown that this bandwidth parameter is also related to the ratio of the expected number of zero upcrossings per unit time, $E[N_+(0)]$ and the total expected number of peaks of maxima per unit time, $E[N(-\infty, 0)]$. The ratio is

$$\frac{E[N_+(0)]}{E[N(-\infty, 0)]} = \frac{m_2}{(m_0 m_4)^{\frac{1}{2}}} = (1 - \varepsilon^2)^{\frac{1}{2}}. \qquad (11.25)$$

In general, for an ergodic random process $\zeta(t)$ of wave elevation, the expected number of upcrossings per unit time at a chosen level ζ_0 is

$$E[N_+(\zeta_0)] = \frac{1}{2\pi} \left(\frac{m_2}{m_0}\right)^{\frac{1}{2}} \exp\left(-\frac{\zeta_0^2}{2m_0}\right), \qquad (11.26)$$

whereas the expected number of upcrossings at the level ζ_0 with velocity $\dot{\zeta}_0$ or more per unit time is given by

$$E[N_+(\zeta_0, \dot{\zeta}_0)] = \frac{1}{2\pi} \left(\frac{m_2}{m_0}\right)^{\frac{1}{2}} \exp\left[-\frac{1}{2}\left(\frac{\zeta_0^2}{m_0} + \frac{\dot{\zeta}_0^2}{m_2}\right)\right]. \qquad (11.27)$$

Finally the expected number of maxima per unit time, regardless of their magnitude, is given by the expression

$$E[N(-\infty, 0)] = \frac{1}{2\pi} \left(\frac{m_4}{m_0}\right)^{\frac{1}{2}}. \qquad (11.28)$$

11.3.2   The ITTC wave spectrum

When the only information available is the significant wave height $h_{\frac{1}{3}}$ (defined as the average of the one-third highest waves recorded), then it is assumed that in equation (11.18) the constants are

$$A = 8.1 \times 10^{-3} g^2 \quad \text{and} \quad B = 3.11/h_{\frac{1}{3}}^2.$$

Table 11.2. *Measured data of wave spectra at Weather Station India (after Hoffman, 1975)*

Group no.	Heights (m)	$h_{\frac{1}{3}}$ (m)	T_1(s)	Measured ε	Approximate value of n giving suitable truncation point
1	<0.91	0.63	6.97	0.5782	3.0
2	0.91–1.83	1.59	7.33	0.5784	3.0
3	1.83–2.74	2.30	8.46	0.6368	3.7
4	2.74–3.66	3.24	8.31	0.6345	3.65
5	3.66–4.88	4.28	9.03	0.6806	4.4
6	4.88–6.40	5.43	8.80	0.6803	4.4
7	6.40–8.23	7.33	9.50	0.7075	5.1
8	8.23–10.36	8.82	9.93	0.7328	5.9
9	10.36–12.80	11.35	11.21	0.7638	7.4
10	12.80>	14.65	11.61	0.7870	9.0

The spectrum is illustrated in fig. 11.5 which has a maximum value of $0.25 h_{\frac{1}{3}}^{\frac{5}{2}} e^{-\frac{5}{4}}$ m^2 s at frequency $\omega_m = 1.26 h_{\frac{1}{3}}^{-\frac{1}{2}}$ or period $T_m = 5.0 h_{\frac{1}{3}}^{\frac{1}{2}}$.

Price & Bishop (1974) show that the significant wave height is related to the area m_0 beneath the wave spectrum defined in the frequency range $0 \leqslant \omega \leqslant \infty$ through the result

$$h_{\frac{1}{3}} = 4.0 \, m_0^{\frac{1}{2}} \tag{11.29}$$

and to the second moment m_2 by

$$h_{\frac{1}{3}} = 5.1 \, m_2. \tag{11.30}$$

Further, the average wave period $\bar{T}$, measured at the level of the calm water surface is

$$\bar{T} = 2\pi \left(\frac{m_0}{m_2} \right)^{\frac{1}{2}}, \tag{11.31}$$

and, in the light of the foregoing properties, this reduces to

$$\bar{T} = 3.55 \, h_{\frac{1}{3}}^{\frac{1}{2}} = 4.47 \, \omega_m^{-1} = 0.71 \, T_m.$$

The variable $\bar{T}$ is dependent only on the significant wave height and is not treated as an independent variable.

Fig. 11.5. A typical ITTC version of the Pierson–Moskowitz wave spectrum. For the curve shown, $h_{\frac{1}{3}}$ has been taken as 7 m, 5 m, 3 m, 2 m.

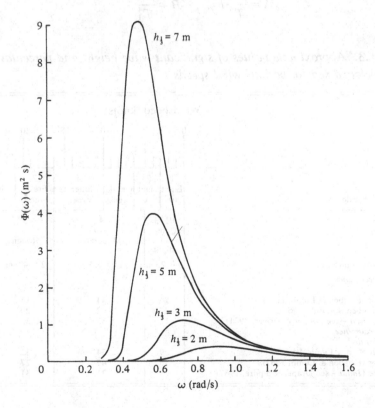

If the truncated spectrum is adopted as discussed in section 11.3.1 then the previous results become

$$
\left.
\begin{aligned}
h_{\frac{1}{3}} &= 4.0 \exp\left(\frac{5}{8n^4}\right) m_0^{\frac{1}{2}}, \\[2mm]
h_{\frac{1}{3}} &= 5.1 m_2 \left\{ 1 - 2\,\mathrm{erf}\left[\left(\frac{5}{2n^4}\right)^{\frac{1}{2}}\right] \right\}^{-1}, \\[2mm]
m_4 &= 2.025 \times 10^{-3} g^2 E_1\!\left(\frac{5}{4n^4}\right)
\end{aligned}
\right\}
\qquad (11.32)
$$

and

$$
\bar{T} = 3.55\, h_{\frac{1}{3}}^{\frac{1}{2}} \left\{ 1 - 2\,\mathrm{erf}\left[\left(\frac{5}{2n^4}\right)^{\frac{1}{2}}\right] \right\}^{-\frac{1}{2}} \exp\left(-\frac{5}{8n^4}\right)
$$

for $n \geqslant 2$.

It is recommended that when only data on wind speed are available, the approximation of Table 11.3 should be used to find suitable values of the all-important quantity $h_{\frac{1}{3}}$.

11.3.3 *The ISSC wave spectrum*

If information is also available on the 'characteristic mean period' T_1, it is assumed that

$$
A = \frac{173}{T_1^4} h_{\frac{1}{3}}^2, \qquad B = \frac{691}{T_1^4}.
$$

Table 11.3. *Approximate values of significant wave height and percentage occurrence of a fully developed sea for various wind speeds*

	Windspeed (knots)				
Beaufort wind scale	1 Light airs / 2 Light breeze / 3 Gentle breeze	4 Moderate breeze	5 Fresh breeze	6 Strong breeze / 7 Near gale	8 Gale
Sea state	0123	4	5	6	
Description	Calm to slight	Moderate	Rough	Very rough	High
Wave height (significant)	0–1.25 m	1.25–2.5 m	2.5–4.0 m	4.0–6.0 m	6.0 m upwards
Percentage occurrence in North Sea					
All seasons (Hogben & Lumb, 1967)	34	44	15	5	2
Winter (Hogben & Lumb, 1967)	27	43	20	8	2
Winter (*Famita* – see Draper & Driver, 1971)	11	33	33	15	8
Percentage occurrence in North Atlantic					
Winter (Hogben & Lumb, 1967)	9	34	32	17	8
All seasons (Hogben & Lumb, 1967)	15	42	27	11	5
All seasons (*India* – see Draper & Squire, 1967)	8	31	31	21	9

The quantity T_1 is defined by

$$T_1 = 2\pi \frac{m_0}{m_1}$$

and can be taken approximately as the value $\bar{T}$ of the random variable of average period T_{av} such that $\bar{T}$ is measured at the level of the calm water surface and is

$$\bar{T} = 2\pi \left(\frac{m_0}{m_2}\right)^{\frac{1}{2}}.$$

(11.33)

This wave spectrum is illustrated in fig. 11.6; it has a maximum value of $0.065 \, h_{\frac{1}{3}}^2 T_1 \, e^{-5/4}$ at a frequency of $\omega_m = 4.85 \, T_1^{-1}$ or period $T_m = 1.30 \, T_1$. The ITTC and ISSC wave spectra have maxima occurring at the same value of frequency only when $T_1 = 3.86 h_{\frac{1}{3}}^{\frac{1}{2}}$.

By substituting the values of the constants A and B into equation (11.18) we discover that in the frequency range $0 \leqslant \omega \leqslant \infty$,

$$h_{\frac{1}{3}} = 4.0 \, m_0^{\frac{1}{2}}$$

(11.34)

and

$$T_1 = (691\pi)^{\frac{1}{4}} \left(\frac{m_0}{m_2}\right)^{\frac{1}{2}} = (691\pi)^{\frac{1}{4}} \frac{\bar{T}}{2\pi} = 1.086 \, \bar{T}$$

$$= 4.85 \, \omega_m^{-1} = 0.77 \, T_m.$$

(11.35)

Fig. 11.6. Typical ISSC wave spectrum for various values of the characteristic wave period $\bar{T}$, with $h_{\frac{1}{3}} = 7$ m.

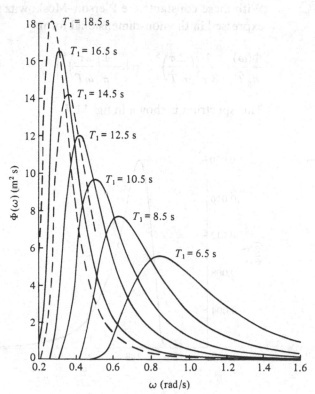

$T_1 = 18.5$ s
$T_1 = 16.5$ s
$T_1 = 14.5$ s
$T_1 = 12.5$ s
$T_1 = 10.5$ s
$T_1 = 8.5$ s
$T_1 = 6.5$ s

$\Phi(\omega) \, (m^2 s)$

$\omega \, (rad/s)$

If the truncated spectrum of section 11.3.1 is adopted then these relationships are modified to

$$
\left.\begin{array}{l}
h_{\frac{1}{3}} = 4.0 \exp\left(\dfrac{5}{8n^4}\right) m_0^{\frac{1}{2}}, \\[3mm]
m_4 = 43.25\, h_{\frac{1}{3}}^2\, E_1\!\left(\dfrac{5}{4n^4}\right)\!\Big/ T_1^4 \\[3mm]
\text{and} \\[3mm]
T_1 = 1.086\, \bar{T}\left\{1 - 2\,\mathrm{erf}\left[\left(\dfrac{5}{2n^4}\right)^{\frac{1}{2}}\right]\right\}^{\frac{1}{2}} \exp\left(\dfrac{5}{8n^4}\right)
\end{array}\right\}
\qquad (11.36)
$$

for $n \geqslant 2$.

11.3.4 Non-dimensional spectra

The values given in equation (11.19) for m_0 and m_2 relate to Pierson–Moskowitz spectra in general. If, in addition, we let

$$
h_{\frac{1}{3}} = 4m_0^{\frac{1}{2}} \quad \text{and} \quad \bar{T} = 2\pi\left(\dfrac{m_0}{m_2}\right)^{\frac{1}{2}},
$$

we find that

$$
A = \frac{h_{\frac{1}{3}}^2 \bar{T}}{8\pi^2}\left(\frac{2\pi}{\bar{T}}\right)^5 \quad \text{and} \quad B = \frac{1}{\pi}\left(\frac{2\pi}{\bar{T}}\right)^4.
$$

With these constants the Pierson–Moskowitz wave spectrum may be expressed in the non-dimensional form

$$
\frac{\Phi(\omega)}{h_{\frac{1}{3}}^2 \bar{T}} = \frac{1}{8\pi^2}\left(\frac{2\pi}{\omega\bar{T}}\right)^5 \exp\left[-\frac{1}{\pi}\left(\frac{2\pi}{\omega\bar{T}}\right)^4\right].
\qquad (11.37)
$$

This spectrum is shown in fig. 11.7.

Fig. 11.7. A non-dimensional form of the Pierson–Moskowitz wave spectrum.

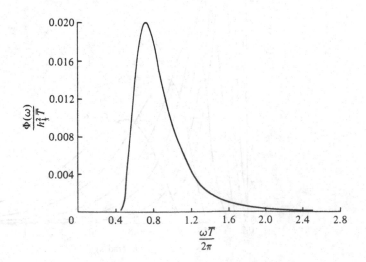

If $\ln \lambda$ is used as the independent variable instead of ω, this spectrum becomes

$$\Phi(\ln \lambda) = \frac{\pi}{2}\left(\frac{\lambda h_{\frac{1}{3}}}{g\overline{T}^2}\right)^2 \exp\left[-4\pi\left(\frac{\lambda}{g\overline{T}^2}\right)^2\right], \tag{11.38}$$

as may be found by applying the transformation techniques of section 11.2.1. This curve is illustrated in fig. 11.8.

In a similar manner, the wave log-slope spectrum is given by

$$\Phi_{\zeta'\zeta'}(\ln \lambda) = \frac{\omega}{2}\Phi_{\zeta'\zeta'}(\omega) = \frac{\omega^5}{2g^2}\Phi(\omega)$$

as derived by Price & Bishop (1974). But the previous results show that this reduces to

$$\Phi_{\zeta'\zeta'}(\ln \lambda) = 2\pi^3\left(\frac{h_{\frac{1}{3}}}{g\overline{T}^2}\right)^2 \exp\left[-4\pi\left(\frac{\lambda}{g\overline{T}^2}\right)^2\right] = 4\pi^2\left[\frac{\Phi(\ln \lambda)}{\lambda^2}\right]$$

or

$$\frac{\Phi_{\zeta'\zeta'}(\ln \lambda)}{(h_{\frac{1}{3}}/g\overline{T}^2)^2} = 4\pi^2\left[\frac{\Phi(\ln \lambda)}{(\lambda h_{\frac{1}{3}}/g\overline{T}^2)^2}\right] \tag{11.39}$$

and the form of the curve is again that of fig. 11.8. The use of such a spectrum has been described by Nordenstrøm, Faltinsen & Pedersen (1971) for the prediction of wave induced motions of ships.

11.3.5 Joint North Sea Wave Project spectrum

The Joint North Sea Wave Project (JONSWAP) was set up in 1967 by a number of bodies in the Netherlands, West Germany, the United States and the United Kingdom as a cooperative experiment on the measurement of waves. The two basic aims of the study were to measure the growth of waves under conditions of limited fetch and the attenuation of waves as they come into shallow water. The necessary conditions are met at the chosen site extending 160 km off the island of Sylt in the German Bight where 13 wave stations were set up. Easterly winds form waves in deep water with limited fetch, while waves arriving mainly from the west and north form a swell which is considerably influenced by the bottom topography.

Fig. 11.8. An alternative non-dimensional form of the Pierson–Moskowitz wave spectrum.

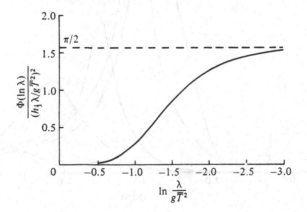

The wave stations were equipped to measure the one-dimensional wave spectrum during wave generation by easterly winds, while a number of stations were also able to measure the directional wave spectrum using pitch–roll buoys and an array of subsurface pressure transducers. During the course of two periods – September 1968 and July 1969 – over 2000 wave spectra were measured, of which about 300 corresponded to well-defined, approximately stationary regimes.

For the simplest conditions of near-uniform offshore winds, Ewing (1975) obtained a set of wave spectra like those shown in fig. 11.9(a). These spectra, and those obtained from other cases in which the conditions of wave generation were ideal, were fitted to the analytical function

$$\Phi(\omega) = \frac{\alpha g^2}{\omega^5} \exp\left[-\frac{5}{4}\left(\frac{\omega}{\omega_m}\right)^{-4}\right] \gamma^{\exp[-(\omega-\omega_m)^2/2\sigma^2\omega_m^2]}$$

by an appropriate least-squares technique. In this expression, ω_m is the frequency at which the maximum of the spectrum occurs, α is the parameter corresponding to the constant in the Phillips (1958a,b)

Fig. 11.9. (a) Family of wave spectra, after Ewing (1975), derived from JONSWAP data. (b) Comparison of the JONSWAP and Pierson–Moskowitz wave spectra ($\gamma = 3.3$, $\sigma_a = 0.07$, $\sigma_b = 0.09$). (c) Variation of the JONSWAP wave spectrum with peak enhancement factor γ ($h_{\frac{1}{3}} = 7.5$ m, $\bar{T} = 8.5$ s).

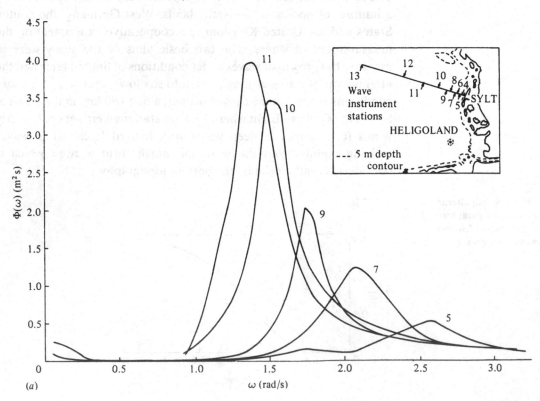

(a)

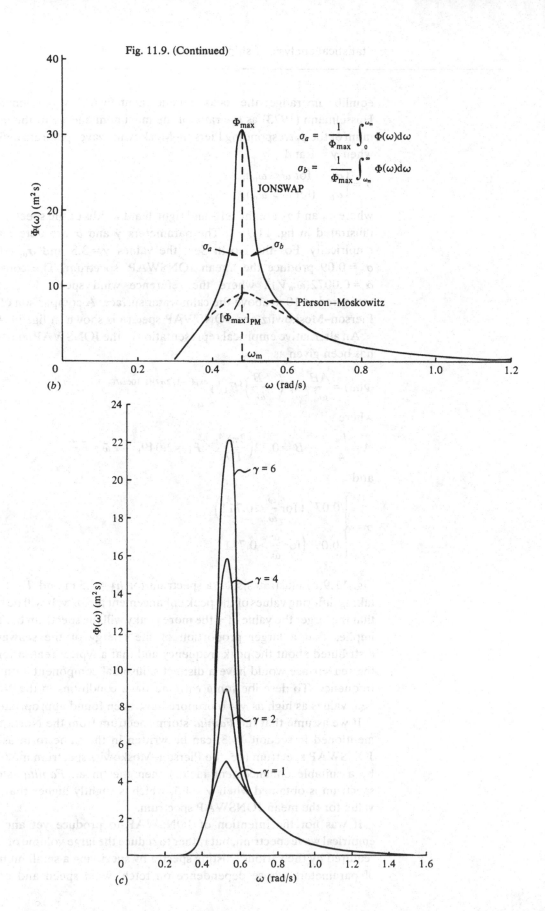

Fig. 11.9. (Continued)

$$\sigma_a = \frac{1}{\Phi_{max}} \int_0^{\omega_m} \Phi(\omega)d\omega$$

$$\sigma_b = \frac{1}{\Phi_{max}} \int_{\omega_m}^{\infty} \Phi(\omega)d\omega$$

JONSWAP

Pierson–Moskowitz

$[\Phi_{max}]_{PM}$

ω_m

equilibrium range; the 'peak enhancement factor' γ is defined by Hasselmann (1973) as the ratio of the maximum energy to the maximum of the corresponding Pierson–Moskowitz wave spectrum defined when $\gamma = 1$ and

$$\sigma = \begin{cases} \sigma_a & \text{(for } \omega \leqslant \omega_m\text{)}, \\ \sigma_b & \text{(for } \omega > \omega_m\text{)}, \end{cases}$$

where σ_a and σ_b are the left- and right-hand widths of the spectrum as illustrated in fig. 11.9(b). The parameters γ and σ are determined empirically. For the North Sea, the values $\gamma = 3.3$ and $\sigma_a = 0.07$, $\sigma_b = 0.09$ produce the 'mean JONSWAP' spectrum. The constant $\alpha = 0.0072(\omega_m V)^{\frac{2}{3}}$, where the reference wind speed V is now measured at 10 m above the calm water surface. A comparison of the Pierson–Moskowitz and JONSWAP spectra is shown in fig. 11.9(b).

An alternative empirical representation of the JONSWAP spectrum has been given as

$$\Phi(\omega) = \frac{AB}{\omega^5} \exp\left(-\frac{B}{\omega^4}\right)\{F_1^{-1} \gamma^{\exp[(-1/2\sigma^2)(1.406\omega/\bar{\omega}-1)^2]}\},$$

where

$$A = \frac{h_{\frac{1}{3}}^2}{4}, \qquad B = 0.32\left(\frac{2\pi}{\bar{T}}\right)^4, \qquad F_1 = 2.049, \qquad \bar{\omega} = \frac{2\pi}{\bar{T}}$$

and

$$\sigma = \begin{cases} 0.07 & \left(\text{for } \dfrac{\omega}{\bar{\omega}} < 0.711\right), \\[2mm] 0.09 & \left(\text{for } \dfrac{\omega}{\bar{\omega}} > 0.711\right). \end{cases}$$

Fig. 11.9(c) illustrates such a spectrum for $h_{\frac{1}{3}} = 7.5$ m and $\bar{T} = 8.5$ s, taking differing values of the peak enhancement factor γ. It will be seen that the larger the value of γ the more peaky will the spectrum be. This implies that a larger proportion of the energy of the seaway is distributed about the peak frequency and that a typical realisation of the sea surface would have a distinct sinusoidal component with this frequency. To describe some extreme wave conditions in the North Sea, values as high as $\gamma = 6$ or more have been found appropriate.

If we assume that the *Famita* storm spectrum from the North Sea, mentioned in section 11.3, can be written in the same form as the JONSWAP spectrum (i.e. the Pierson–Moskowitz spectrum modified by a suitable enhancement factor) then the 'mean *Famita*' storm spectrum is obtained when $\gamma = 4.5$, which is slightly higher than the value for the mean JONSWAP spectrum.

It was not the intention of JONSWAP to produce yet another empirical wave spectrum, but rather to reduce the large volume of data required by the various existing spectra by specifying a small number of parameters whose dependence on fetch, wind speed and other factors could be investigated in a systematic manner. By way of conclusions, it was found by Ewing (1975) that fetch-limited wave spectra, as measured in the North Sea, indicate:

(a) the existence of an 'overshoot phenomenon' by which the energy of a particular frequency component near the peak of the spectrum grows to a level that is approximately double its final value in the asymptotic state;

(b) that the spectra have narrower but higher peaks than are predicted by the appropriate Pierson–Moskowitz formulation (see fig. 11.3)

(c) that the Phillips constant α is fetch-dependent for the high-frequency 'ω^{-5} tails' of spectra. This appears to support the findings of Longuet-Higgins (1969) who suggested that the constant varies very slightly with fetch and wind speed;

(d) that non-linear wave–wave interactions control the overall energy balance in the spectrum and contribute to the growth of waves on the forward face of the wave spectrum, particularly for shorter fetches;

(e) that the introduction of the additional factors, γ, σ_a, σ_b yields a wider variety of spectral shapes than does the basic Pierson–Moskowitz formulation and consequently that it admits a better fit with measured data (e.g. see fig. 11.3).

11.3.6 Directional wave spectra

The total average wave energy per unit horizontal surface area or mean square wave elevation in a confused or short-crested irregular seaway may be obtained from the two-dimensional spectrum $\Phi(\omega, \mu)$. The volume represented by $\Phi(\omega, \mu)\, d\omega\, d\mu$ is proportional to the average wave energy per unit horizontal surface area from waves of frequency ω and bandwidth $d\omega$ travelling at an angle μ to the dominant wind direction and spread over the range $d\mu$. The simplest representation of the directional wave spectrum is

$$\Phi(\omega, \mu) = \Phi(\omega)M(\mu), \tag{11.40}$$

where $\Phi(\omega)$ is the unidirectional wave spectrum and $M(\mu)$ is a function describing the spread of waves about the dominant wind direction. The form of this latter function is usually determined semi-empirically.

For use with the Pierson–Moskowitz spectrum, the ITTC and ISSC recommend a spreading function of the form

$$M(\mu) = A_n \cos^n \mu, \tag{11.41}$$

where $-\pi/2 \leqslant \mu \leqslant \pi/2$ and the normalising factor is given by

$$A_n = \frac{2}{\pi} \qquad (n = 2),$$

for the ITTC spectrum, and

$$A_n = \frac{8}{3\pi} \qquad (n = 4),$$

for the ISSC spectrum. These simple forms are the most widely used and the spreading functions are illustrated in fig. 11.10.

Oceanographers have suggested more elaborate spreading functions of the type

$$M(\mu) = G(s) \cos^{2s}[(\mu - \bar{\mu})/2],$$ (11.42)

where $-\pi/2 \le \mu \le \pi/2$ and the normalising function is given by

$$\frac{1}{G(s)} = \int_{-\pi/2}^{\pi/2} \cos^s \frac{\lambda}{2} \, d\lambda.$$

The angular width of the spectrum depends on s which varies from about 20 at low frequencies to about 2 at high frequencies. The mean wave direction may also vary with wave frequency ω.

11.4   Wave data

In order to specify the wave spectra that we have discussed for the sea areas in which marine vessels or structures operate, information is needed on the significant wave height $h_{\frac{1}{3}}$ and the average period $\bar{T}$. Most of the worldwide long-term data available, however, are presented in terms of visually estimated wave height h_v and period T_v. Hogben & Lumb (1967) collated over one million observations collected from about 500 ships during the period 1953–61 covering fifty different sea areas around the world; these were grouped in ranges

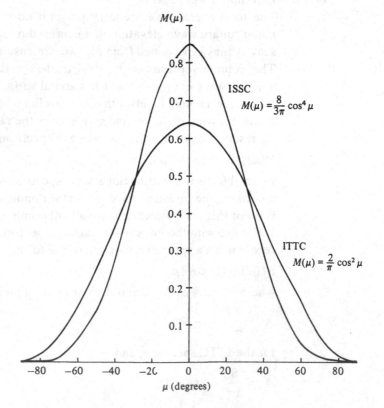

Fig. 11.10. Spreading functions $M(\mu)$ proposed by the ITTC and ISSC for use in directional wave spectra.

ISSC
$$M(\mu) = \frac{8}{3\pi} \cos^4 \mu$$

ITTC
$$M(\mu) = \frac{2}{\pi} \cos^2 \mu$$

Table 11.4(*a*). *Percentage frequency of occurrence of wave heights and characteristic wave periods for northern North Atlantic*

Wave height (m)	Visual wave period in seconds at class midpoints										Total
	2.5	6.5	8.5	10.5	12.5	14.5	16.5	18.5	20.5	Over 21	
0–1	13.7204	3.4934	0.8559	0.3301	0.1127	0.0438	0.0249	0.0172	0.0723	0.3584	19.0291
1–2	11.4889	15.5036	6.4817	1.8618	0.5807	0.1883	0.0671	0.0254	0.0203	0.0763	36.2941
2–3	1.5944	7.8562	8.0854	3.7270	1.1790	0.3713	0.1002	0.0321	0.0091	0.0082	22.9629
3–4	0.3244	2.2487	4.0393	2.9762	1.3536	0.4477	0.1307	0.0428	0.0050	0.0040	11.5724
4–5	0.1027	0.7838	1.6998	1.5882	0.9084	0.3574	0.1443	0.0433	0.0072	0.0049	5.6400
5–6	0.0263	0.1456	0.3749	0.4038	0.2493	0.1200	0.0382	0.0067	0.0027	0.0027	1.3702
6–7	0.0277	0.1477	0.3614	0.4472	0.2804	0.1301	0.0504	0.0113	0.0011	0.0032	1.4605
7–8	0.0084	0.0714	0.1882	0.2199	0.1634	0.0785	0.0353	0.0069	0.0018	0.0034	0.7772
8–9	0.0037	0.0325	0.0856	0.1252	0.1119	0.0558	0.0303	0.0045	0.0027	0.0033	0 4555
9–10	0.0034	0.0204	0.0674	0.1173	0.0983	0.0550	0.0303	0.0173	0.0079	0.0047	0.4220
10–11		0.0005	0.0012	0.0023	0.0031	0.0012		0.0005			0.0088
11+		0.0005	0.0007	0.0019	0.0035	0.0002			0.0005		0.0073
Total	27.3003	30.3043	22.2415	11.8009	5.0443	1.8493	0.6517	0.2080	0.1306	0.4691	100.0000

Table 11.4(*b*). *Number of measured wave heights and periods at Weather Station India based on 2400 records as given by Draper & Squire (1967)*

Measured wave height (m)	Measured wave period in seconds at class midpoint								All periods
	6.5	7.5	8.5	9.5	10.5	11.5	12.5	13.5	
0.00–0.61	1		2	3	1	1	1		9
0.61–1.22	7	15	33	17	1				73
1.22–1.83	4	31	63	36	7	1			142
1.83–2.44	3	20	62	62	18	2			167
2.44–3.05	2	14	47	63	20	4	1		151
3.05–3.66	2	6	31	46	24	8	1		118
3.66–4.27		2	16	33	24	9	3		87
4.27–4.88		1	8	22	20	12	4	1	68
4.88–5.49		1	6	21	16	5	4	2	55
5.49–6.10			3	11	12	7	3		36
6.10–6.71			1	2	10	5	1		19
6.71–7.32		1		4	7	4	2		18
7.32–7.92				1	6	8	1	1	17
7.92–8.53				1	5	2			8
8.53–9.14			1	2	4	3	1		11
9.14–9.75				3	1		1		5
9.75–10.36					3	1			4
10.36–10.97				1	2	2	1		6
10.97–11.58				1		1			2
11.58–12.19					1	1			2
12.19–12.80							1		1
12.80–13.41									
13.41–14.02									
14.02–14.63							1		1
Total	19	91	272	322	174	82	31	9	1000

Table 11.4(c). *Number of visual wave heights and periods at Weather Station India as given by Walden (1964)*

Visual wave height (m) at class midpoint	Visual wave period in seconds at class midpoint								All periods
	<5	6	8	10	12	14	16	>17	
0.75	39	178	50	8				5	280
1.25	303	2086	1716	243	17	3		10	4378
2.25	81	2712	3564	518	33	6	6	15	6935
3.25	12	1381	2924	655	94	7	6	10	5089
4.25	3	716	1765	624	86	16		14	3224
5.25	2	216	791	388	69	3	4	2	1475
6.25	1	162	507	319	60	13		1	1063
7.25	2	78	228	156	48	7		1	520
8.25		63	124	104	64	8	1		364
9.25		59	54	93	22	8	4		240
10.25		1	4	5	1				11
11.25		1	1	2		2			6
12.25		6		2	2	1			11
13.25		4		1	2	4			11
14.25						2	1		3
15.25		4			1	5			10
Total	443	7667	11728	3118	501	83	22	58	23620

of wave height and wave period. A condensed version of these data is given by Price & Bishop (1974), as illustrated in Table 11.4(a) and fig. 11.11. Further data on the North Sea has been given by Roll (1958) and the U.S. Naval Oceanographic Office (1963) while the U.S. Department of Commerce (1959, 1961) has published data for sea areas in the North Atlantic and North Pacific Oceans.

Weather ships and light vessel stations located at fixed positions in the oceans are the main source of measured wave data. Draper (1966) described the various wave-recording devices in use; these include shipborne wave recorders, pressure sensors on the sea bed or on legs of offshore platforms, instrumentation buoys, and so forth. In their survey of sea waves, Cartwright & Draper (1975) described methods of wave prediction and quoted instances where such instrument packages have been installed by the Institute of Oceanographic Sciences at locations around the British coast and in the North Sea.

Wave statistics have been published for Weather Station *Juliette* (52° 30′ N, 20′ W) by Draper & Whitaker (1965), for Ocean Station *India* (59° N, 19° W) by Draper & Squire (1967) and at the rescue and weathership MV *Famita* in position 57° 30′ N, 3° 00′ E by Draper & Driver (1971). Additional measured long-term wave data recorded from light vessel stations have been published by Draper & Fricker (1965) for Land's End, by Draper (1968a) for Morecambe Bay, by

Draper (1968b) for Smith's Knoll in the North Sea, by Draper & Graves (1968) for the Varne light vessel in the Dover Strait and by Draper & Blakey (1969) for the Mersey Bar light vessel. The data collected generally consisted of records of 12 minutes duration taken every 3 hours during a period of 1 year. Each single record is regarded as a short sample from a stationary Gaussian random process and gives at least an estimate of the significant value and mean zero crossing period appropriate to the random process of which the individual record is a sample.

Table 11.4(b) illustrates measured significant wave height and period data at Weather Station *India* obtained by Draper & Squire (1967) whereas Table 11.4(c) shows visual wave data at the same location compiled by Walden (1964) (see also Nordenstrøm, 1969). Similar trends are evident but the measured significant wave height extends over a much wider range of values than the comparable visual range.

Several attempts have been made to relate 'visual' to 'measured' data for specified sea areas but the degree of success achieved has been limited, not least by the quality of observers used. Nordenstrøm (1973) suggests that the significant wave height $h_{\frac{1}{3}}$ (metres) can be estimated

Fig. 11.11. Probability distribution function of exceeding a visual wave height in three sea areas.

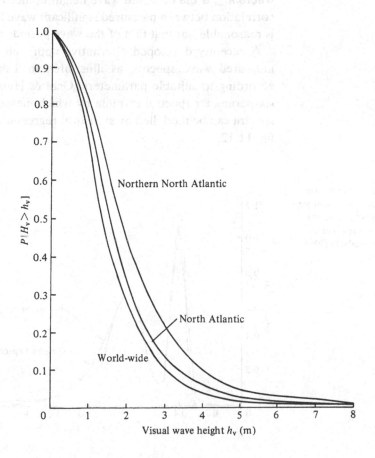

Visual wave height h_v (m)

from the visually estimated values h_v obtained by trained observers (such as the data given by Hogben & Lumb (1967) for the North Atlantic) from the formula

$$h_{\frac{1}{3}} = 1.68 \, h_v^{0.75},$$

and that the average value of the average apparent wave period is

$$\mu_{\bar{T}} = 2.83 \, T_v^{0.44}.$$

Another relationship may be found by adapting results given by Hoffman (1972). The information was obtained by trained observers and relates to the North Atlantic. It is

$$h_{\frac{1}{3}} = 1.78 \, h_v^{0.72},$$
$$\mu_{\bar{T}} = 3.39 \, T_v^{0.41}.$$

A formula of more general application has been suggested for the significant wave heights when the information is reported by untrained observers; it is

$$h_{\frac{1}{3}} = 1.25 + 0.89 \, h_{\text{obs}},$$

where h_{obs} is the observed wave height in metres. It is found that the correlation between measured significant wave height and visual data is reasonable but that that of the wave periods is not as accurate.

A recently developed alternative approach is to use families of measured wave spectra, as illustrated in Table 11.2, categorised according to suitable parameters. Ochi & Hubble (1976) derived a six-parameter spectral formula by which measured families of wave spectra can be modelled by statistical regression analysis as is seen in fig. 11.12.

Fig. 11.12. Comparison between observed and six-parameter wave spectra derived by statistical regression analysis ($h_{\frac{1}{3}} = 2.38$ m).

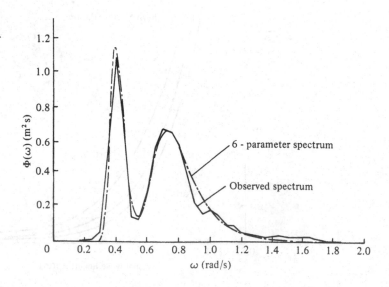

11.4.1 *Hindcasting*

Methods of predicting wave data from information on the wind distribution over a specified sea area may be referred to as 'hindcasting'. Wave forecasting and hindcasting methods involve analysis procedures which track energy flux over large ocean areas and are based on the numerical integration of the energy-balance equation

$$\frac{\partial \Phi}{\partial t} + \mathbf{v} \cdot \operatorname{grad} \Phi = S,$$

where $\Phi(\omega, \mu)$ is the two-dimensional wave energy spectrum, $\mathbf{v}$ is the wave group velocity for deep water and $S(\omega, \mu)$ is the source function. (When $S = 0$ the equation describes the propagation of swell in deep water.)

The main uncertainty in this approach is in the correct specification of the source function. In one possible formulation it is expressed as

$$S = (a + b\Phi)\left[1 - \left(\frac{\Phi}{\Phi_{\text{eqn}}}\right)^{n}\right] + N,$$

where a, b represent the linear and exponential growth terms of Phillips and Miles fitted to experimentally determined growth rates, Φ_{eqn} is the equilibrium wave spectrum, n determines the rapidity of the approach to equilibrium and N is a complicated function of the wave spectrum describing the energy transfer due to non-linear wave–wave interactions. (Its presence implies that all wave components assumed present in the wave spectrum are coupled.)

To solve this partial differential equation, the directional wave spectrum is represented by a finite number of frequencies and wave directions at a suitable number of chosen grid points covering the sea area of interest. As an example (to indicate the enormous size of the problem), in a major hindcasting project covering the North Atlantic, Pierson, Tick & Baer (1966) produced directional spectra four times daily for a 15-month period covering the whole of 1959 at each of 519 grid points situated in the North Atlantic. The results of this analysis were used by Wachnik & Zainick (1965) to investigate the seakeeping performance of a ship in a realistic short-crested seaway.

The approach of forecasting and hindcasting waves using the energy-balance equation is a rational attempt to solve this problem, and as more detailed understanding of the propagation and generation of waves in the ocean is acquired, predictions based on these methods are sure to improve. Though the methods raise problems of data handling they are becoming more widely used for engineering purposes.

11.5 Probability density and statistical properties

Oceanographers have found that over a short period of time (say 20 minutes) wave records can be assumed to represent stationary, relatively narrow band processes. The statistical properties and characteristics of such processes have been fully described elsewhere (e.g. see Price & Bishop, 1974). Only a summary of the relevant results needed for present purposes will be given here. It will be assumed that the mean square spectral density function $\Phi(\omega)$ of the wave elevation random process $\zeta(t)$ has been estimated and its lower order moments determined (see equation (11.17)).

11.5.1 The normal or Gaussian density function

The elevation of the wave surface $\zeta(t)$ measured from its calm water position, at random or equal intervals of time, closely follows a normal or Gaussian distribution with zero mean value and variance m_0. The probability density and distribution functions are

$$
\left.
\begin{aligned}
f_\zeta(x) &= \frac{1}{(2\pi m_0)^{\frac{1}{2}}} \exp\left(\frac{-x^2}{2m_0}\right) \quad (-\infty < x < \infty) \\
&\text{and} \\
F_\zeta(x) &= P[\zeta \leq x] = \int_{-\infty}^{x} f_\zeta(z)\, dz \\
&= \frac{1}{(2\pi m_0)^{\frac{1}{2}}} \int_{-\infty}^{x} \exp\left(\frac{-z^2}{2m_0}\right) dz \\
&= 0.5 + \mathrm{erf}\,(x/m_0^{\frac{1}{2}})
\end{aligned}
\right\}
\tag{11.43}
$$

respectively, x being a measure of the wave elevation. In the latter expression, the error function

$$
\mathrm{erf}\,(x) = \frac{1}{(2\pi)^{\frac{1}{2}}} \int_0^\infty \exp\,(-z^2/2)\, dz
$$

has the properties

$$
\mathrm{erf}\,(-x) = -\mathrm{erf}\,(x), \qquad \mathrm{erf}\,(\infty) = 0.5.
$$

Fig. 11.13 illustrates the form of these functions. Strictly, such functions are associated with a random process having a bandwidth $\varepsilon = 1$, i.e. with a broad-band random process.

Sometimes it is convenient to express the distribution function in a logarithmic form. The log-normal distribution is

$$
P[\zeta > x] = P[\ln \zeta > \ln x] = \frac{1}{(2\pi m_0)^{\frac{1}{2}}} \int_{-\infty}^{\ln x} \exp\left(\frac{-\ln^2 z}{2m_0}\right) d(\ln z).
$$

With suitable logarithmic scales, this function can be represented in graphical form as a straight line.

11.5.2 The Rayleigh density function

If the wind has been blowing for a sufficiently long time, it will produce a fully developed seaway. That is to say the surface has the appearance of a sinusoidal wave of slowly varying period and amplitude and the elevation is a narrow-band process such that $\varepsilon = 0$. It is found that under these conditions the random process of wave height $H(t)$ ($=2\zeta(t)$) measured from troughs to crests can be described by a Rayleigh probability density function. That is to say

$$f_H(h) = \frac{h}{4m_0} \exp\left(-\frac{h^2}{8m_0}\right) \quad (0 \leqslant h \leqslant \infty), \tag{11.44}$$

where h is a measure of the wave height and m_0 is the moment of zero order of the wave amplitude random process $\zeta(t)$ and not $H(t)$. The function is illustrated in fig. 11.14(a). The corresponding Rayleigh distribution function

$$F_H(h) = P[H \leqslant h] = \int_0^h \frac{y}{4m_0} \exp\left(-\frac{y^2}{8m_0}\right) dy$$

$$= 1 - \exp\left(-\frac{h^2}{8m_0}\right) \tag{11.45}$$

is illustrated in fig. 11.14(b); it is such that the probability of obtaining a wave height in the range $h_1 < H \leqslant h_2$ is

$$P[h_1 < H \leqslant h_2] = P[H \leqslant h_2] - P[H \leqslant h_1]$$

$$= \exp\left(-\frac{h_1^2}{8m_0}\right) - \exp\left(\frac{h_2^2}{8m_0}\right).$$

Fig. 11.13. (a) The normal probability density function for values of $x/m_0^{\frac{1}{2}}$ and (b) the corresponding distribution function.

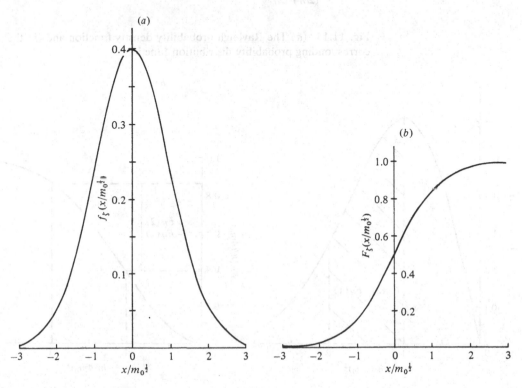

(a)

(b)

The corresponding value may be found from the probability density function of fig. 11.14(a) by integration between the limits $h_1 < h \leq h_2$.

The probability of obtaining a wave height $H > h$ is

$$P[H > h] = 1 - P[H \leq h] = \exp\left(-\frac{h^2}{8m_0}\right).$$

Thus if $h_{1/n}$ is the lowest wave height associated with the $1/n$th highest observations,

$$P[H > h_{1/n}] = \frac{1}{n} = \exp\left(-\frac{h_{1/n}^2}{8m_0}\right)$$

and

$$h_{1/n}^2 = 8m_0 \ln n. \tag{11.46}$$

The average value of these $1/n$th highest observations is

$$\bar{h}_{1/n} = n \int_{h_{1/n}}^{\infty} h f_H(h)\, \mathrm{d}h = n \int_{h_{1/n}}^{\infty} \frac{h^2}{4m_0} \exp\left(-\frac{h^2}{8m_0}\right)\, \mathrm{d}h$$

which on integration reduces to

$$\bar{h}_{1/n} = (8m_0)^{\frac{1}{2}}\{(\ln n)^{\frac{1}{2}} + n\pi^{\frac{1}{2}}[0.5 - \mathrm{erf}\,(2 \ln n)^{\frac{1}{2}}]\}, \tag{11.47}$$

where

$$\mathrm{erf}\,(x) = \frac{1}{(2\pi)^{\frac{1}{2}}} \int_0^x \mathrm{e}^{-y^2/2}\, \mathrm{d}y.$$

Fig. 11.14. (a) The Rayleigh probability density function and (b) the corresponding probability distribution function.

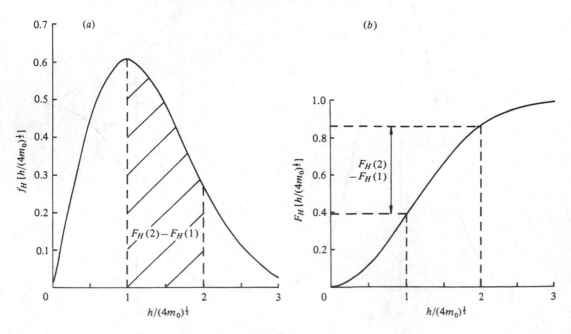

For very large n, the error function tends to 0.5, so that the average wave height of these $1/n$th highest observations is

$$\bar{h}_{1/n} = (8m_0 \ln n)^{\frac{1}{2}}$$

and the average wave amplitude of these observations is

$$\bar{a}_{1/n} = 0.5\,\bar{h}_{1/n} = (2m_0 \ln n)^{\frac{1}{2}}.$$

Values of $\bar{h}_{1/n}$ are given in Table 11.5.

It will be found from Table 11.5 that the average or mean value of the wave height ($n = 1$) is $0.886\,(8m_0)^{\frac{1}{2}}$, whence

$$\bar{h} = 2.51 m_0^{\frac{1}{2}}$$

whilst the average wave amplitude

$$\bar{a} = 0.5\,\bar{h} = 1.25 m_0^{\frac{1}{2}}.$$

Similarly for the 'significant wave height' ($n = 3$),

$$\bar{h}_{\frac{1}{3}} = 4.0 m_0^{\frac{1}{2}} \quad \text{and} \quad \bar{a}_{\frac{1}{3}} = 2.0 m_0^{\frac{1}{2}}$$

whereas the one-tenth highest wave height ($n = 10$) gives

$$\bar{h}_{1/10} = 5.05 m_0^{\frac{1}{2}} \quad \text{and} \quad \bar{a}_{1/10} = 2.55 m_0^{\frac{1}{2}}.$$

Another statistical parameter that is sometimes quoted is the 'most frequent' or 'most probable wave height', h_{mp}. This is the value of the wave height corresponding to the maximum of the probability density function. By differentiating equation (11.44) this is found to give

$$h_{mp} = 2.0 m_0^{\frac{1}{2}}$$

and so the most probable wave amplitude is

$$a_{mp} = m_0^{\frac{1}{2}},$$

m_0 again relating to the wave elevation amplitude spectrum.

Fig. 11.15 illustrates the meaning of these various statistical quantities associated with the Rayleigh probability density function describing wave height.

Another statistical result, based on the Rayleigh probability density function and derived by Longuet-Higgins (1952), is the value of the expected or average value of the maximum wave height $\bar{h}_m$, or wave

Table 11.5. *Average values of $1/n$th largest wave heights in a fully developed seaway*

$1/n$	$\bar{h}_{1/n}(8m_0)^{-\frac{1}{2}}$	$P[H > \bar{h}_{1/n}]$	$1/n$	$\bar{h}_{1/n}(8m_0)^{-\frac{1}{2}}$	$P[H > \bar{h}_{1/n}]$
0.001	2.628	1.00×10^{-3}	0.5	1.256	0.2065
0.01	2.359	3.83×10^{-3}	0.6	1.176	0.2508
0.1	1.801	3.90×10^{-2}	0.7	1.102	0.2969
0.2	1.591	7.96×10^{-2}	0.8	1.031	0.3454
0.3	1.454	1.21×10^{-1}	0.9	0.961	0.3971
0.4	1.347	1.63×10^{-1}	1.0	0.886	0.4561

amplitude $\bar{a}_{\mathrm{m}}$, in a sample of N waves. It may be shown that

$$\bar{h}_{\mathrm{m}} = (8m_0)^{\frac{1}{2}}[(\ln N)^{\frac{1}{2}} + 0.2886(\ln N)^{-\frac{1}{2}}] = 2\bar{a}_{\mathrm{m}}.$$

(It is to be expected that $\bar{h}_{\mathrm{m}}$ is in general less than $\bar{h}_{1/n}$.) A discussion and a full theoretical derivation of this result is also given by Price & Bishop (1974). Similar expressions can be derived for the expected value of the second and third highest wave heights denoted by $\bar{h}_{2\mathrm{m}}$ and $\bar{h}_{3\mathrm{m}}$ respectively. Table 11.6 shows the appropriate numerical results.

11.5.3 Probability density functions of extremes

It is in the nature of things for the engineer to need information on probabilities of extreme conditions. The height of the highest wave a ship is ever likely to encounter, for example, is plainly of interest and will serve as a useful illustration. The ship concerned will be subjected to many sea states in its lifetime, each associated with a random variable such as wave elevation at a point. We assume that each of these random variables is associated with a stationary random process. Together, they define another random process.

Consider a sample of N similarly distributed independent random variables $\{X_1, X_2, \ldots, X_n\}$ which, together, determine a random process $X(t)$; that is to say, all the independent random variables have the same sort of probability density and distribution functions. The probability that the single variable X_1 is less than some chosen value x_{M} is given by the probability distribution function

$$P[X_1 \leqslant x_{\mathrm{M}}] = F_X(x_{\mathrm{M}}),$$

Fig. 11.15. The Rayleigh probability density function of wave height for a narrow band process having bandwidth $\varepsilon = 0$. For this curve $P[H > h_{\mathrm{mp}}] = 0.6065$, $P[H > \bar{h}] = 0.4561$, $P[H > \bar{h}_{\frac{1}{3}}] = 0.1354$, $P[H > \bar{h}_{\frac{1}{10}}] = 0.0387$.

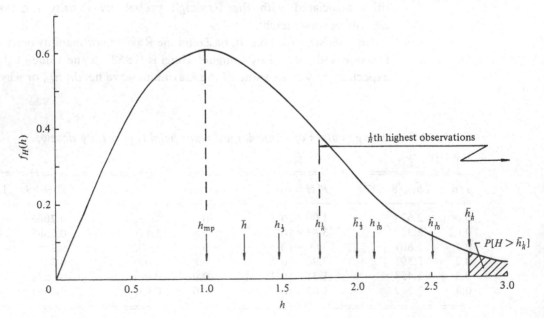

and the probability that every independent random variable in the sample is less than x_M is

$$P[(X_1 \leqslant x_M) \cap (X_2 \leqslant x_M) \cap \ldots \cap (X_N \leqslant x_M)]$$
$$= P[X_1 \leqslant x_M]P[X_2 \leqslant x_M] \ldots P[X_N \leqslant x_M]$$
$$= F_X^N(x_M).$$

Alternatively this condition may be described as the probability that x_M is the largest value achieved by any of the N independent variables. In other words, if the random variable

$$\eta = \max \{X_1, X_2, \ldots, X_N\}, \tag{11.48}$$

the probability that the largest value of η is less than the maximum x_M is

$$P[\eta \leqslant x_M] = F_\eta(x_M) = F_X^N(x_M). \tag{11.49}$$

When it is differentiated, this result gives

$$f_\eta(x_M) = N F_X^{N-1}(x_M) f_X(x_M).$$

This is the probability density function for the maximum or largest value among the sample of N independent variables or observations and $f_X(x_M)$, $F_X(x_M)$ are the probability density and distribution functions respectively of the individual variables, evaluated at the chosen peak value.

In a similar manner, the probability that the variable X_1 is greater than some chosen value x_m is

$$P[X_1 > x_m] = 1 - P[X_1 \leqslant x_m] = 1 - F_X(x_m)$$

and the probability that every independent random variable exceeds x_m is

$$P[(X_1 > x_m) \cap (X_2 > x_m) \cap \ldots \cap (X_N > x_m)] = [1 - F_X(x_m)]^N.$$

This is the probability that x_m is the smallest value among the N observations. Again defining a new random variable

$$\xi = \min \{X_1, X_2, \ldots, X_N\},$$

we see that the probability that ξ is greater than the chosen minimum x_m is

$$P[\xi > x_m] = 1 - P[\xi \leqslant x_m] = 1 - F_\xi(x_m) = [1 - F_X(x_m)]^N.$$

Table 11.6. *The expected values of the first, second and third highest waves in a sample of N wave crests*

N	50	100	200	800
$\bar{h}_m/(8m_0)^{\frac{1}{2}}$	2.12	2.28	2.43	2.70
$\bar{h}_{2m}/(8m_0)^{\frac{1}{2}}$	1.91	2.05	2.23	2.51
$\bar{h}_{3m}/(8m_0)^{\frac{1}{2}}$	1.80	1.98	2.12	2.40

That is
$$F_\xi(x_m) = 1 - [1 - F_X(x_m)]^N. \tag{11.50}$$
The probability density function associated with the minimum value among the N independent observations is
$$f_\xi(x_m) = \frac{dF_\xi(x_m)}{dx_m} = N[1 - F_X(x_m)]^{N-1} f_X(x_m). \tag{11.51}$$
Now $f_X(x_m)$ and $F_X(x_m)$ are the probability density and distribution functions respectively of the individual random variables evaluated at the minimum value.

The probability density and distribution functions of the extreme values x_M and x_m can evidently be determined provided that the values of the initial probability functions at the maximum or minimum are known.

The value of x_M for which $f_\eta(x_M)$ is a maximum is the extreme value which is most likely to be realised in the N independent observations; it is called the 'most probable extreme value' and denoted by x_{Mp}. This means that the value x_{Mp} satisfies the equation
$$\frac{df_\eta(x_M)}{dx_M} = 0 = (N-1)f_X^2(x_M) + F_X(x_M)\frac{df_X(x_M)}{dx_M}$$
for $x_M = x_{Mp}$. If, for example, $f_X(x)$ is the Rayleigh probability density function associated with a narrow-band random process, then for large values of N it may be shown that the most probable extreme maximum value is given by
$$x_{Mp} \simeq (2m_0 \ln N)^{\frac{1}{2}}. \tag{11.52}$$
Ochi (1973) has shown that for large N, the probability of the extreme value random variable exceeding the most probable extreme value x_{Mp} is
$$P[\eta > x_{Mp}] = 1 - e^{-1} \simeq 0.632$$
which is significant. It is sometimes desirable to predict an extreme value, referred to as the 'design extreme value', x_{Md} which gives a far more conservative probability estimate. Thus if β is a small quantity we might wish to ascertain x_{Md} for which
$$P[\eta > x_{Md}] = \beta = 1 - F_\eta(x_{Md}) = 1 - F_X^N(x_{Md}).$$
For large values of N, this approximates to
$$F_X(x_{Md}) = (1 - \beta)^{1/N} \simeq 1 - \frac{\beta}{N} + O(\beta^2).$$

Again using the Rayleigh probability distribution function as an example, we should find that the design extreme value is given by
$$x_{Md} = \left[2m_0 \ln \left(\frac{N}{\beta}\right)\right]^{\frac{1}{2}}.$$

For example, if $\beta = 0.01$, then the extreme value may be estimated which is unlikely to be exceeded with 99% certainty. This implies that one ship in 100 sister ships operating in the same statistical environment may experience an extreme value greater than the predicted design extreme value x_{Md}. Alternatively, a single ship may experience a greater value than the predicted extreme design value once in 100 encounters with the same statistical environment. If greater assurance for safety is required, a smaller value of β may be chosen.

It is interesting to note that the design extreme value exceeds the significant value by the ratio

$$\frac{x_{Md}}{x_{\frac{1}{3}}} = \left[\frac{1}{2}\ln\left(\frac{N}{\beta}\right)\right]^{\frac{1}{2}}$$

which is neither sensitive to the value N nor to the choice of value of β. Again choosing $\beta = 0.01$, we see that for $N = 100$ the ratio is 2.15 whilst for $N = 10^6$ the ratio increases to 2.63. It would thus appear that the design extreme value is far more sensitive to a correct evaluation of the moment m_0 of the random process.

Ochi (1973) and Ochi & Motter (1975) discuss the variation of the parameters x_M, x_{Mp}, x_{Md} and β when the random process has a non-zero bandwidth. They show that in most cases the net reduction in the magnitude of the design extreme value is less than 10%.

11.5.4 Weibull probability density function

Gumbel (1954) discusses several valid types of probability function which may be used empirically for such extreme value problems. One such function, called the 'Weibull' probability distribution or density function, has the form

$$F_\xi(x) = \begin{cases} 1 - \exp(-ax^b) & (x \geq 0), \\ 0 & (x < 0), \end{cases} \qquad (11.53)$$

$$f_\xi(x) = abx^{b-1} \exp(-ax^b),$$

where a and b are positive. The functions are illustrated in fig. 11.16. The probability density function has a maximum value

$$ab(b-1/ab)^{(b-1)/b} e^{-(b-1/b)}$$

at $x = (b - 1/ab)^{1/b}$. Weibull (1951, 1952) used such functions to describe experimentally observed variations in the fatigue resistance of steel, its elastic limit, etc. Although the function carries the name of Weibull, it was used in probability theory before the appearance of his experimental work in the discussion of limiting distributions of maximum and minimum values. In ship response and wave dynamics the function is used to describe long-term trends in responses and wave data.

Gumbel (1954) also discussed the asymptotic distributions of the largest and smallest values of the various functions used in the extreme problem. That is to say he examined how the function and the extreme values become modified if the number of constituent random variables, N, tends to infinity.

Fig. 11.16. Weibull probability distribution and density functions.

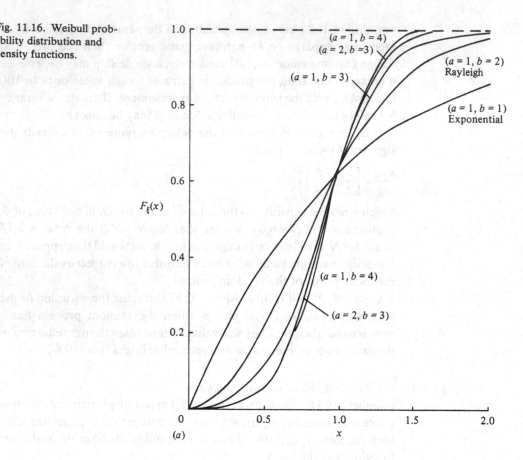

(a)

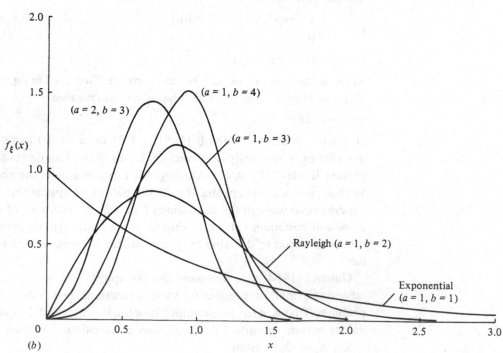

(b)

The asymptotic Weibull probability distribution function for the smallest value is of the form

$$F_\xi(x) = 1 - \exp\left[-\left(\frac{x-\alpha}{\beta-\alpha}\right)^\gamma\right] \quad (x > \alpha, \beta > \alpha),$$

where α, β and γ are constants. Weibull used a function like this in discussing the breaking strength of materials but it will be more helpful to discuss its practical application in terms of waves. The use of a Weibull distribution, as we shall see, is really an exercise in curve fitting in which a probability distribution is derived on the basis of observed data. The attractions of this particular distribution are (a) that having three disposable parameters α, β and γ, it is fairly adaptable, (b) that these parameters can be determined systematically as we shall show and (c) that it has some theoretical justification, in the light of Gumbel's findings, when one is concerned with extreme value problems.

Suppose that it is wished to fit a Weibull distribution function to visually observed heights of waves for which the random variable of visual wave period T_V has values T_v lying within some chosen interval. To be specific suppose we select the range $7.5\,\text{s} < T_v \le 9.5\,\text{s}$; we might then observe the frequency of occurrence of waves whose heights are $0{-}1, 1{-}2, 2{-}3, \ldots, m$ as shown in Table 11.4(a). By this means a histogram could be constructed as given by the probability distribution illustrated in fig. 11.11.

If a Weibull distribution is to be fitted to the observed data, the probability that the visual wave height random variable H_V does not exceed a value h_v is

$$P[H_V \le h_v] = F_{H_V}(h_v) = 1 - \exp\left[-\left(\frac{h_v-h_0}{h_c-h_0}\right)^\gamma\right]. \tag{11.54}$$

In this expression, h_0, h_c and γ are parameters whose values are to be found from the observed data and we can see immediately that h_0 is a cut-off value since it is necessary that $h_v > h_0$ if this function is to be a valid distribution. Moreover we shall require that $h_c > h_0$. According to Battjes (1970) the constant h_0 can be loosely described as a measure of background noise (which may be due, for example, to swells) and it appears to be related to the degree of exposure of the location at which measurements are taken. The positive constant $(h_c - h_0)$ is a scale parameter and γ is a non-dimensional shape parameter; i.e. the distribution becomes steeper with increasing values of γ. When $h_0 = 0$ and $\gamma = 2$ the Rayleigh distribution is again achieved.

The values of the three disposable parameters may be determined as follows:

(a) If $h_v = h_c$,

$$F_{H_V}(h_c) = 1 = e^{-1} = 0.63212 = P[H_V \le h_c].$$

An estimate of h_c may therefore be obtained from a sample of visual wave height observations by choosing the wave height corresponding to this value of the probability. It may readily be found from a histogram constructed for the given data.

(b) The mean value of the visual wave height sample is given by

$$\mu_{H_V} = \int_{h_0}^{\infty} h_V f_{H_V}(h_V)\, dh_V.$$

On substitution and integration, this result is found to give

$$\frac{\mu_{H_V} - h_0}{h_c - h_0} = \Gamma\left(1 + \frac{1}{\gamma}\right).$$

But the observed mean of the sample of visual wave heights is such that

$$F_{H_V}(\mu_{H_V}) = 1 - \exp\left[-\left(\frac{\mu_{H_V} - h_0}{h_c - h_0}\right)^{\gamma}\right],$$

and so

$$\left(\frac{\mu_{H_V} - h_0}{h_c - h_0}\right)^{\gamma} = \left[\Gamma\left(1 + \frac{1}{\gamma}\right)\right]^{\gamma} = \ln\left[\frac{1}{1 - F_{H_V}(\mu_{H_V})}\right] = \ln\left[T(\mu_{H_V})\right],$$

where

$$\ln\left[T(\mu_{H_V})\right] = \left[\Gamma\left(1 + \frac{1}{\gamma}\right)\right]^{\gamma}.$$

The quantity $T(\mu_{H_V})\{ = 1/P[H_V > \mu_{H_V}]\}$ is the 'recurrence interval' or 'return period'; it is the average number of observations (or average length of time) in which μ_{H_V} will be equalled or exceeded once when a large number of observations is made (or a long period of record used). If $T(\mu_{H_V})$ is known from the sample of visual wave height observations, it gives a relationship for γ only.

(c) To obtain an estimate of the lower limit h_0 of the sample, the variance must be determined, i.e.

$$\sigma_{H_V}^2 = \int_{h_0}^{\infty} (h_V - \mu_{H_V})^2 f_{H_V}(h_V)\, dh_V = \int_{h_0}^{\infty} h_V^2 f_{H_V}(h_V)\, dh_V - \mu_{H_V}^2$$

and when this integration is performed it is found that

$$\sigma_{H_V}^2 = (h_c - h_0)^2\left[\Gamma\left(1 + \frac{2}{\gamma}\right) - \Gamma^2\left(1 + \frac{1}{\gamma}\right)\right].$$

Since h_c and γ are determined previously the only unknown is h_0, for $\sigma_{H_V}^2$ is determined from the sample of visual wave height observations. If h_0 is known to be zero, then the distribution of equation (11.54) becomes simplified accordingly and there remain only two disposable parameters.

It will be seen, by taking the natural logarithm of equation (11.54) twice, that

$$\ln\left(\ln\{1 - P[H_V \leqslant h_V]\}^{-1}\right) = \gamma \ln(h_V - h_0) - \gamma \ln(h_c - h_0).$$

A plot of the Weibull distribution is a straight line with $\ln(\ln\{1 - P[H_v \leqslant h_v]\}^{-1})$ as one coordinate and $\ln(h_v - h_0)$ as the other. This was pointed out by Gumbel (1954). The slope of the line is γ, while h_0 is given by the intersection of the curve with the $\ln(\ln\{1 - P[H_v \leqslant h_v]\}^{-1})$ axis. It is usually found that the best fit of the wave height data is achieved when $h_0 \neq 0$.

Nordenstrøm *et al.* (1971) determined values of h_0, h_c and γ for the North Atlantic and the North Sea. These are given in Tables 11.7(*a*) and (*b*). Fig. 11.17 shows the Weibull distribution (on log–log scale) of visual wave height for the chosen time intervals in the North Atlantic.

Table 11.7(*a*). *Parameters of Weibull distributions of visual wave heights at Weather Stations A, B, C, D, E, I, J, K and M in the North Atlantic*

Visual wave period intervals (s)	h_0 (m)	h_c (m)	γ	Probability of T_v falling within designated wave period interval
$T_v \leqslant 5.5$	1.00	1.25	0.63	0.1190
$5.5 < T_v \leqslant 7.5$	1.35	2.20	0.85	0.3453
$7.5 < T_v \leqslant 9.5$	1.10	3.15	1.13	0.3586
$9.5 < T_v \leqslant 11.5$	0.75	4.05	1.56	0.1385
$11.5 < T_v \leqslant 13.5$	0.35	5.35	1.82	0.0290
$13.5 < T_v \leqslant 15.5$	0.20	6.35	2.02	0.0056
$15.5 < T_v \leqslant 17.5$	0.00	6.30	1.86	0.0010
$17.5 < T_v$	0.35	2.35	0.85	0.0028

Table 11.7(*b*). *Parameters of Weibull distribution of visual wave heights on the North Sea according to data obtained by Roll (1958)*

Visual wave period interval (s)	h_0 (m)	h_c (m)	γ	Probability of T_v falling within designated interval
$T_v \leqslant 3.5$	0.0	0.72	3.06	0.0335
$3.5 < T_v \leqslant 4.5$	0.0	0.89	2.27	0.0496
$4.5 < T_v \leqslant 5.5$	0.0	0.80	1.15	0.1517
$5.5 < T_v \leqslant 6.5$	0.0	1.25	1.54	0.4319
$6.5 < T_v \leqslant 7.5$	0.0	2.03	2.27	0.1785
$7.5 < T_v \leqslant 8.5$	0.0	2.40	2.63	0.1415
$8.5 < T_v \leqslant 9.5$	0.0	2.40	2.17	0.0100
$9.5 < T_v$	0.0	2.95	2.90	0.0033

The probability of the significant wave height random variable $H_{\frac{1}{3}}$ not exceeding a value $h_{\frac{1}{3}}$ may be determined from equation (11.54) by using the relationship

$$h_{\frac{1}{3}} = 1.68 h_v^{0.75}$$

which was discussed in section 11.4, i.e.

$$P[H_{\frac{1}{3}} \leqslant h_{\frac{1}{3}}] = 1 - \exp\left\{-\left[\frac{(h_{\frac{1}{3}}/1.68)^{\frac{4}{3}} - h_0}{h_c - h_0}\right]^\gamma\right\}.$$

By way of illustration, we may note from Table 11.7(b) that if the visual wave period lies in the region $6.5 < T_v \leqslant 7.5$ s then $h_0 = 0.0$ m, $h_c = 2.03$ m and $\gamma = 2.27$ so that

$$P[H_{\frac{1}{3}} \leqslant h_{\frac{1}{3}}] = 1 - \exp\left(-\frac{h_{\frac{1}{3}}^3}{24}\right).$$

The probability of $H_{\frac{1}{3}}$ exceeding 6 m (say) is 1.2×10^{-4} provided that the visual wave period is in the given interval. The probability that the visual wave period lies within the chosen interval is given in Table 11.7(b) as 0.1785 and if also $H_{\frac{1}{3}}$ exceeds 6 m then the combined probability of these independent events occurring together is $0.1785 \times 1.2 \times 10^{-4} = 2.2 \times 10^{-5}$ which corresponds to $2.2 \times 10^{-5} \times 8760 = 0.193$ hours/year.

11.6   Short-term response

Ever since the introduction of spectral techniques by St Denis & Pierson (1953) probabilistic techniques have been used in studies of

Fig. 11.17. Weibull distribution (on log–log scale) of visual wave height for chosen time intervals.

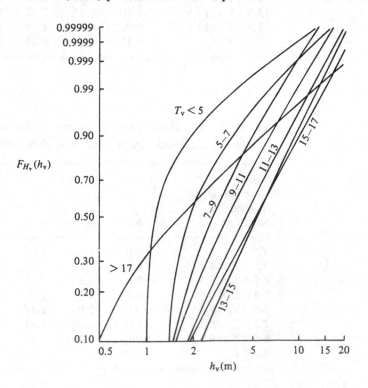

ship dynamics in a seaway. A short-term response of a ship may be determined from a knowledge of the appropriate receptance and the relevant wave spectrum. The basic assumptions made in this probabilistic approach are that

(a) both the responses (motions, velocity, acceleration, bending moment, shear force, relative motions, etc.) and the irregular seaway providing the excitation are Gaussian random processes with zero mean value,
(b) ship response to waves is linear so that superposition is valid and
(c) the input and output spectra are essentially narrow band (i.e. $\varepsilon \to 0$).

Price & Bishop (1974) show that under these conditions the peak value of ship response may be described by a Rayleigh probability density function and the statistical averages derived in section 11.4.2 may be adopted. For example, the peak to trough variation of any chosen ship response, $X(t)$, is given by the Rayleigh probability density function

$$f_X(x) = \frac{x}{4m_0} \exp\left(-\frac{x^2}{8m_0}\right) \qquad (x \geqslant 0),$$

where m_0 is the area under the mean square spectral density curve of the response amplitude. On the other hand the Rayleigh probability density function of the amplitude of this response, $Y(t) = X(t)/2$ is

$$f_Y(y) = \frac{y}{m_0} \exp\left(-\frac{y^2}{2m_0}\right) \qquad (y \geqslant 0, x = 2y).$$

Sometimes in the literature the quantity m_0 is replaced by $E/8$, where E is defined as the mean square value of the peak to trough response variations in the random process.

Consider a ship travelling at a speed $\bar{U}$ and heading angle χ with respect to the dominant wave direction. Price & Bishop (1974) show that the amplitude of any ship response $Y(t)$ – i.e. of distortion, bending moment, shear force, etc. – may be expressed in the form

$$\Phi_{YY}(\omega_e, \chi) = |H_Y(\omega_e, \chi)|^2 \Phi(\omega_e, \chi). \tag{11.55}$$

The quantities $\Phi_{YY}(\omega_e, \chi)$ and $\Phi(\omega_e, \chi)$ are the response and wave directional spectral density function respectively; these are dependent on the heading angle χ and the frequency of encounter

$$\omega_e = \omega - \bar{U}\frac{\omega^2}{g} \cos \chi.$$

The quantity $|H_Y(\omega_e, \chi)|$ is the modulus of the appropriate receptance; it is the response amplitude operator (RAO), defined as the amplitude of the response to a sinusoidal wave of unit amplitude when the ship speed is $\bar{U}$ and the heading χ. Such a function is discussed at length in chapter 8. The result given in equation (11.55) is of cardinal importance and it may be compared with the input–output relationship given in equation (11.9).

In a long-crested sea no wave spreading occurs so that all the wave crests are parallel to one another and all advance in the same direction. In this case χ is a constant and equation (11.55) may be simplified to

$$\Phi_{YY}(\omega_e) = |H_Y(\omega_e)|^2 \Phi(\omega_e). \tag{11.56}$$

That is to say, we may discuss the response of the ship in a uni-directional head sea ($\chi = 180°$), a beam sea ($\chi = 90°$), or in a sea with any other fixed direction. The moments of the response spectrum are then given by

$$m_n = \int_0^\infty \omega_e^n \Phi_{YY}(\omega_e)\, d\omega_e = \int_0^\infty \omega_e^n |H_Y(\omega_e)|^2 \Phi(\omega_e)\, d\omega_e. \tag{11.57}$$

In a 'short-crested sea' in which spreading occurs about the dominant wind direction, the ship experiences waves approaching it from a range of directions and the relative heading angle must now be regarded as a variable. If χ_0 is the constant angle between the ship and dominant wave direction, the ship suffers waves approaching over the range $\chi = \chi_0 + \mu$, where μ $(-\pi/2 \leqslant \mu \leqslant \pi/2)$ is the assumed spreading of the seaway as discussed in section 11.3.6. By integrating over this range of χ the one-dimensional response function is found; it is

$$\Phi_{YY}(\omega_e) = \int_\chi |H_Y(\omega_e, \chi)|^2 \Phi(\omega_e, \chi)\, d\chi,$$

where $\Phi(\omega_e, \chi)$ is expressed in the form given in section 11.3.6. It must be remembered, however, that ω_e and χ are not always independent of each other. The moments of the response are now given by

$$m_n = \int_0^\infty \int_\chi \omega_e^n |H_Y(\omega_e, \chi)|^2 \Phi(\omega_e, \chi)\, d\chi\, d\omega_e, \tag{11.58}$$

where the mean square value is

$$\langle Y^2(t) \rangle = m_0,$$

while the root mean square value is $m_0^{\frac{1}{2}}$.

In an actual sea spectrum the largest concentration of wave energy per unit wavelength corresponds to the region of small values of frequency (or as explained in chapter 8 to the region of small values of l/λ). The wave energy decreases with increasing l/λ. It will be seen from the response operators given in chapter 8 that although the magnitudes of the resonant responses to waves of unit amplitude for large l/λ may be large, the contributions that they represent to the overall energy content of the response will be small. Only resonances at small values of l/λ will contribute significantly to the total energy of the response.

Figs. 11.18(a) and (b) illustrate the mean square spectral density of amidships bending moment $\Phi_{BM}(\omega_e)$ and shearing force $\Phi_{SF}(\omega_e)$ respectively for the ballasted tanker travelling at 9 m/s in a uni-directional head sea with a significant wave height of 6 m and average period 9.46 s. Since structural damping only alters the magnitudes of

Fig. 11.18. Amidships mean square spectral density response of (a) bending moment and (b) shearing force for the 250 000 DWT tanker in ballast travelling at 9 m/s in a long-crested head sea for which the significant wave height is $h_{\frac{1}{3}} = 6$ m and the average period is $T_1 = 9.46$ s.

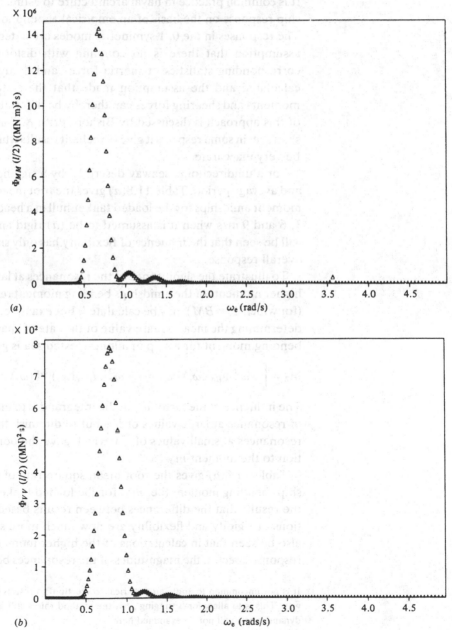

the resonance peaks at large values of l/λ, its effect on the total energy of the response is small. For this reason, the influence of fluid damping is much greater than that of structural damping in deciding the overall energy of the response in an irregular seaway. (Note that this is not always true for the evaluation of the higher moments m_2, m_4, etc., where the influence of contributions from large l/λ values are significant and structural damping is important.)

11.6.1   *The assumption of hull rigidity*

It is common practice in naval architecture to estimate the statistics of ship response on the basis of an empirical assumption of hull rigidity. The responses in the 0, 1 symmetric modes are determined under the assumption that there is no coupling with distortion modes. The corresponding statistics of inertia forces and fluid actions are then calculated and the assumption made that the statistics of bending moments and shearing forces can thereby be determined. The efficacy of this approach is discussed by Bishop, Price & Tam (1977);† as we shall see, in some respects it gives excellent results while in others it can be very inaccurate.

For a unidirectional seaway described by a significant wave height and average period, Table 11.8(a) gives the root mean square bending moment amidships for the loaded tanker hull in a head sea travelling at 3, 6 and 9 m/s when it is assumed to be (a) rigid and (b) flexible. It will be seen that the influence of flexibility has only small effect on the overall response.

To illustrate the significance of the resonances at large values of l/λ, higher moments of the amidships bending moment response spectrum (for which $Y \equiv BM$) may be calculated. For example, the moment m_2, determining the mean square value of the rate of change of amidships bending moment for a ship in a long-crested sea is given by

$$m_2 = \int_0^\infty \omega_e^2 \Phi_{BM}(\omega_e)\, d\omega_e = \int_0^\infty \omega_e^2 |H_{BM}(\omega_e)|^2 \Phi(\omega_e)\, d\omega_e.$$

The influence of the factor ω_e^2 in the integrand is to magnify the effects of resonance at large values of l/λ but to diminish the significance of resonances at small values of l/λ which gave the dominant contribution to the moment m_0.

Table 11.8(b) gives the root mean square rate of change of amidships bending moment (i.e. $m_2^{\frac{1}{2}}$) for the loaded tanker. It is seen from the results that the differences between results based on the assumptions of rigidity and flexibility are now much more significant. It will also be seen that in calculations of the higher moments m_n ($n \geq 2$) of response spectra, the magnitudes of the resonances become important

† It is not uncommon to apply an empirical 'correction' to results obtained in this way. This is to allow for 'springing', but the method raises still more questions of dynamics and will not be examined here.

Table 11.8. *Statistical data on bending moment amidships for the loaded tanker in head seas:* (a) *RMS value* $m_0^{\frac{1}{2}}$ (MN m), (b) *RMS value of rate of change* $m_2^{\frac{1}{2}}$ (MN m/s), (c) *upcrossing period* (s)

Table	Parameter	Significant wave height (m)	Average wave period (s)	$\bar{U} = 3$ m/s			$\bar{U} = 6$ m/s			$\bar{U} = 9$ m/s		
					Flexible			Flexible			Flexible	
				Rigid	Theory A	Theory B	Rigid	Theory A	Theory B	Rigid	Theory A	Theory B
a	RMS value (MN m)	2.0	5.46	66.3	86.4	84.0	67.3	138.0	133.4	70.7	239.6	239.6
		4.0	7.72	451.3	458.4	448.6	448.1	476.4	456.7	440.0	528.8	506.7
		6.0	9.46	1072	1074	1079	1055	1073	1083	1036	1090	1093
		8.0	10.92	1705	1699	1725	1663	1675	1741	1623	1665	1751
b	RMS value of rate of change (MN m/s)	2.0	5.46	64.1	147.5	147.5	81.5	330.0	322.5	108.0	670.8	673.8
		4.0	7.72	300.9	334.5	326.8	345.4	495.4	476.9	389.2	844.2	835.9
		6.0	9.46	617.7	635.1	632.1	694.8	789.0	775.4	768.3	1092	1080
		8.0	10.92	903.7	913.5	919.5	1001	1070	1082	1095	1352	1365
c	Upcrossing period (s)	2.0	5.46	6.50	3.68	3.58	5.19	2.63	2.60	4.11	2.24	2.23
		4.0	7.72	9.42	8.61	8.62	8.20	6.04	6.02	7.10	3.94	3.81
		6.0	9.46	10.90	10.36	10.79	9.54	8.54	8.78	8.47	6.27	6.36
		8.0	10.92	11.85	11.69	11.79	10.44	9.84	10.11	9.31	7.74	8.06

and so the influence of the structural damping (which influences them greatly) becomes more marked.

Table 11.8(c) gives the average period of crossing the zero mean value of the amidships bending moment response, i.e. $\bar{T} = 2\pi(m_0/m_2)^{\frac{1}{2}}$; the results are based on those quoted in Tables 11.8(a) and (b). Naturally, the variations between the results which flow from the assumptions of rigidity and that of flexibility are again marked.

It is noticeable in Table 11.8 that the largest discrepancies in the results found under the two assumptions occur with the lower sea states. In section 11.3 it was shown that increases in the significant wave height and average wave period imply that the maximum of the wave spectrum shifts to lower frequencies. Since the modulus of the receptance for a given forward speed is unchanged, the shape of the wave spectrum is critical in determining the overall value of the statistical parameter under investigation.

The sketches in fig. 11.19 illustrate this last point. Sketch (a) is a typical response amplitude operator with resonance peaks, A and B

Fig. 11.19. Sketches illustrating how the severity of a sea state can influence a response spectrum.

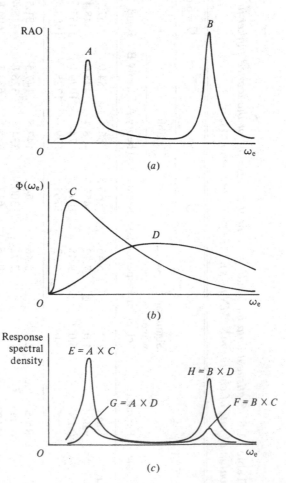

(say). Sketch (b) shows two possible wave spectra; C represents the condition when the wave height is great and D is the spectrum for a lower significant wave height. The resultant response spectra are sketched in diagram (c). It is seen that in the rougher sea the shape of the response spectrum is dominated by the peak E whereas peak F is comparatively insignificant. In determining the statistics of the response the major contribution will come from the portion E of the curve and not from F. By contrast, the regions around the response peaks G and H may produce contributions to the overall response of comparable magnitude.

If the ship is assumed rigid there is effectively only one peak in the response amplitude operator; this approximates to A in fig. 11.19(a). The absence of peak B has a greater effect when the encounter spectrum is of the form D than when the spectrum resembles C. That is to say the assumption of rigidity is the less dependable for low sea states. This effect is clearly visible in the results of Table 11.8(a).

If in addition the effects of the higher peak B are accentuated by multiplication by ω_e^2 in the calculation of m_2, the absence of such a peak is made even more plain. This is illustrated in Table 11.8(b) where the results depend on m_2, greater differences between the results based on the assumptions of rigidity and flexibility appearing than occur in Table 11.8(a), where m_0 is adduced. These discrepancies will be even more marked in moments of higher order, such as m_4.

11.6.2 Generalised gamma functions

The Rayleigh and Weibull distributions discussed in section 11.5 represent only a small step in the development of a probability function that is sufficiently versatile to cater for all sea states and ship motions. When these distributions are used, assumptions have to be made which may be unjustified in practice. For example, short-term ship motions are usually assumed to be linear, to be narrow band and to be described by the single-parameter Rayleigh distribution. But it has been found by Voznesenkiy (1967) that the two-parameter Weibull distribution agrees more closely with some observed amplitudes of rolling motion. Moreover analysis of data recorded by Bledsoe, Bussemaker & Cummins (1960) during seakeeping trials of a Dutch destroyer revealed several instances where the Rayleigh distributions failed to describe observed amplitudes of heave accelerations, roll and pitch angle.

The three-parameter gamma distribution function discussed by Ochi (1976) is a generalisation of the two previous distributions. It is more flexible and opens up the possibility of adequately describing a wider range of measured or observed data. The generalised gamma density function for the random variable X has the form

$$f_X(x; \lambda, m, c) = c\lambda^{cm} x^{cm-1} \exp\left[-(\lambda x)^c\right]/\Gamma(m) \qquad (x \geq 0).$$

The equivalent probability distribution function is

$$F_X(x; \lambda, m, c) = P[X \le x] = \frac{1}{\Gamma(m)} \int_0^{(\lambda m)^c} y^{m-1} \exp(-y)\, dy.$$

The quantities λ, m and $c > 0$ are constant parameters and Γ is the gamma function.

By assigning certain specific values to λ, m and c some well-known probability density and distribution functions are obtained, e.g.

(a) if $m = 1$ and $c = 1$, the exponential probability density and distribution functions are obtained

(b) if $m = 1$ and $c = 2$, the Rayleigh probability density and distribution functions are found

(c) if $m = 1$, the Weibull probability density and distribution functions are found.

Table 11.9 illustrates some of the well-known different types of distribution functions derived from the generalised gamma function for differing values of m, λ and c.

In using such a function to describe a sample of data (of waves, ship responses, etc.) no initial assumption need be made as to which particular distribution best describes the sample. In other words, the best fit for the data is determined from a free choice of m, λ and c; by contrast, if it is initially assumed that the data are described by a Weibull distribution (say) then m is necessarily equal to unity. But while such a gamma function gives greater versatility in fitting data it has the disadvantage of being more cumbersome to use.

Fig. 11.20 shows a histogram for the observed amplitude of pitch angle for the Dutch destroyer referred to by Bledsoe *et al.* (1960) and the corresponding Rayleigh probability density function ($\lambda = 0.510$,

Table 11.9. *Special cases of the generalised gamma probability density function*

Type	Probability density function ($x \ge 0$)	
Exponential	$f_X(x; \lambda, 1, 1)$	
Weibull	$f_X(x; \lambda, 1, c)$	
Rayleigh	$f_X(x; (2m_0)^{-\frac{1}{2}}, 1, 2)$	m_0 = zeroth moment of spectrum
Gamma	$f_X(x; \lambda, m, 1)$	
Truncated normal	$f_X\left(x; \frac{1}{2^{\frac{1}{2}}}, \frac{1}{2}, 2\right)$	
Chi-squared	$f_X\left(x; \frac{1}{2}, \frac{n}{2}, 1\right)$	n degrees of freedom
Chi	$f_X\left(x; \frac{1}{2^{\frac{1}{2}}}, \frac{n}{2}, 2\right)$	n degrees of freedom
Hydrograph	$f_X(x; \lambda, m, 2)$	

$m = 1$, $c = 2$). Using these data, Andrew & Price (1979) have determined the equivalent Weibull function ($\lambda = 0.489$, $m = 1$, $c = 2.832$) and generalised gamma probability density function ($\lambda = 0.413$, $m = 0.853$, $c = 2.748$). These are also shown in fig. 11.20 and it is seen that this latter function describes the observed data adequately – markedly better than either the Rayleigh or Weibull distribution.

11.7 'Long-term' description of waves

Wave data as illustrated in Table 11.4(a)–(c) are usually tabulated in terms of the significant wave height random variable $H_{\frac{1}{3}}$ with values $h_{\frac{1}{3}}$ and the average apparent or zero upcrossing period random variable T_{av} with values $\bar{T}$. Thus the joint probability distribution function

$$F_{H_{\frac{1}{3}}T_{\mathrm{av}}}(h_{\frac{1}{3}}, \bar{T}) = P[(H_{\frac{1}{3}} \leqslant h_{\frac{1}{3}}) \cap (T_{\mathrm{av}} \leqslant \bar{T})]$$

describes the probability of having a significant wave height of magnitude less than $h_{\frac{1}{3}}$ and with an average period less than a value $\bar{T}$. Price & Bishop (1974) show that there exists the corresponding joint probability density function $f_{H_{\frac{1}{3}}T_{\mathrm{av}}}(h_{\frac{1}{3}}, \bar{T})$ satisfying the relationship

$$f_{H_{\frac{1}{3}}T_{\mathrm{av}}}(h_{\frac{1}{3}}, \bar{T}) \, \mathrm{d}h_{\frac{1}{3}} \, \mathrm{d}\bar{T} = P[(h_{\frac{1}{3}} < H_{\frac{1}{3}} \leqslant h_{\frac{1}{3}} + \mathrm{d}h_{\frac{1}{3}}) \cap (\bar{T} < T_{\mathrm{av}} \leqslant \bar{T} + \mathrm{d}\bar{T})].$$

$$(11.59)$$

The probability density function of the significant wave height is

$$f_{H_{\frac{1}{3}}}(h_{\frac{1}{3}}) = \int f_{H_{\frac{1}{3}}T_{\mathrm{av}}}(h_{\frac{1}{3}}, \bar{T}) \, \mathrm{d}\bar{T}$$

Fig. 11.20. Probability density function of destroyer's pitch amplitude.

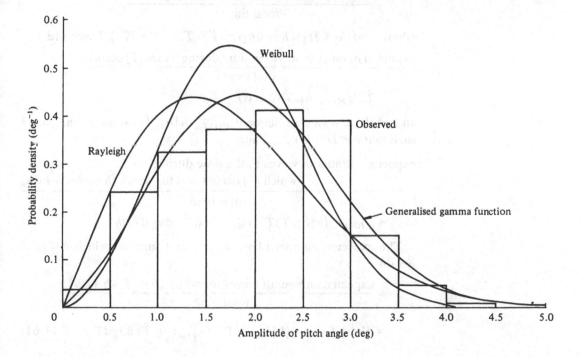

while that of the average period

$$f_{T_{av}}(\bar{T}) = \int f_{H_{\frac{1}{3}}T_{av}}(h_{\frac{1}{3}}, \bar{T}) \, dh_{\frac{1}{3}},$$

the integrations being over the complete range of values of $\bar{T}$ and $h_{\frac{1}{3}}$ respectively.

The short-term probability density function of the individual wave height random variable H with values h can be obtained from these functions, conditional on a knowledge of $H_{\frac{1}{3}}$ and T_{av}. In practice, this is assumed to be given by the Rayleigh probability density and distribution functions:

$$\left.\begin{array}{l} f_{H|H_{\frac{1}{3}}T_{av}}(h|h_{\frac{1}{3}}, \bar{T}) = \dfrac{4h}{h_{\frac{1}{3}}^{2}} \exp\left[-2\left(\dfrac{h}{h_{\frac{1}{3}}}\right)^{2}\right] \qquad (h \geqslant 0), \\[4mm] \text{and} \\[2mm] F_{H|H_{\frac{1}{3}}T_{av}}(h|h_{\frac{1}{3}}, \bar{T}) = 1 - \exp\left[-2\left(\dfrac{h}{h_{\frac{1}{3}}}\right)^{2}\right]. \end{array}\right\} \qquad (11.60)$$

These functions are dependent only on the significant wave height and are independent of the average period. Note that since $h_{\frac{1}{3}} = 4m_0^{\frac{1}{2}}$ these equations reduce to the forms given in section 11.5.2.

The long-term probability density function of the individual wave height $f_H(h)$ may be derived as a weighted sum of the short-term Rayleigh probability densities. Following the approach of Battjes (1970), equation (11.59) may be expressed as

$$f_{H_{\frac{1}{3}}T_{av}}(h_{\frac{1}{3}}, \bar{T}) \, dh_{\frac{1}{3}} \, d\bar{T}$$

$$= \frac{\text{expected time during which event } [I] \text{ occurs}}{\text{total time}},$$

where $I = [(h_{\frac{1}{3}} < H_{\frac{1}{3}} \leqslant h_{\frac{1}{3}} + dh_{\frac{1}{3}}) \cap (\bar{T} < T_{av} \leqslant \bar{T} + d\bar{T})]$. Therefore

$$\frac{\text{expected number of waves in time during which } [I] \text{ occurs}}{\text{total time}}$$

$$= \bar{T}^{-1} f_{H_{\frac{1}{3}}T_{av}}(h_{\frac{1}{3}}, \bar{T}) \, dh_{\frac{1}{3}} \, d\bar{T}$$

and of these waves a fraction $f_{H|H_{\frac{1}{3}}T_{av}}(h|h_{\frac{1}{3}}, \bar{T}) \, dh$ have a height H such that $h < H \leqslant h + dh$. Thus,

expected number of waves in the time during

$$\frac{\text{which } [I] \text{ occurs and for which } [h < H \leqslant h + dh]}{\text{total time}}$$

$$= f_{H|H_{\frac{1}{3}}T_{av}}(h|h_{\frac{1}{3}}, \bar{T})\bar{T}^{-1} f_{H_{\frac{1}{3}}T_{av}}(h_{\frac{1}{3}}, \bar{T}) \, dh_{\frac{1}{3}} \, d\bar{T} \, dh.$$

The expected number of waves per unit time for which $h < H \leqslant h + dh$

$$= \frac{\text{expected number of waves for which } [h < H \leqslant h + dh]}{\text{total time}}$$

$$= dh \iint f_{H|H_{\frac{1}{3}}T_{av}}(h|h_{\frac{1}{3}}, \bar{T})\bar{T}^{-1} f_{H_{\frac{1}{3}}T_{av}}(h_{\frac{1}{3}}, \bar{T}) \, dh_{\frac{1}{3}} \, d\bar{T} \qquad (11.61)$$

and the expected total number of waves per unit time

$$= \iint \bar{T}^{-1} f_{H_{\frac{1}{3}}T_{av}}(h_{\frac{1}{3}}, \bar{T}) \left[\int f_{H|H_{\frac{1}{3}}T_{av}}(h | h_{\frac{1}{3}}, \bar{T}) \, dh \right] dh_{\frac{1}{3}} \, d\bar{T}$$
$$= \overline{T^{-1}} \qquad\qquad (11.62)$$

because the expression in the square brackets is unity. Finally

$$\frac{\text{expected number of waves per unit time for which } h < H \le h + dh}{\text{expected total number of waves per unit time}}$$

$$= \text{fraction of the waves for which } h < H \le h + dh$$
$$= f_H(h) \, dh \qquad\qquad (11.63)$$

in which $f_H(h)$ is the long-term probability density function for the individual wave height random variable H.

Equations (11.61)–(11.63) give the probability density function

$$f_H(h) = \frac{\iint f_{H|H_{\frac{1}{3}}T_{av}}(h | h_{\frac{1}{3}}, \bar{T}) \bar{T}^{-1} f_{H_{\frac{1}{3}}T_{av}}(h_{\frac{1}{3}}, \bar{T}) \, dh_{\frac{1}{3}} \, d\bar{T}}{\iint \bar{T}^{-1} f_{H_{\frac{1}{3}}T_{av}}(h_{\frac{1}{3}}, \bar{T}) \, dh_{\frac{1}{3}} \, d\bar{T}} \qquad (11.64)$$

and the distribution function

$$F_H(h) = P[H \le h]$$
$$= \frac{\int_0^h dh^* \iint \bar{T}^{-1} f_{H_{\frac{1}{3}}T_{av}}(h_{\frac{1}{3}}, \bar{T}) f_{H|H_{\frac{1}{3}}T_{av}}(h^* | h_{\frac{1}{3}}, \bar{T}) \, dh_{\frac{1}{3}} \, d\bar{T}}{\iint \bar{T}^{-1} f_{H_{\frac{1}{3}}T_{av}}(h_{\frac{1}{3}}, \bar{T}) \, dh_{\frac{1}{3}} \, d\bar{T}} .$$

Substitution from equation (11.60) now shows that

$$1 - P[H \le h] = P[H > h]$$
$$= \frac{\iint \bar{T}^{-1} f_{H_{\frac{1}{3}}T_{av}}(h_{\frac{1}{3}}, \bar{T}) \exp[-2(h/h_{\frac{1}{3}})^2] \, dh_{\frac{1}{3}} \, d\bar{T}}{\iint \bar{T}^{-1} f_{H_{\frac{1}{3}}T_{av}}(h_{\frac{1}{3}}, \bar{T}) \, dh_{\frac{1}{3}} \, d\bar{T}} . \qquad (11.65)$$

If it is assumed that the significant wave height random variable $H_{\frac{1}{3}}$ and average period random variable T_{av} are statistically independent so that

$$f_{H_{\frac{1}{3}}T_{av}}(h_{\frac{1}{3}}, \bar{T}) = f_{H_{\frac{1}{3}}}(h_{\frac{1}{3}}) f_{T_{av}}(\bar{T}),$$

it follows that

$$P[H > h] = \int f_{H_{\frac{1}{3}}}(h_{\frac{1}{3}}) \exp[-2(h/h_{\frac{1}{3}})^2] \, dh_{\frac{1}{3}}. \qquad (11.66)$$

Although there is in general some correlation between $H_{\frac{1}{3}}$ and T_{av} so that the assumption of statistical independence is not strictly justified, equation (11.66) has been used by Lewis (1967), Nordenstrøm (1969) and others. If the variation of T_{av} is ignored, H is overestimated because the number of large waves occurring in a given length of time will on average be less than the number of small waves.

Wave data are in discrete form and the equations used hitherto have thus to be slightly modified. Let the midpoints of the class intervals of $H_{\frac{1}{3}}$ and T_{av} be $h_{\frac{1}{3}i}$ ($i = 1, 2, \ldots$) and $\bar{T}_j$ ($j = 1, 2, \ldots$) respectively, let the class widths be $\Delta h_{\frac{1}{3}}$ and $\Delta \bar{T}$ and let the joint probability associated with $h_{\frac{1}{3}i}$ and $\bar{T}_j$ be p_{ij}. These numbers represent estimates of the joint probability at the point (i, j); i.e.

$$p_{ij} = f_{H_{\frac{1}{3}}T_{av}}(h_{\frac{1}{3}i}, \bar{T}_j) \Delta h_{\frac{1}{3}} \Delta \bar{T}$$

and all integrals become summations. Hence equation (11.65) becomes

$$P[H > h] = \frac{\sum_{\text{all } i} \sum_{\text{all } j} \exp\left[-2(h/h_{\frac{1}{3}i})^2\right]\bar{T}_j^{-1} p_{ij}}{\sum_{\text{all } i} \sum_{\text{all } j} \bar{T}_j^{-1} p_{ij}}$$

and equation (11.66) becomes

$$P[H > h] = \sum_{\text{all } i} \exp\left[-2(h/h_{\frac{1}{3}i})^2\right]p_i,$$

where

$$p_i = f_{H_{\frac{1}{3}}}(h_{\frac{1}{3}i})\,\Delta h_{\frac{1}{3}}.$$

Jasper (1956) initially suggested that the logarithm of the significant wave height data is Gaussian-distributed, thereby permitting a long-term wave description to be made. It has since been found, however, that this distribution function fails to give an adequate fit to the data, especially at high values of significant wave height. Nordenstrøm (1969) found that a better long-term description of significant wave height data could be obtained by using the modified Weibull distribution function

$$P[H_{\frac{1}{3}} > h_{\frac{1}{3}}] = 1 - P[H_{\frac{1}{3}} \leq h_{\frac{1}{3}}] = \exp\left[-\left(\frac{h_{\frac{1}{3}} - h_0}{h_c - h_0}\right)^{\gamma}\right].$$

Fig. 11.21 illustrates a histogram of measured wave data, $h_{\frac{1}{3}}$, at Weather Station *India* in Table 11.4(*b*) for apparent wave period $\bar{T} = 8.5$ s. The parameters in the Weibull distribution derived by

Fig. 11.21. Histogram of wave height data at Weather Station *India* for average wave period $\bar{T} = 8.5$ s (see Table 11.4(*b*)).

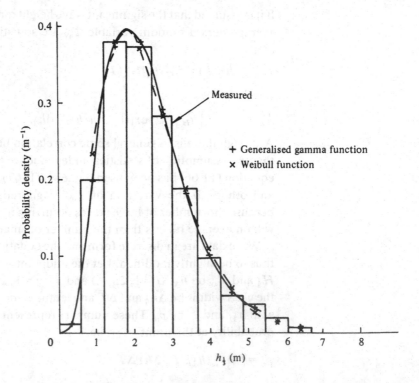

Nordenstrøm (1969) are given in Table 11.10. The data were also fitted by Andrew & Price (1979) by a generalised gamma function as defined in section 11.6.2, and the estimated values of the parameters are also included in Table 11.10. It is seen from fig. 11.21 that both distribution functions fit the data very satisfactorily. The Weibull distribution is defined only in the range $h_{\frac{1}{3}} > 0.6$ m and shows the greatest error between measured and estimated values for $h_{\frac{1}{3}} \leqslant 1$ m. The gamma distribution provides a fit over the whole range of wave height data.

11.7.1 *The design wave*

Nowadays many marine structures are designed to withstand the hydrodynamic loading actions associated with some particular wave height that is predicted to occur once in a given period – e.g. the '100-year wave', or the '50-year wave'. The only way in which the requisite information on the wave can be obtained is by extrapolation of existing wave height data and this, in its turn, depends on the analytical description of those data.

If N^{-1} is the fraction of wave measurements in a great number of assumed independent measurements for which $H > h_N$, then N is defined as the dimensionless return period corresponding to the probability of exceeding the wave height h_N; i.e.

$$N = \frac{1}{P[H > h_N]} = \frac{1}{1 - P[H \leqslant h_N]}. \tag{11.67}$$

For example, for waves of mean wave period $\bar{T}$ s, the expected number of waves in the return period of y years (say) or $3.1536y \times 10^7$ s is

$$N = 3.1536y \times 10^7 \bar{T}^{-1}. \tag{11.68}$$

As we have already pointed out, the long-term expected number of waves per unit time $\overline{T^{-1}}$ should strictly be used instead of $\bar{T}^{-1}$. However, the differences between these values are found to be small and the value $\bar{T}^{-1}$ appears to be a very satisfactory approximation.

The extreme long-term statistical wave heights with return periods of 1, 10 and 100 years associated with a mean period of 8.5 s in Table 11.4(b) were calculated by Andrew & Price (1979) using both a Weibull and a generalised gamma probability density function to

Table 11.10. *Parameters of distribution functions for the measured wave data in Table 11.4(b) ($\bar{T} = 8.55$ s)*

Weibull distribution				Generalised gamma distribution			
h_0(m)	h_c(m)	$(h_c - h_0)^{-1}$	m	$\gamma (\equiv c)$	λ	m	c
0.6	2.59	0.5025	1	1.67	2.306	4.976	0.946

describe the significant wave height. By substituting these functions
into equation (11.66) and combining it with equations (11.67) and
(11.68), the required extreme wave heights are found by interpolation.
The values are given in Table 11.11.

It will immediately be seen that the 'design wave' depends
significantly on the quality of the representation with which the data
are used. Unfortunately, because of the lack of data the correct value
of these extreme values is unknown and one can only assume that the
better the description of the existing data, the more correct will the
estimated value be. Implicitly the estimated value of extreme wave
height is based on the assumption that the distribution which is valid
for measured sea conditions holds good for all sea conditions.

11.7.2   Long-term ship response

The purpose of seeking a long-term distribution of ship response –
especially of pressure variation on the hull, bending moment and
stresses amidships – is to predict what may be expected in the lifetime
of the ship (see e.g. St Denis, 1975). Cummins (1969) proposed a
convenient method for evaluating long-term ship response for design
purposes in which the long-term variance of moment $\tilde{m}_0$ of any
arbitrary response $Y(t)$ is given by

$$\tilde{m}_0 = \int_0^\infty |H_Y(\omega)|^2 \tilde{\Phi}(\omega) \, d\omega.$$

In this equation, $\tilde{\Phi}(\omega)$ is the 'long-term average wave spectrum'
defined by

$$\tilde{\Phi}(\omega) = \int_0^\infty \int_0^\infty \Phi(\omega, h_{\frac{1}{3}}, \bar{T}) f_{H_{\frac{1}{3}} T_{av}}(h_{\frac{1}{3}}, \bar{T}) \, dh_{\frac{1}{3}} \, d\bar{T},$$

where $\Phi(\omega, h_{\frac{1}{3}}, \bar{T})$ is the wave spectrum which, as shown in section
11.3, may be defined in terms of the significant wave height $h_{\frac{1}{3}}$, average
period $\bar{T}$ (and, if need be, wave spreading as well). The quantity
$f_{H_{\frac{1}{3}} T_{av}}(h_{\frac{1}{3}}, \bar{T})$ is the joint probability density function of the significant

Table 11.11. *Comparison of long-term statistical wave
data*

Return period (years)	Wave height (m)	
	Weibull distribution	Generalised gamma distribution
1	14.4	15.6
10	16.7	18.6
100	19.3	21.7

wave height random variable, $H_{\frac{1}{3}}$, and the average wave period random variable, T_{av} as defined in the previous section. Thus the long-term average spectrum which directly determines the long-term m_0 is obtained by integrating the short-term wave spectrum Φ over the significant wave height – average period plane along with the appropriate joint probability density function. Information on the joint probability density function, or on the long-term probabilities of occurrence of different sea conditions, can be obtained in various atlases of wave statistics, as discussed in section 11.4.

Several methods exist to evaluate the long-term distribution of the response. For example, when a ship travels at constant speed at a constant heading angle χ in a unidirectional seaway the short-term statistics of the response may be discussed in terms of the Rayleigh probability density function. Thus the probability that the response $Y(t)$ will exceed an arbitrary level y is given by

$$P[Y > y] = \exp(-y^2/2m_0) \quad (y \geqslant 0),$$

where m_0 is the zero moment of the amplitude of the response spectrum and is a function of the significant wave height and average period describing the seaway. If service conditions are assumed never to change, the long-term probability is

$$\tilde{P}[Y > y] = \int_0^\infty \int_0^\infty P[Y > y] f_{H_{\frac{1}{3}}T_{av}}(h_{\frac{1}{3}}, \bar{T}) \, dh_{\frac{1}{3}} \, d\bar{T},$$

where $f_{H_{\frac{1}{3}}T_{av}}(h_{\frac{1}{3}}, \bar{T})$ is the joint probability density function of the random variables $H_{\frac{1}{3}}$ and T_{av}.

Another approach, developed by Nordenstrøm (1973) in a series of papers, is to use the square root of the zero moment as a random variable denoted by M_0 with values $m_0^{\frac{1}{2}}$ (or in his notation $(0.5E_X)^{\frac{1}{2}}$) since the short-term distribution of the response is completely characterised statistically by this function. Further, values of m_0 may be determined for any weather condition defined by the significant wave height and average period. Most wave statistics are given in terms of their visually observed values. It was found that the long-term statistical distribution of the visually observed significant wave height may be well described by a Weibull distribution as discussed in section 11.5.4 while the long-term distribution of the average period random variable T_{av} with values $\bar{T}$ may be described in terms of the observed period T_v by means of a normal (or log normal) probability density function having the form

$$f_{T_{av}}(\bar{T}) = \frac{1}{(2\pi)^{\frac{1}{2}}} \exp[-(\bar{T} - \mu_{\bar{T}})^2/2],$$

where, as we saw in section 11.4, the parameter $\mu_{\bar{T}} = 2.83T_v^{0.44}$ for a given value of the visual wave period T_v. The corresponding probability distribution function is given by

$$F_{T_{av}}(\bar{T}) = P[T_{av} \le \bar{T}]$$

$$= \frac{1}{(2\pi)^{\frac{1}{2}}} \int_{-\infty}^{\bar{T}} \exp\left[-(x - \mu_{\bar{T}})^2/2\right] dx$$

$$= 0.5 + \text{erf}\left(\frac{\bar{T} - 2.83 T_v^{0.44}}{2}\right)$$

and

$$P[T_{av} > \bar{T}] = 0.5 - \text{erf}\left(\frac{\bar{T} - 2.83 T_v^{0.44}}{2}\right),$$

where

$$\text{erf}(-x) = -\text{erf}(x) \quad \text{and} \quad \text{erf}(\infty) = 0.5.$$

In the example of section 11.5.4, if we take the visual wave period $T_v = 7.0$ s so that $\mu_{\bar{T}} = 6.66$ s, then for an average apparent wave period of $\bar{T} = 9$ s (say) the probability of exceeding this value is 0.12. Thus the joint probability that the significant wave height exceeds 6 m and the average period exceeds 9 s is $1.2 \times 10^{-4} \times 0.12 = 1.44 \times 10^{-5}$. (The random variables of wave height and average wave period are assumed to be statistically independent.)

Thus the long-term statistical distribution of the random variable M_0 may be found by combining short-term response characteristics and long-term weather statistics. This combination produces a Weibull distribution of the root mean square value of the response random variable. That is to say the distribution is of the form

$$F_{M_0}(m_0^{\frac{1}{2}}) = P[M_0 \le m_0^{\frac{1}{2}}] = 1 - \exp\left[-\left(\frac{m_0^{\frac{1}{2}}}{\beta}\right)^{\gamma}\right],$$

where $m_0^{\frac{1}{2}}$ denotes some value of the root mean square of the response variable and where the constants β and γ are determined in the manner discussed in section 11.5.4.

Since each value of the random variable M_0 represents a short-term Rayleigh distribution of the response Y, the long-term distribution of Y may be found by a summation of all these short-term distributions with due regard to their different probabilities of occurrence as discussed previously. That is to say the long-term probability that the response Y does not exceed some arbitrary value y is

$$\tilde{P}[Y \le y] = 1 - \tilde{P}[Y > y] = 1 - \int_0^{\infty} \exp\left(-\frac{y^2}{2m_0}\right) f_{M_0}(m_0^{\frac{1}{2}}) \, dm_0^{\frac{1}{2}}$$

and this has also been shown to be approximately of the form

$$P[Y \le y] = 1 - \exp\left[-\left(\frac{y^2}{a}\right)^b\right],$$

where the constants a and b are functions of α and β of the Weibull distribution. Nordenstrøm *et al.* (1971) determined values of these constants which are given in Table 11.12.

It is thus possible to find the long-term distribution of the response Y from the long-term Weibull distribution of the root mean square random variable M_0. In practice it is common to assign a limit to some response Y, like amidships bending moment, such that a ship life of 20 years corresponds approximately to the probability

$$\tilde{P}[Y > y] = 10^{-8}.$$

The corresponding value y is the most probable largest value of the response to occur during the lifetime of the ship.

Table 11.12. *Values of the parameters associated with the long-term probability distribution of the response* Y

$2/\gamma$	b	a/β^2	$2/\gamma$	b	a/β^2
0.0	1.00	1.00	1.4	0.45	0.42
0.5	0.72	0.61	1.5	0.43	0.41
0.6	0.68	0.57	1.6	0.42	0.40
0.7	0.64	0.54	1.7	0.40	0.39
0.8	0.60	0.52	1.8	0.39	0.39
0.9	0.57	0.50	1.9	0.37	0.38
1.0	0.54	0.48	2.0	0.36	0.37
1.1	0.52	0.46	3.0	0.27	0.34
1.2	0.49	0.45	4.0	0.21	0.31
1.3	0.47	0.43	5.0	0.18	0.30

12 Responses of other marine structures to waves

For duty, duty must be done;
The rule applies to everyone,
And painful though that duty be,
To shirk the task were fiddle-de-dee!
Ruddigore

12.1 Resonance of marine structures

The purpose of this book has been to show how structural, hydro-dynamic and oceanographic theories can be used in the dynamics of ships. We have considered the responses of ships to sinusoidal waves and discovered that they display sharply defined resonances. These resonances account for the narrow-band responses to random excitation by waves that are actually observable. One might reasonably expect that marine vehicles and structures other than ships will also behave in this way. It is also probable that this behaviour is most easily examined in terms of modal theory. In this chapter we shall briefly examine the matter. In doing so, however, we shall not pretend to present a fully developed theory and we shall come even less close than hitherto to writing a handbook on what procedure to adopt.

To investigate the resonances of a marine vehicle or structure in waves it is helpful to know the relevant principal modes and natural frequencies. But whereas the modes of a wave-excited ship were adequately derived by the use of Timoshenko beam theory, those of an oil platform (say) are likely to be more complicated. Even so the general approach for an oil platform is the same as that for the ship – the differences are of detail only.

Throughout this book we have dealt mainly with matters of principle, using specific ships for the purposes of numerical illustrations. Where we needed to know the principal modes of a hull, for instance, we were careful to point out that there are several ways of finding them. (For the sake of definiteness we have in fact used the Prohl–Myklestad technique, but there are others.) Similarly we employed two forms of strip theory for finding the generalised fluid actions although, again, other approaches could be used. This is the attitude we shall adopt in the following introduction to modal analysis of structures other than ships.

It will be convenient to refer to a particular type of marine structure for which the Timoshenko beam theory can be adapted, namely a tower buoy. Such structures are normally used in water more than 100 m deep and fig. 12.1 shows a typical design. The column is essentially a non-uniform buoyant beam, attached to the sea bed at its lower end by means of a universal joint. Its purpose is to provide support for a pipe which conveys crude oil and which comes to the installation lying on the ocean floor. The column allows the crude oil to be brought up and fed to a tanker that is moored to the column. In heavy weather tankers may not be allowed access to the column which then moves freely in the waves, and it is this condition that we shall investigate in this chapter by way of illustration, using a simplified model of a tower buoy.

Again, as in the rest of this book, a linear analysis is proposed and only conservative axial loadings are admitted – tangential, follower and partial follower axial forces are outside the scope of this chapter.

Fig. 12.1. A typical design of tower buoy. [By courtesy of David Brown (Offshore) Ltd.]

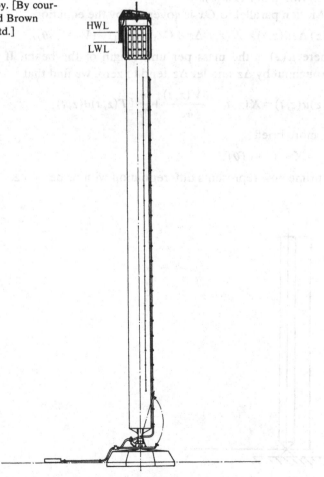

HWL

LWL

12.2   Equations of motion of a buoyant column

The axes $Oxyz$ are placed with the plane Oxy horizontal, as shown in fig. 12.2. The point O lies at the hinged attachment of the beam to the sea bed, and for the moment we shall leave open the boundary conditions. The beam occupies the region $0 \leqslant z \leqslant l$ and we shall assume that a tensile force T exists at any section. The lateral displacement at any section $u(z, t)$ in the plane Oxz, is caused by a distributed applied force $X(z, t)$ per unit length.

Consider the motion of an element of thickness Δz. This element suffers a lateral displacement of its centre of mass C, the magnitude of which is $u(z, t)$ as shown in fig. 12.3(a). The element also suffers two types of distortion and it is convenient to examine these separately.

First consider the shear strain $\gamma(z, t)$ that is imparted to the element by the shear forces $V(z, t)$. This is illustrated in fig. 12.3(b) and it will be seen that, while the axial tensions T_1 and T_2 remain parallel to Oz, they now tend to rotate the element about an axis which passes through C and is perpendicular to the plane of the paper. The bending moment rotates the faces of the element through slightly different angles $\theta(z, t)$ as shown in fig. 12.3(c).

Motion parallel to Ox is governed by the equation

$$\mu(z)\,\Delta z \ddot{u}(z, t) = X(z, t)\Delta z + (V_1 + T_1\theta_1) - (V_2 + T_2\theta_2),$$

where $\mu(z)$ is the mass per unit length of the beam. If we divide throughout by Δz and let Δz tend to zero, we find that

$$\mu(z)\ddot{u}(z, t) = X(z, t) + \frac{\partial V(z, t)}{\partial z} + \frac{\partial}{\partial z}[T(z, t)\theta(z, t)],$$

or, more briefly,

$$\mu\ddot{u} = X + V' + (T\theta)'. \tag{12.1}$$

A prime now represents differentiation with respect to z.

Fig. 12.2. The tower buoy is a column which may be represented as a non-uniform Timoshenko beam, shown here in its equilibrium position relative to fixed axes $Oxyz$. The end $z = 0$ is pinned while the end $z = l$ is free. At any section there is an axial tension T.

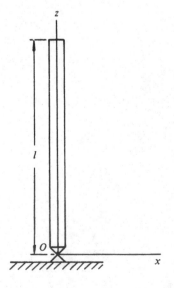

In discussing rotation of the element it is convenient to refer to fig. 12.3(d). Moments taken about an axis parallel to Oy and passing through C give

$$I_y(z)\,\Delta z \ddot{\theta}(z, t) = M_1 - M_2 - (T_1 - V_1\theta_1)\frac{\Delta z}{2}(\gamma + \theta_1)$$

$$- (T_2 - V_2\theta_2)\frac{\Delta z}{2}(\gamma + \theta_2)$$

$$+ (T_1\theta_1 + V_1)\frac{\Delta z}{2} + (T_2\theta_2 + V_2)\frac{\Delta z}{2},$$

where $I_y(z)$ is the moment of inertia per unit length. It follows that
$$I_y\ddot{\theta} = M' - T(\gamma + \theta) + T\theta + V$$
and, since

$$u' = \gamma + \theta, \tag{12.2}$$

$$I_y\ddot{\theta} = M' - Tu' + T\theta + V. \tag{12.3}$$

The equation of motion parallel to Oz is essentially one of static equilibrium since, to the first order, the element has no acceleration in that direction. That is to say
$$(T_1 - V_1\theta_1) - (T_2 - V_2\theta_2) + Z(z, t)\,\Delta z = 0,$$
whence

$$T' + Z = 0 \tag{12.4}$$

Fig. 12.3. An element of the column suffers lateral displacement (a). In addition it is sheared as in (b), its centre of mass C remaining fixed. In addition it rotates as in (c). The forces shown in (c) may be resolved parallel to the axes as in (d). Notice that the axial force T and shearing force V remain, respectively, perpendicular and parallel to the faces of the element.

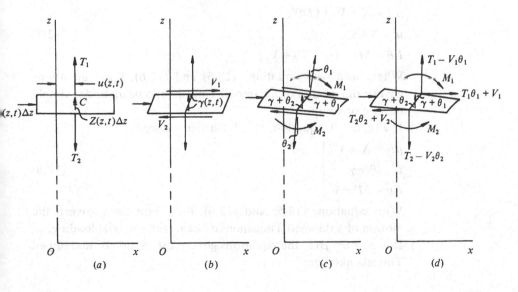

or

$$T(z, t) = -\int_0^z Z(q, t)\, dq,$$

$Z(z, t)$ being the upward force per unit length, to which there may in general be localised contributions.

As with ship dynamics we shall assume that the shearing force and bending moment relations are

$$V(z, t) = kAG(z)[\gamma(z, t) + \alpha(z)\dot{\gamma}(z, t)] \tag{12.5}$$

and

$$M(z, t) = EI(z)[\theta'(z, t) + \beta(z)\dot{\theta}'(z, t)], \tag{12.6}$$

where $\alpha(z)$ and $\beta(z)$ represent the distributed structural damping. The equations (12.1)–(12.6) will allow us to make a modal analysis of the tower buoy response, but before proceeding it is worthwhile to consider the standing of equations (12.1)–(12.4).

Equations (12.1)–(12.4) are adaptations of the Timoshenko beam theory that we introduced in chapter 3. They make allowance for the axial force $T(z, t)$ and their form collapses to that of the equations we used previously if we let $T(z, t) = 0$. It must be accepted, however, that this present use of the equations contains an extra element of arbitrariness. The basic assumptions of beam theory have not been discarded (e.g. plane sections are still assumed to remain plane) and yet we have made further demands on the theory. At best we can achieve only a logical development of the underlying empirical theory, such as it is.

In these circumstances it is as well to check that the foregoing equations conform to simpler existing theory for special cases.

(a) If $Z(z, t) = 0$, T is constant and so equations (12.1)–(12.3) become

$$\left.\begin{aligned} \mu\ddot{u} &= X + V' + (T\theta)', \\ u' &= \theta + \gamma, \\ I_y\ddot{\theta} &= M' - Tu' + T\theta + V. \end{aligned}\right\} \tag{12.7}$$

When used with equations (12.5) and (12.6), these equations govern the motion of a damped Timoshenko beam under simple tension or compression.

(b) If $T(z, t) = 0$, equations (12.7) further reduce to

$$\left.\begin{aligned} \mu\ddot{u} &= X + V', \\ u' &= \theta + \gamma, \\ I_y\ddot{\theta} &= M' + V. \end{aligned}\right\} \tag{12.8}$$

With equations (12.5) and (12.6), these equations govern the motion of a damped Timoshenko beam without axial loading.

(c) If $\alpha(z) = 0 = \beta(z)$ the equations govern the motion of undamped Timoshenko beams.

(d) If, in addition, $I_y = 0$, $\gamma(z, t) \to 0$ and $kAG(z) \to \infty$, equations (12.6) and (12.8) reduce further to

$$\left.\begin{aligned}\mu\ddot{u} &= X + V', \\ 0 &= M' + V, \\ M &= EIu''.\end{aligned}\right\} \tag{12.9}$$

These results form the basis of the Bernoulli–Euler beam theory.

(e) If $Z(z) = 0$ as in (a) and, in addition, $EI(z) = 0 = kAG(z) = \gamma(z, t) = I_y(z)$, equation (12.7) reduces further to

$$\mu\ddot{u} = X + Tu''. \tag{12.10}$$

This is the equation of motion of a taut string.

(f) If $T(z, t) = T(t)$ and the conditions appropriate to the Bernoulli–Euler beam theory as stated in (d) are met, equations (12.7) reduce to

$$\left.\begin{aligned}\mu\ddot{u} &= X + V' + T(t)u'', \\ 0 &= M' + V, \\ M &= EIu''.\end{aligned}\right\} \tag{12.11}$$

Bolotin (1964) discusses such equations in great detail in connection with the dynamic stability of elastic systems such as, for example, a simply supported beam loaded by a periodic longitudinal force of the form $T(t) = T_0 + T \cos \omega t$.

A word has also to be said about our use of these equations as if they can strictly govern the motion of the *dry* column. For positive contributions to the tensile force $T(z, t)$ will usually be of hydrostatic origin and, therefore, non-existent if the column is assumed to move *in vacuo*. Here we shall make further assumptions. They are that the tensile force $T(z, t)$ and body force $Z(z, t)$

(a) are present whether or not the hydrodynamic action $X(z, t)$ exists, i.e. even if the column is assumed to move *in vacuo*;

(b) are essentially constant quantities of the form $T(z)$ and $Z(z)$ when the column moves *in vacuo*.

12.3 Principal modes

Were there no structural damping in the buoyant column, no lateral fluid force $X(z, t)$ and if the axial tension were time-independent, the equation of motion (12.1) would reduce to

$$\mu\ddot{u} = V' + (T\theta)', \tag{12.12}$$

while equations (12.5) and (12.6) would be

$$V = kAG\gamma, \tag{12.13}$$

$$M = EI\theta'. \tag{12.14}$$

Equations (12.2) and (12.3) would remain the same. Under these conditions (i.e. for the dry column) we may seek a free motion described by

$$
\left.
\begin{aligned}
u &= u_r(z) \sin \omega_r t, \\
\theta &= \theta_r(z) \sin \omega_r t, \\
\gamma &= \gamma_r(z) \sin \omega_r t,
\end{aligned}
\right\}
\tag{12.15}
$$

in which $u_r(z)$, $\theta_r(z)$ and $\gamma_r(z)$ define the rth principal mode whose natural frequency is ω_r.

Equation (12.12) shows that

$$-\omega_r^2 \mu u_r = V_r' + (T\theta_r)',$$

$V_r(z)$ being the distribution of shearing force in the rth mode. Multiply this result throughout by $u_s(z)$ and integrate over the length of the column; it is then found that

$$-\omega_r^2 \int_0^l \mu u_r u_s \, dz = \int_0^l V_r' u_s \, dz + \int_0^l (T\theta_r)' u_s \, dz.$$

Integrating the terms on the right-hand side by parts, we find that

$$-\omega_r^2 \int_0^l \mu u_r u_s \, dz = [V_r u_s]_0^l - \int_0^l V_r u_s' \, dz + [T\theta_r u_s]_0^l - \int_0^l T\theta_r u_s' \, dz.$$

It will be seen that the first integrated term vanishes because $u_s(0) = 0 = V_r(l)$. The second integrated term vanishes if $T(z) = 0$ at the upper extremity $z = l$. In other words, if we ignore this second integrated term and wish to apply the theory to a column which carries a concentrated mass or a buoyant chamber of enlarged diameter at the upper end, the mass of the chamber must be treated as part of the beam. With this proviso, then, we have

$$-\omega_r^2 \int_0^l \mu u_r u_s \, dz = -\int_0^l V_r u_s' \, dz - \int_0^l T\theta_r u_s' \, dz.$$

By reversing the subscripts it is established that

$$-\omega_s^2 \int_0^l \mu u_r u_s \, dz = -\int_0^l V_s u_r' \, dz - \int_0^l T\theta_s u_r' \, dz,$$

whence

$$(\omega_r^2 - \omega_s^2) \int_0^l \mu u_r u_s \, dz = \int_0^l (V_r u_s' - V_s u_r') \, dz + \int_0^l T(\theta_r u_s' - \theta_s u_r') \, dz.$$

$$\tag{12.16}$$

Turning next to equation (12.3) we note that

$$-\omega_r^2 I_y \theta_r = M_r' - Tu_r' + T\theta_r + V_r.$$

Multiply by θ_s and integrate; this gives

$$-\omega_r^2 \int_0^l I_y \theta_r \theta_s \, dz = [M_r \theta_s]_0^l - \int_0^l M_r \theta_s' \, dz - \int_0^l Tu_r' \theta_s \, dz$$

$$+ \int_0^l T\theta_r \theta_s \, dz + \int_0^l V_r \theta_s \, dz.$$

It will be seen that the integrated term vanishes provided $M_r(0) = 0 = M_r(l)$, a condition that is usually met by a tower buoy. Therefore

$$-\omega_r^2 \int_0^l I_y \theta_r \theta_s \, dz = -\int_0^l M_r \theta_s' \, dz - \int_0^l Tu_r' \theta_s \, dz$$

$$+ \int_0^l T\theta_r \theta_s \, dz + \int_0^l V_r \theta_s \, dz$$

and, by reversal of subscripts,

$$-\omega_s^2 \int_0^l I_y \theta_r \theta_s \, dz = -\int_0^l M_s \theta_r' \, dz - \int_0^l Tu_s' \theta_r \, dz$$

$$+ \int_0^l T\theta_r \theta_s \, dz + \int_0^l V_s \theta_r \, dz.$$

If one of these two equations is subtracted from the other and it is noted that, in accordance with equation (12.14),

$$M_r = EI(z)\theta_r',$$

then it is found that

$$(\omega_r^2 - \omega_s^2) \int_0^l I_y \theta_r \theta_s \, dz = \int_0^l T(u_r' \theta_s - u_s' \theta_r) \, dz - \int_0^l (V_r \theta_s - V_s \theta_r) \, dz.$$

$$(12.17)$$

Equations (12.16) and (12.17) now provide orthogonality conditions for the principal modes. By adding them together we find that

$$(\omega_r^2 - \omega_s^2) \int_0^l (\mu u_r u_s + I_y \theta_r \theta_s) \, dz = \int_0^l (V_r \gamma_s - V_s \gamma_r) \, dz$$

and, since

$$V_r = kAG\gamma_r,$$

the right-hand integral vanishes. We may thus write

$$\int_0^l (\mu u_r u_s + I_y \theta_r \theta_s) \, dz = a_{rs} \delta_{rs},$$ (12.18)

where δ_{rs} is the Kronecker delta function.

A second orthogonality relationship is found by returning to the equations containing only ω_r^2 and adding them together. This gives

$$-\omega_r^2 a_{rs} \delta_{rs} = -\int_0^l [M_r \theta_s' + T(u_r' \theta_s + u_s' \theta_r) - T\theta_r \theta_s + V_r(u_s' - \theta_s)] \, dz$$

which simplifies to

$$\int_0^l [EI\theta_r' \theta_s' + T(u_r' \theta_s + u_s' \theta_r - \theta_r \theta_s) + kAG\gamma_r \gamma_s] \, dz = \omega_r^2 a_{rs} \delta_{rs}.$$ (12.19)

or

$$\int_0^l \left[\frac{M_r M_s}{EI} + T\left(\frac{V_r \theta_s}{kAG} + \frac{V_s \theta_r}{kAG} + \theta_r \theta_s \right) + \frac{V_r V_s}{kAG} \right] \, dz = \omega_r^2 a_{rs} \delta_{rs} = c_{rs},$$

where T is a function of z only.

12.3.1   The tension T

It will be seen that we have chosen to define $T(z)$ as the tension in the column due to the excess of buoyancy over weight. That is to say

$$T(z) = \int_z^l [Z_B(\xi) - \mu(\xi)g]\,d\xi \tag{12.20}$$

in accordance with equation (12.4), in which $Z \equiv Z_B(z)$ is the buoyancy force per unit length of the column. Being independent of time, this tension partially determines the principal modes which must be extracted numerically in some way.

If, due to the presence of waves, there is a time dependent component of $Z(z)$ (equal to $Z_w(z, t)$, say), this will augment the tension; that is to say it will add a time-dependent component to the axial tension. We then have

$$T(z, t) = \int_z^l [Z_B(\xi) - \mu(\xi)g]\,d\xi + \int_z^l Z_w(\xi, t)\,d\xi. \tag{12.21}$$

The first component is independent of time and partially determines the principal modes as before, while the second integral represents an

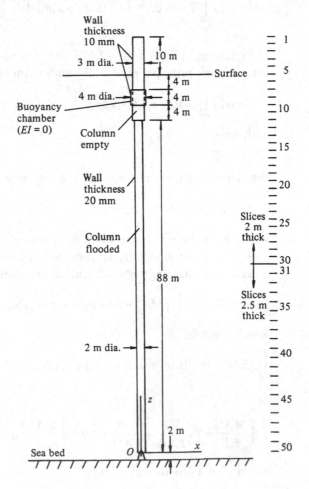

Fig. 12.4. An 'ideal' tower buoy whose characteristics have been computed for the purposes of illustration.

'excitation' and plays a part in producing a forced oscillation. Inclusion of the second integral brings us close to the work of Bolotin (1964), but removes the problem from *linear* dynamics as may be seen by inspection of the equations in which products of time-dependent terms then appear.

12.3.2   *Extraction of the principal modes*

The axial tension $T(z)$ demands a modification of the Prohl–Myklestad method for extracting the principal modes. A suitable numerical technique can readily be formulated on the basis of equations (12.12)–(12.15). It will suffice, here to show some typical results.

Fig. 12.4 shows an idealised tower buoy. The material is steel with $E = 207\,070\,\text{MN/m}^2$ and $G = 79\,600\,\text{MN/m}^2$.

For the circular cross-section, the value of k used is 0.5.

The variation of the axial tension $T(z)$ is as shown in fig. 12.5 and it will be seen that the structure is almost neutrally buoyant. The curve

Fig. 12.5. The curve of axial tension for the tower buoy of fig. 12.4. Note that no allowance has been made for the weight of ends, stiffeners or fittings.

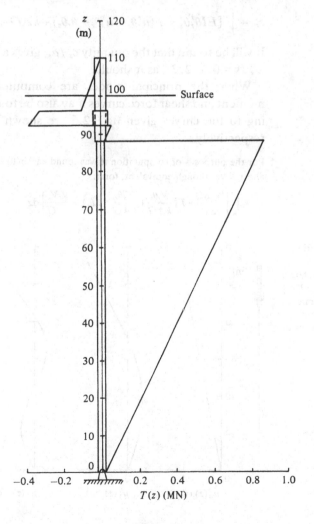

shows that the universal joint at the sea bed has apparently to apply a holding-down force of about 0.02 MN, but if a suitable allowance were made for the weight of ends, stiffeners and fittings this would be transformed into a modest upward force. The principal modes shown in fig. 12.6, which are all scaled to give a lateral displacement $u(l) = 1$ m at the top, have the associated natural frequencies:

$\omega_0 = 0.18$ rad/s or 0.03 Hz,

$\omega_1 = 2.73$ rad/s or 0.47 Hz,

$\omega_2 = 8.74$ rad/s or 1.39 Hz,

$\omega_3 = 17.30$ rad/s or 2.75 Hz.

Orthogonality of the modes should be checked when they are computed. By the method employed in arriving at the present results, the values quoted in Table 12.1 were found for the integrals†

$$a_{rs} = \int_0^l (\mu u_r u_s + I_y \theta_r \theta_s)\, dz,$$

$$c_{rs} = \int_0^l [EI\theta_r'\theta_s' + T(u_r'\theta_s + u_s'\theta_r - \theta_r\theta_s) + kAG\gamma_r\gamma_s]\, dz.$$

It will be found that the quantity c_{ss}/a_{ss} gives a close approximation to ω_s^2 ($s = 0, 1, 2, 3$), as it should.

When the principal modes are computed the modal bending moment and shear force curves may also be found. Those corresponding to the curves given fig. 12.6 are shown in figs. 12.7 and 12.8 respectively.

† For the purposes of computation it was found easier to calculate c_{rs} in the alternative, though equivalent, form:

$$c_{rs} = \int_0^l \left[\frac{M_r M_s}{EI} + T\left(\frac{V_r\theta_s}{kAG} + \frac{V_s\theta_r}{kAG} + \theta_r\theta_s\right) + \frac{V_r V_s}{kAG}\right] dz.$$

Fig. 12.6. The four principal modes of lowest order of the column shown in fig. 12.4. These modes refer to the dry column but allow for hydrostatic actions and for the mass of water in the flooded portion.

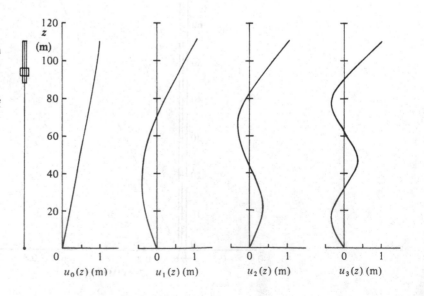

Table 12.1. *Values of a_{rs} and c_{rs} for the column shown in fig. 12.4 using the method devised for determining the principal modes*

(a) Values of a_{rs} in tonne m²:

s \ r	0	1	2	3
0	98 000	−36	21	−6
1	−36	38 000	9	14
2	21	9	26 000	2
3	−6	14	2	26 000

(b) Values of c_{rs} in kN m:

s \ r	0	1	2	3
0	3100	−4	0	−3
1	−4	320 000	48	210
2	0	48	2 000 000	−1200
3	−3	210	−1200	7 800 000

Fig. 12.7. Modal bending moment curves corresponding to the principal modes shown in fig. 12.5.

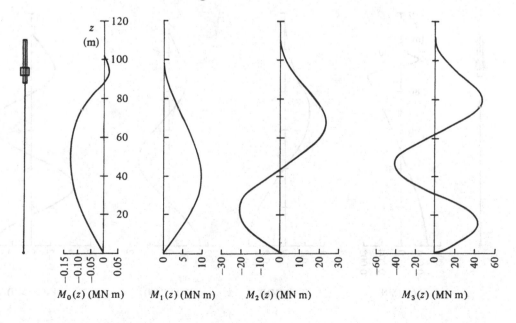

$M_0(z)$ (MN m) $M_1(z)$ (MN m) $M_2(z)$ (MN m) $M_3(z)$ (MN m)

12.4 Forced oscillation

Forced oscillation of the tower buoy by wave excitation can now be investigated using the principal coordinates of the dry structure. The approach is similar to the one we used for ships if linearity of the analysis is preserved by admitting only time independent tension $T(z)$ and excluding the $Z_w(x, t)$ term in equation (12.21) which otherwise produces a non-linear analysis.

The horizontal distortion of the column in a seaway may be expressed as the sum of distortions in its principal modes; i.e. we may write

$$
\left.
\begin{aligned}
u(z, t) &= \sum_{r=0}^{\infty} u_r(z)p_r(t), \\
\theta(z, t) &= \sum_{r=0}^{\infty} \theta_r(z)p_r(t), \\
\gamma(z, t) &= \sum_{r=0}^{\infty} \gamma_r(z)p_r(t),
\end{aligned}
\right\}
\tag{12.22}
$$

where $p_r(t)$ is the rth principal coordinate which can conveniently be scaled so that unit deflection at the end $z = l$ corresponds to $p_r = 1$.

Substitution of equations (12.22) into (12.1) and (12.5) shows that

$$
\mu \sum_{r=0}^{\infty} u_r\ddot{p}_r - kAG \sum_{r=0}^{\infty} (\gamma_r p_r + \alpha\gamma_r\dot{p}_r)' - \sum_{r=0}^{\infty} (T\theta_r)'p_r = X(z, t).
$$

Fig. 12.8. Modal shearing force curves corresponding to the principal modes shown in fig. 12.5.

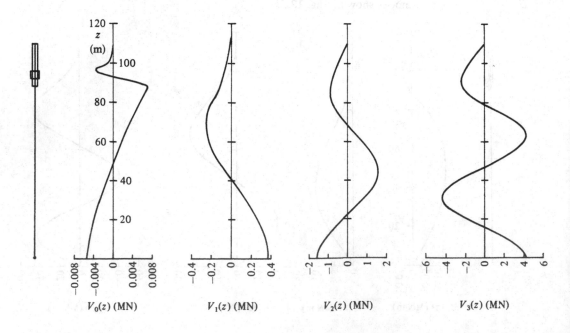

If this equation is multiplied by $u_s(z)$ and integrated it is found that

$$\sum_{r=0}^{\infty} \ddot{p}_r \int_0^l \mu u_r u_s \, dz - \sum_{r=0}^{\infty} p_r [kAG\gamma_r u_s]_0^l$$

$$+ \sum_{r=0}^{\infty} p_r \int_0^l kAG\gamma_r u_s' \, dz - \sum_{r=0}^{\infty} \dot{p}_r [kAG\alpha\gamma_r u_s]_0^l$$

$$+ \sum_{r=0}^{\infty} \dot{p}_r \int_0^l kAG\alpha\gamma_r u_s' \, dz - \sum_{r=0}^{\infty} p_r [T\theta_r u_s]_0^l$$

$$+ \sum_{r=0}^{\infty} p_r \int_0^l T\theta_r u_s' \, dz = \int_0^l X(z, t)u_s(z) \, dz.$$

The three integrated terms vanish since $u_s(0) = 0$ and $V_r(l) = 0 = T(l)$. We are thus left with

$$\sum_{r=0}^{\infty} \ddot{p}_r \int_0^l \mu u_r u_s \, dz + \sum_{r=0}^{\infty} p_r \int_0^l (kAG\gamma_r + T\theta_r)(\theta_s + \gamma_s) \, dz$$

$$+ \sum_{r=0}^{\infty} \dot{p}_r \int_0^l kAG\alpha\gamma_r(\theta_s + \gamma_s) \, dz = \int_0^l X(z, t)u_s(z) \, dz, \qquad (12.23)$$

where the result (12.2) has been used.

Similarly if equations (12.22) are substituted into equation (12.3) and the result is multiplied throughout by $\theta_s(z)$ and integrated, it is found that

$$\sum_{r=0}^{\infty} \ddot{p}_r \int_0^l I_y \theta_r \theta_s \, dz + \sum_{r=0}^{\infty} p_r \int_0^l EI\theta_r' \theta_s' \, dz$$

$$+ \sum_{r=0}^{\infty} \dot{p}_r \int_0^l \beta EI\theta_r' \theta_s' \, dz - \sum_{r=0}^{\infty} p_r \int_0^l kAG\gamma_r \theta_s \, dz$$

$$- \sum_{r=0}^{\infty} \dot{p}_r \int_0^l \alpha kAG\gamma_r \theta_s \, dz$$

$$+ \sum_{r=0}^{\infty} p_r \int_0^l Tu_r' \theta_s \, dz - \sum_{r=0}^{\infty} p_r \int_0^l T\theta_r \theta_s \, dz = 0 \qquad (12.24)$$

when the integrated terms are removed by reason of the boundary conditions $EI\theta_r'(0) = 0 = EI\theta_r'(l)$. When added together, equations (12.23) and (12.24) give

$$a_{ss}\ddot{p}_s + \omega_s^2 a_{ss}p_s + \sum_{r=0}^{\infty} \dot{p}_r(\alpha_{rs} + \beta_{rs})$$

$$= \int_0^l X(z, t)u_s(z) \, dz \qquad (s = 0, 1, 2, \ldots), \qquad (12.25)$$

where

$$\alpha_{rs} = \int_0^l \alpha kAG\gamma_r \gamma_s \, dz, \qquad \beta_{rs} = \int_0^l \beta EI\theta_r' \theta_s' \, dz. \qquad (12.26)$$

In arriving at this result, we have used the orthogonality relations (12.18) and (12.19). It will be seen that equation (12.25) has the form with which we have already become familiar.

12.4.1   Steady state deflection

The equation of motion admits a steady state solution for which the only fluid action aside from buoyancy is that of a steady current. That is to say there are no fluctuating fluid forces due to the presence of waves. Let

$$X(z, t) = \bar{X}(z), \qquad p_r(t) = \bar{p}_r,$$

where the overbars indicate constant values. The variation of fluid force $\bar{X}(z)$ per unit length in the steady flow may be found from experiments if need be. With these values, equation (12.25) gives

$$\bar{p}_s = \frac{1}{\omega_s^2 a_{ss}} \int_0^l \bar{X}(z) u_s(z) \, dz \qquad (s = 0, 1, 2, \ldots). \tag{12.27}$$

We thus reach a result that is entirely comparable with those encountered previously in chapter 5.

The static deflection, bending moment and shearing force distributions can now be found by adding the modal contributions; i.e.

$$\left. \begin{aligned} \bar{u}(z) &= \sum_{s=0}^{n} \bar{p}_s u_s(z), \\ \bar{M}(z) &= \sum_{s=0}^{n} \bar{p}_s M_s(z), \\ \bar{V}(z) &= \sum_{s=0}^{n} \bar{p}_s V_s(z). \end{aligned} \right\} \tag{12.28}$$

12.4.2   Motion about a steady deflection

If the column assumes a steady deflection and a time-dependent motion due to the presence of waves, we may use the fresh coordinates $q_r(t)$, where

$$p_r(t) = \bar{p}_r + q_r(t)$$

and also let

$$X(z, t) = \bar{X}(z) + X_w(z, t).$$

Equations (12.25) and (12.26) now show that

$$a_{ss}\ddot{q}_s + \omega_s^2 a_{ss} q_s + \sum_{r=0}^{\infty} \dot{q}_r(\alpha_{rs} + \beta_{rs}) = \int_0^l X_w u_s \, dz \quad (s = 0, 1, 2, \ldots). \tag{12.29}$$

This is the simplest form of the equations of motion. A more complete form may be found by including a term embodying a time-dependent component of the tension T on the right-hand side.

A steady state solution of equation (12.29) may be found by the methods employed previously when $X_w(z, t)$ varies sinusoidally with waves of unit amplitude. And when the quantities q_r have been found, the displacements, bending moment and shearing force may be found from modal addition in the usual way (cf. equation (12.28)).

12.4.3   *Equivalent linearisation of fluid loading*
The horizontal fluid force per unit length, $X(z, t)$, may be represented by the non-linear Morison equation; that is to say

$$X(z, t) = m(z)\dot{v}(z, t) + b(z)[\bar{U}(z) + v(z, t)]|\bar{U}(z) + v(z, t)|, \qquad (12.30)$$

where

$$v(z, t) = \dot{u}_w(z, t) - \dot{u}(z, t).$$

The quantity $v(z, t)$ is the relative velocity between the water particles, whose velocity is $\dot{u}_w(z, t)$ in the absence of a steady current on the one hand and the structure, whose local velocity is $\dot{u}(z, t)$ on the other. The steady current has the speed $\bar{U}(z)$. The added mass coefficient $m(z)$ and drag coefficient $b(z)$ determine the distributed fluid action.

The steady state deflection discussed previously was one in which current speed $\bar{U}(z)$ is non-zero whilst $\dot{u}_w = 0 = \dot{u}$. The horizontal fluid loading then reduces to

$$X(z, t) = \bar{X}(z) = b(z)\bar{U}(z)|\bar{U}(z)|$$

and the steady state solution given in equation (12.27) becomes

$$\bar{p}_s = \frac{1}{\omega_s^2 a_{ss}} \int_0^l u_s(z)b(z)\bar{U}(z)|\bar{U}(z)| \, dz \qquad (s = 0, 1, 2, \dots). \qquad (12.31)$$

For motions about this steady state the change in the horizontal loading is

$$\begin{aligned}
X_w(z, t) &= X(z, t) - \bar{X}(z) \\
&= m\dot{v} + b(\bar{U} + v)|\bar{U} + v| - b\bar{U}|\bar{U}|. \qquad (12.32)
\end{aligned}$$

If this expression is introduced into equation (12.29) we arrive at a non-linear equation of motion whose solution is a matter of some difficulty. We shall therefore examine a technique by which the equation may be linearised.

In the method of linearisation to be employed the non-linear horizontal loading expression is replaced by the 'equivalent' linear expression

$$X_w(z, t) = m\dot{v} + b_e v$$

where the error due to the linearisation is

$$\varepsilon(z, t) = b(\bar{U} + v)|\bar{U} + v| - b\bar{U}|\bar{U}| - b_e v.$$

For this error to be a minimum, its mean square value $\langle \varepsilon^2 \rangle$ must be made as small as possible. This is accomplished when

$$\frac{\partial \langle \varepsilon^2 \rangle}{\partial b_e} = -2\langle [b(\bar{U} + v)|\bar{U} + v| - b\bar{U}|\bar{U}| - b_e v]v \rangle = 0,$$

giving the equivalent linearised drag coefficient

$$b_e = \frac{b\langle [(\bar{U} + v)|\bar{U} + v| - \bar{U}|\bar{U}|]v \rangle}{\langle v^2 \rangle}.$$

Were we to introduce this value into the equations of motion we should thereby introduce time averages which are partially determined by the solution of the equation. Now if the motions of the water particles at all sections z are sinusoidal with zero mean values, the approximate values of the responses of the tower buoy (about the steady reference position), determined from the equivalent linearised equations of motion, are also sinusoidal. Under these conditions the equivalent linearised drag coefficient becomes

$$b_e = \frac{2b\langle v(\bar{U}+v)|\bar{U}+v|\rangle}{v_0^2},$$

where v_0 is the amplitude of the relative motion between structure and water particles and

$$\langle \bar{U}|\bar{U}|v\rangle = \bar{U}|\bar{U}|\langle v\rangle = 0$$

because $\langle \dot{u}_w \rangle = 0 = \langle \dot{u} \rangle$. The multiplier of b in the numerator of this expression after integration reduces to

$$\langle v(\bar{U}+v)|\bar{U}+v|\rangle = \frac{2v_0^3}{3\pi}\cos\alpha\,(3-\cos^2\alpha)+\frac{2}{\pi}\bar{U}v_0^2\alpha$$

if during a portion of a cycle $(\bar{U}+v)\leq 0$ such that $\alpha = \sin^{-1}(\bar{U}/v_0)$. Alternatively it reduces to

$$\langle v(\bar{U}+v)|\bar{U}+v|\rangle = \bar{U}v_0^2$$

if during the whole cycle $(\bar{U}+v)\geq 0$ provided that $\bar{U}>0$. For $\bar{U}=0$, the integration gives

$$\langle v^2|v|\rangle = \frac{4}{3\pi}v_0^3.$$

The drag coefficient is therefore given by

$$b_e = \begin{cases} \dfrac{4}{3\pi}v_0\cos\alpha\,(3-\cos^2\alpha)+\dfrac{4\bar{U}\alpha}{\pi} & (\bar{U}\leq|v_0|), \\[2mm] 2\bar{U} & (\bar{U}\geq|v_0|). \end{cases}$$

When the equivalent linearised horizontal loading $X_w(z,t)$ is substituted into equation (12.29) the equation of motion takes the form

$$a_{ss}\ddot{q}_s + \omega_s^2 a_{ss}q_s + \sum_{r=0}^{\infty}\dot{q}_r(\alpha_{rs}+\beta_{rs})$$
$$= \int_0^l u_s\left[m\left(\ddot{u}_w - \sum_{r=0}^{\infty}u_r\ddot{q}_r\right) + b_e\left(\dot{u}_w - \sum_{r=0}^{\infty}u_r\dot{q}_r\right)\right]\mathrm{d}z.$$

This may be written:

$$a_{ss}\ddot{q}_s + \omega_s^2 a_{ss}q_s + \sum_{r=0}^{\infty}(\ddot{q}_r A_{rs}+\dot{q}_r B_{rs}) = \int_0^l u_s(m\ddot{u}_w + b\dot{u}_w)\,\mathrm{d}z, \qquad (12.33)$$

where the coefficients are

$$A_{rs} = \int_0^l m u_r u_s\,\mathrm{d}z,$$

$$B_{rs} = \alpha_{rs} + \beta_{rs} + \int_0^l b_e u_r u_s\,\mathrm{d}z.$$

If the motions are random, rather than deterministic as hitherto, the method of equivalent linearisation may be again employed. For example, if the motions of the water particles at all sections z are represented by stationary Gaussian random processes with zero mean values, the approximate values of the equivalent linearised responses of the tower buoy are also stationary Gaussian random processes. Therefore the probability density function for the random process of relative velocity V, with values v as defined previously, is given by

$$f_V(v) = \frac{1}{\sigma_V(2\pi)^{\frac{1}{2}}} \exp\left(\frac{-v^2}{2\sigma_V^2}\right) \qquad (-\infty < v < \infty).$$

In this expression σ_V is the standard deviation of the relative velocity between structure and fluid particles and so the variance is

$$\sigma_V^2 = \langle v^2 \rangle.$$

The equivalent linearised drag coefficient becomes

$$b_e = \frac{b\langle v(\bar{U}+v)|\bar{U}+v|\rangle}{\sigma_V^2},$$

where

$$\langle v(\bar{U}+v)|\bar{U}+v|\rangle = \int_{-\infty}^{\infty} v(\bar{U}+v)|\bar{U}+v|f_V(v)\,\mathrm{d}v.$$

And after integration this takes the form

$$\langle v(\bar{U}+v)|\bar{U}+v|\rangle = \left(\frac{8}{\pi}\right)^{\frac{1}{2}} \sigma_V^3 \left(1 + \frac{\bar{U}^2}{\sigma_V^2}\right) \exp\left(-\frac{\bar{U}^2}{2\sigma_V^2}\right).$$

The drag coefficient is therefore given by

$$b_e = b\left(\frac{8}{\pi}\right)^{\frac{1}{2}} \sigma_V \left(1 + \frac{\bar{U}^2}{\sigma_V^2}\right) \exp\left(-\frac{\bar{U}^2}{2\sigma_V^2}\right)$$

which depends upon the current velocity and the standard deviation of the relative velocity. When $\bar{U} = 0$, this value simplifies to

$$b_e = b\left(\frac{8}{\pi}\right)^{\frac{1}{2}} \sigma_V$$

and this expression agrees with that determined by Penzien, Kaul & Berge (1972).

The equation of motion is similar in form to equation (12.33) but it is now of a stochastic nature. In this case, it has been suggested by Malhotra & Penzien (1970) and by Penzien et al. (1972) that an iterative form of solution must be adopted in solving the equations of motion since the equivalent linearised drag coefficient depends upon the responses. If the convergence rate is rapid, however, only a few cycles of iteration are necessary.

12.4.4   *Forced oscillation with time-dependent tension*

When allowance is made for time-dependence of the axial tension, so that

$$T(z, t) = T(z) + Z_w(z, t),$$

where $T(l) = 0 = Z_w(l, t)$, then equation (12.23) takes the modified form

$$\sum_{r=0}^{\infty} \ddot{p}_r \int_0^l \mu u_r u_s \, dz + \sum_{r=0}^{\infty} p_r \int_0^l [kAG\gamma_r + T(z)\theta_r](\theta_s + \gamma_s) \, dz$$

$$+ \sum_{r=0}^{\infty} \dot{p}_r \int_0^l kAG\gamma_r(\theta_s + \gamma_s) \, dz + \sum_{r=0}^{\infty} p_r \int_0^l Z_w(z, t)\theta_r(\theta_s + \gamma_s) \, dz$$

$$= \int_0^l X(z, t)u_s(z) \, dz$$

and equation (12.24) becomes

$$\sum_{r=0}^{\infty} \ddot{p}_r \int_0^l I_y \theta_r \theta_s \, dz + \sum_{r=0}^{\infty} p_r \int_0^l EI\theta_r' \theta_s' \, dz + \sum_{r=0}^{\infty} \dot{p}_r \int_0^l \beta EI\theta_r' \theta_s' \, dz$$

$$- \sum_{r=0}^{\infty} p_r \int_0^l kAG\gamma_r \theta_s \, dz - \sum_{r=0}^{\infty} \dot{p}_r \int_0^l \alpha kAG\gamma_r \theta_s \, dz$$

$$+ \sum_{r=0}^{\infty} p_r \int_0^l [T(z) + Z_w(z, t)](u_r' \theta_s - \theta_r \theta_s) \, dz = 0.$$

If these last two equations are added together and the orthogonality relationships are invoked it is found that

$$a_{ss}\ddot{p}_s + \omega_s^2 a_{ss}p_s + \sum_{r=0}^{\infty} \dot{p}_r(\alpha_{rs} + \beta_{rs}) + \sum_{r=0}^{\infty} p_r Z_{rs}(t)$$

$$= \int_0^l X(z, t)u_s(z) \, dz \qquad (s = 0, 1, 2, \dots), \tag{12.34}$$

where the time-dependent coefficient

$$Z_{rs}(t) = \int_0^l Z_w(z, t)(u_r'\theta_s + \theta_r\gamma_s) \, dz.$$

For sinusoidal waves, equation (12.34) is a generalisation of the type of equations discussed by Bolotin (1964). The horizontal loading $X(z, t)$ takes the form we discussed in the previous section.

12.5   Concluding remarks

Our purpose in this chapter has been the limited one of demonstrating the versatility of those techniques of linear dynamics which we have developed primarily for use with ships, in accordance with a suggestion made by Bishop & Price (1975). The tower buoy which we have used for the purposes of illustration does not possess any rigid body modes when it moves freely in the absence of hydrodynamic actions. The possession of a stable upright configuration under the influence of an axial tension ensures that even the $s = 0$ mode entails some *lateral* distortion. But despite this considerable difference between the modes and those of a ship, there is a complete (and quite obvious) analogy between the theory we have outlined and that developed previously for ships.

In practice, considerable interest centres on the force transmitted at the hinge of a tower buoy. In the horizontal direction this is represented as the shearing force $V(0, t)$ in the theory and it may readily be estimated by the usual method. Normally, of course, it would be necessary to examine the motion in two orthogonal vertical planes if the hinge is of the conventional cruciform type.

Bibliography

Note: *Entries marked with an asterisk are not referred to in the text but contain relevant material.*

Abramowitz, M. & Stegun, I. A. (1964). *Handbook of mathematical functions.* National Bureau of Standards Mathematics Series 55, Washington, DC.

Aertssen, G. (1963). Service performance and seakeeping trials on M.V. *Lukuga. Trans. RINA*, **105**, 293–335.

Aertssen, G. (1966). Service performance and seakeeping trials on M.V. *Jordaens. Trans. RINA*, **108**, 303–43.

Aertssen, G. (1968). Labouring of ships in rough seas with emphasis on the fast ship. *Proc. SNAME*, Diamond Jubilee Spring Meeting.

Aertssen, G. (1971). Discussion of the paper by R. A. Goodman. *Trans. RINA*, **113**, 167–84.

Aertssen, G. & de Lembre, R. (1971). A survey of vibration damping factors found from slamming experiments on four ships. *Trans. NECIES*, **87**, 83–6.

Aertssen, G. & van Sluijs, M. F. (1972). Service performance and seakeeping trials on a large container ship. *Trans. RINA*, **114**, 429–47.

Andrew, N. J. & Price, W. G. (1979). Application of generalised Gamma functions in ship dynamics. *Trans. RINA*, **121**, 137–43.

Bai, K. J. & Yeung, R. W. (1974). Numerical solutions to free-surface flow problems. *Tenth Symposium on Naval Hydrodynamics*, pp. 609–47.

Battjes, J. A. (1970). *Long-term wave height distribution at seven stations around the British Isles.* Nat. Inst. Oceanogr. Report A44.

Bendat, J. S. (1958). *Principles and applications of random noise theory.* New York: John Wiley and Sons.

Betts, C. V. (1975). On the damping of ship hulls. M. Phil. Thesis, University of London.

Betts, C. V., Bishop, R. E. D. & Price, W. G. (1977a). A survey of hull damping. *Trans. RINA*, **119**, 125–42.

Betts, C. V., Bishop, R. E. D. & Price, W. G. (1977b). The symmetric generalised fluid forces applied to a ship in a seaway. *Trans. RINA*, **119**, 265–78.

Bishop, R. E. D. (1956). Myklestad's method for non-uniform vibrating beams. *Engineer*, **202**, 838–40 and 874–5.

* Bishop, R. E. D. (1971). On the strength of large ships in heavy seas. *S. African Mechanical Engineer*, **21**, 12, 2–17.

* Bishop, R. E. D. & Eatock Taylor, R. (1973). On wave induced stress in a ship executing symmetric motions. *Phil. Trans. Roy. Soc. London.* **A275**, 1–32.

Bishop, R. E. D. & Gladwell, G. M. L. (1963). An investigation into the theory of resonance testing. *Phil. Trans. Roy. Soc.* **A255**, 241–80.

Bishop, R. E. D. & Johnson, D. C. (1960). *The mechanics of vibration.* Cambridge University Press.

* Bishop, R. E. D. & Price, W. G. (1974). On modal analysis of ship strength. *Proc. Roy. Soc. London*, **A341**, 121–34.

Bishop, R. E. D. & Price, W. G. (1975). A general method of structural analysis for marine structures and vehicles. *J. Mech. Eng. Sci.* **17**, 363–5.

* Bishop, R. E. D. & Price, W. G. (1975). Ship strength as a problem of structural dynamics. *Naval Architect*, **2**, 61–3.

* Bishop, R. E. D. & Price, W. G. (1976). Antisymmetric response of a box-like ship. *Proc. Roy. Soc. London*, **A349**, 157–67.

* Bishop, R. E. D. & Price, W. G. (1976). On the transverse strength of ships with large deck openings. *Proc. Roy. Soc. London*, **A349**, 169–82.

* Bishop, R. E. D. & Price, W. G. (1976). On modal analysis of ship distortion in still water. *Trans. RINA*, **119**, 151–60.

* Bishop, R. E. D. & Price, W. G. (1976). On the relationship between 'dry modes' and 'wet modes' in the theory of ship response. *J. Sound Vib.* **45**, 157–64.

* Bishop, R. E. D. & Price, W. G. (1976). Allowance for shear distortion and rotatory inertia of ship hulls. *J. Sound Vib.* **47**, 303–11.

* Bishop, R. E. D. & Price, W. G. (1977). Coupled bending and twisting of a Timoshenko beam. *J. Sound Vib.* **50**, 469–77.

Bishop, R. E. D. & Price, W. G. (1977). The generalised antisymmetric fluid forces applied in a seaway. *Int. Shipbldg. Prog.* **24**, 3–14.

* Bishop, R. E. D. & Price, W. G. (1978). On the truncation of spectra. *Int. Shipbldg. Prog.* **25**, 3–6.

* Bishop, R. E. D. & Price, W. G. (1978). A note on structural damping of ship hulls. *J. Sound Vib.* **56**, 495–9.

* Bishop, R. E. D. & Price, W. G. (1978). The vibration characteristics of a beam with an axial force. *J. Sound Vib.* **59**, 237–44.

* Bishop, R. E. D. & Price, W. G. (1979). An investigation into the linear theory of ship response to waves. *J. Sound Vib.* **62**, 353–63.

* Bishop, R. E. D., Burcher, R. K. & Price, W. G. (1973). The representation of unsteady fluid forces acting on flexible structures. *Proc. Int. Symposium on Vibration Problems in Industry.* UK Atomic Energy Authority/National Physical Laboratory, Keswick, paper 327.

* Bishop, R. E. D., Burcher, R. K. & Price, W. G. (1976). Some suggestions concerning linear theory of aero- and hydroelasticity. *Strojnicky Casopis*, **27**, 14–22.

Bishop, R. E. D., Gladwell, G. M. L. & Michaelson, S. (1965). *The matrix analysis of vibration.* Cambridge University Press.

* Bishop, R. E. D., Price, W. G. & Tam, P. K. Y. (1977). The dynamical characteristics of some dry hulls. *J. Sound Vib.* **54**, 29–38.

Bishop, R. E. D., Price, W. G. & Tam, P. K. Y. (1977). A unified dynamic analysis of ship response to waves. *Trans. RINA*, **119**, 363–90.

* Bishop, R. E. D., Price, W. G. & Tam, P. K. Y. (1977). Wave-induced response of a flexible ship. *Int. Shipbldg. Prog.* **24**, 284–95.

* Bishop, R. E. D., Price, W. G. & Tam, P. K. Y. (1978). On the dynamics of slamming. *Trans. RINA*, **120**, 259–80.

Bishop, R. E. D., Price, W. G. & Tam, P. K. Y. (1978a). Hydrodynamic coefficients of some heaving cylinders of arbitrary shape. *Int. J. Numer. Methods Eng.* **13**, 17–33.

Bishop, R. E. D., Price, W. G. & Tam, P. K. Y. (1978b). The representation of hull sections and its effects on estimated hydrodynamic actions and wave responses. *Trans. RINA*, **121**, 115–26.

Bishop, R. E. D., Price, W. G., Tam, P. K. Y. & Temarel, P. (1978). On polar plots of ship response to wave excitation. *Trans. RINA*, **121**, 103–13.

Bishop, R. E. D., Price, W. G. & Temarel, P. (1979). Antisymmetric vibration of ship hulls. *Trans. RINA*, Paper W5.

Bisplinghoff, R. L. & Doherty, C. S. (1952). Some studies of the impact of vee-wedges on a water surface. *J. Franklin Inst.* **253**, 547–61.

Bledsoe, M. D., Bussemaker, O. & Cummins, W. E. (1960). Seakeeping trials on three Dutch destroyers. *Trans. SNAME*, **68**, 39–137.

Bolotin, V. V. (1964). *The dynamic stability of elastic systems*. San Francisco: Holden-Day.

Borg, S. F. (1957). Some contributions to the wedge water-entry problem. *Proc. ASCE*, paper 1214 (EM2, April), 1–28.

Borg, S. F. (1960). The analysis of ship structures subjected to slamming loads, *J. Ship Res.* **4**, 11–27.

Bretschneider, C. L. (1961). A one dimensional gravity wave spectrum. In *Ocean wave spectra*, pp. 41–56. New Jersey: Prentice-Hall.

Bretschneider, C. L., Crutcher, H. L., Darbyshire, J., Neumann, G., Pierson, W. J., Walden, H. & Wilson, B. W. (1962). Data for high wave conditions observed by the OWS *Weather Reporter* in December 1959. *Deutsche Hydrograph. Z.* **15**, 243–55.

Burrill, L. C., Robson, W. & Townsin, R. L. (1962). Ship vibration: entrained water experiments. *Trans. RINA*, **104**, 415–35.

Cartwright, D. E. & Draper, L. (1975). The science of sea waves after 25 years. In *Dynamics of marine vehicles and structures in waves*, ed. R. E. D. Bishop & W. G. Price, pp. 1–7, London: Mech. Eng. Publ.

Cartwright, D. E. & Longuet-Higgins, M. S. (1956). The statistical distribution of the maxima of a random function. *Proc. Roy. Soc. London*, **A237**, 212–32.

Chu, W. H. & Abramson, H. N. (1961). Hydrodynamic theories of ship slamming – review and extension. *J. Ship Res.* **4**, 9–21.

Chuang, S. L. (1970). *Investigation of impact of rigid and elastic bodies with water*. NSRDC Report 3248.

Chuang, S. L. & Milne, D. T. (1971). *Drop test of cones to investigate the three-dimensional effects of slamming*. NSRDC Report 3543.

Church, J. W. (1962). Computer modelling of the elastic response of ship to sea loads. *Fourth Symposium on Naval Hydrodynamics*, 947–91.

Conolly, J. E. (1969). Rolling and its stabilisation by active fins. *Trans. RINA*, **111**, 21–48.

Cowper, G. R. (1966). The shear coefficients in Timoshenko's beam theory. *J. Appl. Mech.* **33**, 335.

Crandall, H. S. & Mark, W. D. (1963). *Random vibration in mechanical systems*. New York: Academic Press.

Cummins, W. E. (1969). A proposal on the use of multiparameter standard wave spectra. *Proc. Twelfth ITTC*, Rome, pp. 772–5.

Darbyshire, J. (1955). An investigation of storm waves in the North Atlantic Ocean. *Proc. Roy. Soc. London*, A**230**, 560–9.

Darbyshire, J. (1959). A further investigation of wind generated waves. *Deutsche Hydrograph. Z.* **12**, 1–13.

De, S. C. (1955). Contributions to the theory of Stokes waves. *Proc. Camb. Phil. Soc.* **51**, 713–36.

de Jong, B. (1973). *Computation of the hydrodynamic coefficients of oscillating cylinders.* Neth. Res. Centre TNO Report 145a.

Draper, L. (1966). Principles of wave recording. *Proc. IERE Conf. on Electronic Engineering in Oceanography*, paper 3, pp. 1–4.

Draper L. (1968a). *Waves at Morecambe Bay Light Vessel, Irish Sea.* Nat. Inst. Oceanogr. Report A32.

Draper, L. (1968b). *Waves at Smith's Knoll, North Sea.* Nat. Inst. Oceanogr. Report A33.

Draper, L. & Blakey, A. (1968). *Waves at the Mersey Bar Light Vessel.* Nat. Inst. Oceanogr. Report A37.

Draper, L. & Driver, T. (1971). Winter waves in the Northern North Sea at 57° 30′ N 3° 00′ E recorded by M.V. *Famita. Proc. First Int. Conf. on Port and Ocean Eng. under Arctic Conditions*, pp. 966–78.

Draper, L. & Fricker, H. S. (1965). Waves off Land's End. *J. Inst. Navig.* **18**, 180–7.

Draper, L. & Graves, R. (1968). *Waves at Varne Light Vessel, Dover Strait.* Nat. Inst. Oceanogr. Report A34.

Draper, L. & Squire, E. M. (1967). Waves at ocean weather-ship station *India* (59° N 19° W). *Trans. RINA*, **109**, 85–93.

Draper, L. & Whitaker, M. A. B. (1965). Waves at ocean weather ship station *Juliette* (52° 30′ N 20° 00′ W). *Deutsche Hydrogr. Z.* **18**, 25–30.

Eatock Taylor, R. (1972). On problems in ship structural dynamics. Dept. of Mech. Eng. Report Nav. Arch. 3/72. University College London.

* Eatock Taylor, R. (1975). Analysis of the flexural vibrations of variable density spheroids immersed in an ideal fluid, with application to ship structural dynamics. *Phil. Trans. Roy. Soc. London.* A**277**, 623–48.

Ewing, J. A. (1975). Some results from the Joint North Sea Wave Project of interest to engineers. In *Dynamics of marine vehicles and structures in waves*, ed. R. E. D. Bishop & W. G. Price, pp. 41–6. London: Mech. Eng. Publ.

Faltinsen, O. (1969). *A study of the two-dimensional added mass and damping coefficients by the Frank's close fit method.* Det Norske Veritas Report 69–10–S.

Faulkner, D. (1973). Compression strength of welded grillages In *Ship Structural Design Concepts*, ed. J. H. Evans, ch. 21. MIT Sea Grant Report.

Fawzy, I. & Bishop, R. E. D. (1976). On the dynamics of linear non-conservative systems. *Proc. Roy. Soc. London*, A**352**, 25–40.

Fawzy, I. & Bishop, R. E. D. (1977). On the nature of resonance in non-conservative systems. *J. Sound Vib.* **55**, 475–85.

Ferdinande, V. (1966). Theoretical considerations on the penetration of a wedge into the water. *Internat. Shipbldg. Prog.* **13**, 102–16.

Flokstra, C. (1974). Comparison of ship motion theories with experiments for a container ship. *Int. Shipbldg. Prog.* **21**, 168–89.

Flügge, W. & Marguerre, K. (1950). Wölbkröfte in dunwandigen profilstäben. *Ing.-Arch.* **18**, 23–38.

Foxwell, J. H. & Madden, P. E. (1969). *Pressure distribution on a sphere during impact with a water surface.* Admiralty Underwater Establishment Tech. Note 362/69.

Frank, W. (1967). *Oscillations of cylinders in or below the free surface of deep fluids.* NSRDC Report 2375.

Frank, W. & Salvesen, N. (1970). *The Frank close fit ship-motion computer program*, NSRDC Report 3289.

Froude, W. (1861). On the rolling of ships. *Trans. INA*, **2**, 180–229.

Gere, J. M. (1954). Torsional vibration of beams of thin-walled open section. *J. Appl. Mech. Trans. ASME* **76**, 381–7.

Gerritsma, J. & Beukelman, W. (1964). The distribution of the hydrodynamic forces on a heaving and pitching ship model in still water. *Fifth Symposium on Naval Hydrodynamics*, 219–51.

Gerstner, F. J. v. (1809). Theorie der wellen. *Ann. der Physik*, **32**, 412–40.

Goodier, J. N. (1962). In *Handbook of engineering mechanics*, ed. W. Flügge, ch. 36. New York: McGraw-Hill.

Goodman, R. A. (1971). Wave-excited main hull vibration in large tankers and bulk carriers. *Trans. RINA*, **113**, 167–84.

Grim, O. (1959). *Oscillation of buoyant two-dimensional bodies and the calculation of the hydrodynamic forces.* Hamburgische Schiffbau-Versuchsanstalt Report 1171.

Gumbel, E. J. (1954). *Statistical theory of extreme values and some practical applications.* Appl. Math. Series 33. National Bureau of Standards.

Hagiwara, K. & Yuhara, T. (1974). Study of wave impact load on ship bow. *Japan Shipbldg. and Mar. Eng.* **8**, 5–14.

Hasselmann, K. (1973). Measurements of wind wave growth and swell decay during the Joint North Sea Wave Project (JONSWAP). *Deutsche Hydrograph. Z.* A8.15, 1–95.

Hicks, A. N. (1972). *The theory of explosion induced ship whipping motions.* Naval Construction Research Establishment Report R579.

Hirowatari, T. (1963). Magnification factors in the higher-modes of ship vibration. *Trans. SNAJ*, **113**, 156–68.

Hoffman, D. (1972). Analysis of ship structural loading. *J. Mar. Tech.* **9**, 173–94.

Hoffman, D. (1975). Analysis of measured and calculated spectra. Int. Symposium on the *Dynamics of marine vehicles and structures in waves*, ed. R. E. D. Bishop & W. G. Price, pp. 8–18. London: Mech. Eng. Publ.

Hoffman, D. & van Hooff, R. (1973). Feasibility study of springing model tests of a Great Lake bulk carrier. *Int. Shipbldg. Prog.* **20**, 72–86.

Hoffman, D. & van Hooff, R. (1976). Experimental and theoretical evaluation of springing on a Great Lake bulk carrier. *Int. Shipbldg. Prog.* **23**, 173–93.

Hogben, N. & Lumb, F. E. (1967). *Ocean statistics.* London: HMSO.

Ince, E. L. (1956). *Ordinary differential equations.* New York: Dover Publications Inc.

Inglis, C. E. (1929). Natural frequencies and modes of vibration in beams of non-uniform mass and section. *Trans. INA*, **72**, 145–66.

Jasper, N. H. (1956). Statistical distribution patterns of ocean and wave induced ship stresses and motions with engineering application. *Trans. SNAME*, **64**, 375–432.

John, F. (1949). On the motion of floating bodies I. *Comm. Pure Appl. Maths.* **2**, 13–57.

John, F. (1950). On the motion of floating bodies II, *Comm. Pure Appl. Maths.* **3**, 45–101.

Johnson, A. J., Ayling, P. W. & Couchman, A. J. (1962). On the vibration amplitudes of ships' hulls. *Trans. IESS*, **105**, 301–87.

Kaplan, P. (1969). *Development of mathematical models for describing ship structural responses in waves.* US Coast Guard Ship Struct. Comm. Report SSC-193.

Kaplan, P. & Sargent, T. P. (1972). *Further studies of computer simulation of slamming and other wave-induced vibratory structural loadings on ships in waves.* US Coast Guard Ship Struct. Comm. Report SSC-231.

Kaplan, P., Sargent, T. P. & Raff, A. I. (1969). *An investigation of the utility of computer simulation to predict ship structural response in waves.* US Coast Guard Ship Struct. Comm. Report SSC-197.

Kawakami, M., Michimoto, J. & Kobayashi, K. (1977). Prediction of long term whipping vibration stress due to slamming of large full ships in rough seas. *Int. Shipbldg. Prog.* **24**, 83–110.

Kennedy, C. C. & Pancu, C. D. P. (1947). Use of vectors in vibration measurement and analysis. *J. Aero. Sci.* **14**, 603–25.

Kinsman, B. (1965). *Wind waves.* New Jersey: Prentice-Hall.

Korvin-Kroukovsky, B. V. (1961). *Theory of seakeeping.* New York: SNAME.

Krylov, A. (1896). A new theory of the pitching motion of ships on waves and the stresses produced by this motion. *Trans. INA*, **37**, 326–68.

Kumai, T. (1958). Damping factors in the higher modes of ship vibrations. *European Shipbldg.* **1**, 29–34.

Kumai, T. (1975). On the three-dimensional entrained water in vibration of Lewis' section cylinder with finite length. *Trans. West JSNA*, **50**, 173–80.

Kumai, T. (1977). Virtual inertia coefficients and response factors in the higher modes of main hull vibration. *Trans. RINA*, **120**, 187–99.

Kuo, C. (1963). Results of vertical vibration experiments with an 11 ft. xylonite model. *European Shipbldg.* **12**, 2–14.

Lamb, H. (1932). *Hydrodynamics*, 6th edn. Cambridge University Press.

Landweber, L. (1968). Parametric equations of ship forms by conformal mapping of ship sections. *Seventh Symposium on Naval Hydrodynamics*, 1619–28.

Landweber, L. & Macagno, M. C. (1957). Added mass of two dimensional forms oscillating in a free surface. *J. Ship Res.* **1**, 20–9.

Landweber, L. & Macagno, M. C. (1959). Added mass of a 3 parametei family of two dimensional forms oscillating in a free surface. *J. Ship Res.* **2**, 36–48.

Landweber, L. & Macagno, M. C. (1967). Added mass of two dimensional forms by conformal mapping. *J. Ship Res.* **11**, 109–16.

Landweber, L. & Macagno, M. C. (1975). Accurate parametric representation of ship sections by conformal mapping. *Proc. First Int. Conf. on Numerical Ship Hydrodynamics*, pp. 665–82.

Lee, Y. W. (1960). *Statistical theory of communication.* New York: John Wiley and Sons.

Leibowitz, R. C. (1962). *Comparison of theory and experiment for slamming of a Dutch destroyer.* DTMB Report 1511.

Leibowitz, R. C. (1963). *A method for predicting slamming forces and response of a ship hull.* DTMB Report 1691.

Leibowitz, R. C. & Kennard, E. H. (1961). *Theory of freely vibrating non-uniform beams, including methods of solution and application to ships.* DTMB Report 1317 S-F013 1101.

Lewis, E. V. (1967). Predicting long-term distributions of wave-induced bending motions on ship hulls. *Proc. SNAME,* Spring Meeting, Montreal.

Lewis, E. V. (1976). The motions of ships in waves. In *Principles of Naval Architecture,* ed. J. P. Comstock, pp. 607–715, New York: SNAME.

Lewis, F. M. (1929). The inertia of water surrounding a vibrating ship. *Trans. SNAME,* **27,** 1–20.

Lewison, G. R. G. (1970a). On the reduction of slamming pressures, *Trans. RINA,* **112,** 285–306.

Lewison, G. R. G. (1970b). Ship slamming. *Proc. Conf. on the Application of Ship Motion Research to Design.* Southampton Univ. Paper 6.

Lloyd, A. R. J. M. (1976). *Seakeeping criteria for warship design.* Admiralty Experiment Works Tech. Report 76007.

Lockwood Taylor, J. (1930). Vibration of ships. *Trans. INA,* **72,** 162–96.

Longuet-Higgins, M. S. (1952). On the statistical heights of sea waves. *J. Mar. Res.* **11,** 245–66.

Longuet-Higgins, M. S. (1953). Mass transport in water waves. *Phil. Trans. Roy. Soc.* A**245,** 535–81.

Longuet-Higgins, M. S. (1969). A non-linear mechanism for the generation of sea waves. *Proc. Roy. Soc. London,* A**310,** 469–78.

McGoldrick, R. T. (1960). *Ship vibration.* DTMB Report 1451.

Madden, P. E. (1970). *Pressure distribution on a rigid ellipsoidal/cylindrical body during impact with a water surface.* Admiralty Underwater Establishment Tech. Note 407/70.

Maeda, H. (1975). Hydrodynamic forces on a cross section of a stationary hull. In *Dynamics of Marine Vehicles and Structures in Waves,* ed. R. E. D. Bishop & W. G. Price, pp. 80–90, London: Mech. Eng. Publ.

Mahalingam, S. & Bishop, R. E. D. (1974). The response of a system with repeated natural frequencies to force and displacement excitation. *J. Sound Vib.* **36,** 285–95.

Malhotra, A. K. & Penzien, J. (1970). Nondeterministic analysis of offshore structures. *Proc. Amer. Soc. Civ. Eng.* **96,** 985–1001.

Mansour, A. & d'Oliveira, J. M. (1975). Hull bending moment due to ship bottom slamming in regular waves. *J. Ship Res.* **19,** 80–92.

Meek, M., Adams, R., Chapman, J. C., Reibel, H. & Wieske, P. (1972). The structural design of the OCL container ships. *Trans. RINA,* **114,** 241–92.

Michell, J. H. (1893). The highest waves in water. *Phil. Mag.* Series 5, **36,** 430–7.

Middleton, D. (1960). *An introduction to statistical communication theory.* New York: McGraw-Hill.

Miles, J. W. (1957). On the generation of surface waves by shear flows. *J. Fluid. Mech.* **3,** 185–204.

Miles, J. W. (1960). On the generation of surface waves by turbulent shear flow. *J. Fluid Mech.* **7,** 469–78.

Myklestad, N. O. (1944). A new method for calculating natural modes of uncoupled bending vibrations of airplane wings and other types of beams. *J. Aero. Sci.* **11,** 153–62.

Neumann, G. (1954). Zur Charakteristik des Seeganges. *Arch. Meteorol. Geophys. Bioklimatol,* A**7,** 352–77.

Newland, D. E. (1975). *An introduction to random vibrations and spectral analysis*. London: Longman.

Newman, J. N. (1971). *Marine hydrodynamics lecture notes*. MIT Dept. of N. Arch. and Mar. Eng.

Newman, J. N. (1977). *Marine hydrodynamics*. Cambridge (Mass.): MIT Press.

Nordenstrøm, N. (1969). *Methods for predicting long term distributions of wave loads probability of failure for ships*. Appendix I. Long term distribution of wave height and period. Det Norske Veritas Report 69-21-S.

Nordenstrøm, N. (1973). *A method to predict long term distributions of waves and wave-induced motions and loads on ships and other floating structures*. Det Norske Veritas Publication 81.

Nordenstrøm, N., Faltinsen, O. & Pedersen, B. (1971). *Predictions of wave-induced motions and loads for catamarans*. Det Norske Veritas Publication 77.

Ochi, M. K. (1958). Model experiments on ship strength and slamming in regular waves. *Trans. SNAME*, **66**, 345–83.

Ochi, M. K. (1961). See reference to model experiments reported in B. V. Korvin-Kroukovsky (1961, p. 720).

Ochi, M. K. (1964). Prediction of occurrence and severity of ship slamming at sea. *Fifth Symposium on Naval Hydrodynamics*, Bergen, pp. 545–59.

Ochi, M. K. (1973). On prediction of extreme values. *J. Ship Res.* **17**, 29–37.

Ochi, M. K. (1976). *Extreme values of surface effect ship (SES) responses in a seaway*. Part I. Estimation of extreme values for SES design consideration. DTNSRDC Report SPD-690-01.

Ochi, M. K. & Bledsoe, M. D. (1960). *Theoretical consideration of impact pressure during ship slamming*. DTMB Report 1321.

Ochi, M. K. & Hubble, E. N. (1976). On six parameter wave spectra. *Coastal Engineering*, **15**, 301–28.

Ochi, M. K. & Motter, L. E. (1971). A method to estimate the slamming characteristics for ship design. *Mar. Tech.* **8**, 219–32.

Ochi, M. K. & Motter L. E. (1973). Prediction of slamming characteristics and hull responses for ship design. *Trans. SNAME*, **81**, 144–90.

Ochi, M. K. & Motter, L. E. (1975). Prediction of extreme ship responses in rough seas of the North Atlantic. *Dynamics of marine vehicles and structures in waves*, ed. R. E. D. Bishop & W. G. Price, pp. 187–97. London: Mech. Eng. Publ.

Ohtaka, K., Kumai, T., Ushijima, M. & Ohji, M. (1967). *On the coupled torsional-horizontal vibrations of ships*. Mitsubishi Tech. Bull. 54.

Papoulis, A. (1965). *Probability, random variables and stochastic processes*. New York: McGraw-Hill.

Peckham, S. G. (1970). A new method for minimising a sum of squares without calculating gradients. *Computer J.* **13**, 418–20.

Penzien, J., Kaul, M. J. & Berge, B. (1972). Stochastic response of offshore towers to random sea waves and strong motion earthquakes. *Comp. and Struct.* **2**, 733–56.

Phillips, O. M. (1958a). The equilibrium range in the spectrum of wind-generated waves. *J. Fluid Mech.* **4**, 426–34.

Phillips, O. M. (1958b). On some properties of the spectrum of wind-generated ocean waves. *J. Mar. Res.* **16**, 231–45.

Phillips, O. M. (1966). *The dynamics of the upper ocean*. Cambridge University Press.

Pierson, W. J. & Moskowitz, L. (1963). *A proposed spectral form for fully developed wind seas based on the similarity theory of S. A. Kitaigorodsku*. Tech. Report US Naval Oceanographical Office Contract 62306-1042.

Pierson, W. J., Tick, L. J. & Baer, L. (1966). Computer based procedures for preparing global wave forecasts and wind field analyses capable of using wave data obtained by spacecraft. *Sixth Symposium on Naval Hydrodynamics*, paper 20, 1–42.

Porter, W. (1960). *Pressure distributions, added mass and damping coefficients for cylinders oscillating in a free surface*. Univ. of California Eng. Publ. Series 82-16.

Price, W. G. & Bishop, R. E. D. (1974). *Probabilistic theory of ship dynamics*. London: Chapman and Hall.

Prohaska, C. W. (1947). Vibrations verticales du navire. *Bull. Assoc. Tech. Mar. Aeronaut.* 171–215.

Raff, A. I. (1972). *Program SCORES – ship structural response in waves*. US Coast Guard Ship Struct. Comm. Report SSC-230.

Rayleigh, Lord. (1894). *The theory of sound*, 2nd edn, art. 94. London: Macmillan.

Reissner, E. (1956). On torsion with variable twist. *J. Appl. Mech.* **23**, 315–16.

Robson, J. D. (1963). *An introduction to random vibration*. Edinburgh: University Press.

Roll, H. U. (1958). Height, length and steepness of seawaves in the North Atlantic. *SNAME Tech. and Res. Bull.* 1–19. (Translation by M. St Denis.)

Saetre, H. J. (1974). *On the high wave conditions in the Northern North Atlantic*. Inst. Oceanogr. Sci. Report 3.

St Denis, M. (1975). On the statistical techniques for predicting the extreme dimensions of ocean waves and of amplitudes of ship response. *First Ship Tech. Res. (STAR) Symposium*. SNAME paper 3, 1–31.

St Denis, M. & Pierson, W. J. (1953). On the motions of ships in confused seas. *Trans. SNAME*, **61**, 280–357.

Salvesen, N., Tuck, E. O. & Faltinsen, O. (1970). Ship motions and sea loads. *Trans SNAME*, **78**, 250–87.

Saunders, H. E. (1957). *Hydrodynamics in ship design*. New York: SNAME.

Sezawa, K. & Watanabe, W. (1936). Damping forces in vibration of a ship. *Trans. JSNA*, **59**, 99–119.

Skjelbreia, L. & Hendrickson, J. (1961). Fifth order gravity wave theory. *Coastal Engineering*, **7**, 184–96.

Smith, W. E. (1883). Hogging and sagging strains in a seaway as influenced by wave structure. *Trans. INA*, **24**, 135–53.

Smith, W. E. (1967). *Computation of pitch and heave motions for arbitrary ship forms*. Neth. Res. Centre TNO Report 90S.

Stoker, J. J. (1957). *Water waves*. New York: Interscience.

Stokes, G. G. (1880). On the theory of oscillatory waves. *Mathematical and Physical Papers*, **1**, 197–229, 314–26.

Szebehely, V. G. (1956). *On slamming*. DTMB Report 995.

Tasai, F. (1960a). *On the damping force and added mass of ships heaving and pitching*. Univ. of California Eng. Publ. Series 82.

Tasai, F. (1960b). *Formula for calculating hydrodynamic force of a cylinder heaving on a free surface* (*N-parameter family*). Res. Inst. for Appl. Mech., Kyushu Univ. Report 31, pp. 71–4.

Tick, L. J. (1958). Certain probabilities associated with bow submergence and ship slamming in irregular seas. *J. Ship Res.* **2**, 30–7.

Timman, R. & Newman, J. N. (1962). The coupled damping coefficients of a symmetric ship. *J. Ship. Res.* **5**, 1–7.

Timoshenko, S. P. (1945). Theory of bending, torsion and buckling of thin-walled members of open cross section. *J. Franklin Inst.* **239**, 209–19, 249–68, 343–61.

Timoshenko, S. P., Young, D. H. & Weaver, W. (1974). *Vibration problems in engineering*, 4th edn. New York: Wiley.

Tomita, T. (1960). Allowable exciting force or moment of diesel marine engine. *Trans. SNAJ*, **108**, (see also Committee 9, *Proc. Second Int. Ship. Struct. Congr.* Delft, 1964).

Townsin, R. L. (1969). Virtual mass reduction factors J values for ship vibration calculations derived from tests with beams including ellipsoids and ship models. *Trans. RINA*, **111**, 385–97.

Ursell, F. (1949a). On the heaving motion of a circular cylinder in the surface of a fluid. *Quart. J. Mech. Appl. Maths.* **2**, 218–31.

Ursell, F. (1949b). On the rolling motion of cylinders in the surface of a fluid. *Quart. J. Mech. Appl. Maths.* **2**, 335–53.

US Department of Commerce (1959). *Climatological and oceanographical atlas for mariners*, vol. 1, *North Atlantic Ocean*. Washington, DC.

US Department of Commerce (1961). *Climatological and oceanographical atlas for mariners*. vol. 2. *North Pacific Ocean*. Washington, DC.

US Naval Oceanographic Office (1963). *Oceanographic atlas of the North Atlantic Ocean*, sect. IV. *Sea and Swell*. Publ. 700, Washington, DC.

Vlasov, V. Z. (1961). *Thin-walled elastic beams*. Israel Program for Scientific Translations.

von Kerczek, C. & Tuck, E. O. (1969). The representation of ship hulls by conformal mapping functions. *J. Ship Res.* **13**, 284–98.

Voznesenkiy, A. I. (1967). Investigation of amplitude distribution of stationary processes as applied to the problems of irregular rolling of a ship. *Sbornik Statey po Gidromekhanike i Dinamike Sudna.*

Vugts, J. H. (1968). The hydrodynamic coefficients for swaying, heaving and rolling cylinders in a free surface. *Int. Shipbldg. Prog.* **15**, 251–76.

Vugts, J. H. (1971). *The hydrodynamic forces and ship motions in oblique waves*. Neth. Res. Centre TNO Report 150S.

Wachnik, Z. G. & Zainick, E. E. (1965). Ship motion predictions in realistic short-crested seas. *Trans. SNAME*, **73**, 100–34.

Wagner, H. (1932). Uber stoss- und gleitvorgänge an der oberfläche von flüssigkeiten. *ZAMM*, **12**, 193–215.

Wahab, R. & Vink, J. H. (1973). *The hydrodynamic forces and ship motions in oblique waves*. Neth. Res. Centre TNO Report 193S.

Wahed, I. F. A. & Bishop, R. E. D. (1976). On the equations governing the free and forced vibrations of a general non-conservative system. *J. Mech. Eng. Sci.* **18**, 6–10.

Walden, H. (1964). *Die Eigenschaften der Meereswellen im Nordatlantischen Ozean; Statistik 10-jähriger Seebeobachtungen der Nordatlantischen Ozean-Wettershiffe*. Deutscher Welterdienst, Seewetteramt, Einzelveröffentlichunger Nr. 41, Hamburg.

Ward, G. & Willshare, G. T. (1976). Propeller-excited vibration with particular reference to full scale measurements. *Trans. RINA*, **118**, 97–120.

Wehausen, J. V. & Laitone, E. V. (1960). Surface waves. *Handbuch der Physik*, **9**, 446–778.

Weibull, W. (1951). A statistical distribution function of wide applicability. *J. Appl. Mech.* **18**, 293–7.

Weibull, W. (1952). Statistical design of fatigue experiments. *J. Appl. Mech.* **19**, 109–13.

Wereldsma, R. (1972). Statistical approach to the analysis of longitudinal stresses in a simplified ship's girder due to a long-crested irregular oblique sealoading. *Symposium on the Development in Merchant Shipbuilding*, Delft.

Wiegel, R. L. (1964). *Oceanographical engineering.* New Jersey: Prentice-Hall.

Index